PAINTS:

TYPES, COMPONENTS AND APPLICATIONS

CHEMISTRY RESEARCH AND APPLICATIONS

PAINTS:

TYPES, COMPONENTS AND APPLICATIONS

STEPHANIE M. SARRICA

EDITOR

Nova Science Publishers, Inc.

New York

For permission to use material from this book please contact us:
Telephone 631-231-7269; Fax 631-231-8175
Web Site: http://www.novapublishers.com

NOTICE TO THE READER

Additional color graphics may be available in the e-book version of this book.

LIBRARY OF CONGRESS CATALOGING-IN-PUBLICATION DATA

Paints : types, components, and applications / editor, Stephanie M. Sarrica.
p. cm.
  Includes bibliographical references and index.
  ISBN 978-1-61761-813-0 (hardcover)
  1. Paint. I. Sarrica, Stephanie M. II. Title.
  TP936.P225 2010
  667'.6--dc22
    2010033999

Published by Nova Science Publishers, Inc. † New York

CONTENTS

PREFACE

The term "paint" and "surface coating" are often used interchangeably. Surface coating is the more general description of any material that may be applied as a thin continuous layer to a surface. Paints are not only used to color surfaces and make them attractive, they can protect and extend the lifetime of the coated substrate. Paints are complex mixtures of organic and inorganic ingredients of polymer particles, pigments, extenders, diluents and additives. This book presents topical research data in the study of paints, including paint analysis for forensic purposes and heritage objects; cool paint technologies for urban heat buildings; selection of coating techniques for sustainability; environmentally friendly paints; intumescent powder coatings; as well as studying the antimicrobial properties of silver nanoparticle-embedded paints.

Chapter 1 - This chapter shows the potentialities of some prominent analytical techniques for painting evaluations considering performance and formulation. Other topics mentioned are the uses of paint analyses for forensic purposes and heritage objects. The analytical techniques discussed in this chapter are described in the literature from 2004 up to 2010. Spectroanalytical techniques like Atomic Absorption Spectrometry (AAS), Infrared Spectroscopy (IR), Mass Spectrometry (MS), Gas Chromatography and Liquid Chromatography, Electroanalytical methods and X-ray Fluorescence (XRF) are some examples of techniques that are discussed. In addition new trends are highlighted, such as, Chemometrics and image applications. In order to amplify the information about paints, an exhaustive bibliographic searching is done in the main data bases, looking for the many aspects of studied methods and their applications. The foremost intention of this chapter is to show the relevance of paints-related studies focused on (i) forensic analysis, (ii) heritage importance, (iii) performance of the products and (iv) formulation. The principal expected audiences for this publication are graduate students and professionals dedicated to paints quality control or even manufacturing. This chapter can foster the field of paints by means of fresh scientific information.

Chapter 2 - Grime and dirt are hazards to oil-paint-surfaces. To remove these impurities, paintings are usually cleaned dry or wet with surfactants in aqueous medium. Historic paint material (oil-wax colours produced by Schoenfeld Lukas, Düsseldorf) used by the Rhenish painter August Deusser (1870-1942) were obtained and studied. To examine the effects of different cleaning methods, numerous paint surfaces were treated dry with a latex sponge and wet with (individually) water; demineralized water; saliva; methyl cellulose; carboxymethyl cellulose; Marlipal®1618/25; and SDS. The surfaces of the treated paints were examined by

3D-measuring technology based on micro mirrors. This new, transportable technology provides measurements in seconds during the cleaning process and produces measurable images which show changes on the surface and craquelé.

Some aqueous cleaning systems can provoke craquelés with different width up to five times as much as dry cleaning methods on oil paint surfaces. However, dry methods are not sufficient to completly clean the surfaces. Therefore modification of aqueous cleaning methods are necessary and include using mild non-ionic surfactants, thickening of the solutions used, reduction of contact humidity, and increasing temperature and pH.

Chapter 3 - Due to the complex mixtures of acrylic paints it is obviously that grime and dirt are greater hazards to acrylic-paint then to others paint-surfaces. To remove these impurities, paintings are usually cleaned dry or wet with surfactants in aqueous medium. Modern acrylic paints produced by Schoenfeld Lukas and Hermann Schmincke, Düsseldorf were studied. To examine the effects of different cleaning methods, numerous acrylic paint surfaces were treated dry with a latex sponge and wet with (individually) water; demineralised water; magnetised water; saliva; methyl cellulose; carboxymethyl cellulose; Marlipal®1618/25; and SDS. The surfaces of the treated paints were examined by 3D-measuring technology based on micro mirrors. This technology provides measurements in seconds during the cleaning process and produces measurable images which can show changes on the surface.

Impurities on acrylic paintings cannot be easily removed. If they are removed dry, or wet with surfactants in aqueous media, it is very difficult to show the changes of the treated surface, even if abrasions of the surface have taken place. Particles of dirt and grime were absorbed in the paint film. Wet cleaning methods cause swelling of the paint surface. After drying the not visible changed surface contains absorbed particles and appeares quite similar to its previous state.

Due to the properties of the examined paints it can be concluded that 3D-exminations of oilpaints show more differences in the surfaces and are therefore more useful than 3-examinations of acrylic paints.

In some cases dry methods are not sufficient to clean surfaces completely. Therefore modifications of aqueous cleaning methods (such as: using mild non-ionic surfactants, thickening of the solutions used, reduction of contact humidity, and increasing temperature and pH) are possible but should be avoided.

Chapter 4 - Urban heat island measure technologies are classified roughly improvement of the building and pavement coating, promotion of ventilation in the urban area, reduction of the exhaust heat from building and vehicle. Cool paint is one of the technologies of improvement of the building and pavement coating, with green roof, water-holding material, etc. Because the performance of cool paint depends on quantity of reflective solar radiation and those of green roof and water-holding material depend on quantity of evaporation, it is hard to compare their performance directly. They are compared based on the surface heat budget model, and surface temperature on each material is the index for the performance. The effect by the improvement of building and pavement coating depends on surface heat budget of each material, the effect by the promotion of ventilation in the urban area depends on heat flux from nearby surface to upper air, and the effect by the reduction of the exhaust heat from the heat exchange device depends on the heat budget of each machine. So, it is hard to compare their performance directly. They are compared based on surface boundary layer model, and air temperature nearby ground surface is the index for the performance by each heat island measure technology. Japanese engineers have developed several improvement technologies related as above. Through various technical evaluation, the effect by cool paint as heat island measure technology is compared with the other technologies.

Chapter 5 - Since the integration of new techniques affects the total process of the industrial production, an isolated view of an individual emission reduction measure is not reasonable. Rather it requires an integrated technique assessment, which considers conflicting objectives between economic, technical, social and ecological aspects adequately with respect to a sustainable development. Therefore, it is not only important that the material and energy flows of different techniques are classified in an environmental impact assessment, but also that they are evaluated in a techno-economic assessment.

This paper describes two case studies of Multi-Criteria Decision Support for the selection of a coating technique by companies, aiming for a long-term orientation towards sustainability. Coating techniques are of importance because the conventionally used materials have a high solvent content and are thus of environmental relevance. Therefore, a changeover to low-solvent painting processes to avoid VOC emissions is currently under discussion, while at the same time low-solvent aqueous or powder coating systems are being developed. A new challenge is the use of renewable resources for which an outlook is given.

Chapter 6 - The terms 'paint' and 'surface coating' are often used interchangeably. Surface coating is the more general description of any material that may be applied as a thin continuous layer to a surface. Paints do not only used to color surfaces and make them attractive, they can protect and extend lifetime of the coated substrate. Paints are complex mixtures of organic and inorganic ingredients. Among these ingredients that might be present in paint formulations are polymer particles (binder), pigments and extenders, diluents (solvents or water) and additives. The composition of paints is simplified with quantitative percent of ingredients.

The adhesion of paints to substrate like metal (steel) is purely physically due to hydrogen bonds that develop when two surfaces are brought closely together. Binders with polar groups like epoxies and alkyds have good wetting properties and show excellent physical adhesion characteristics. Adhesion takes place when the coating and substrate separation is not more than approximately 0.5 nm. Any contaminant on the substrate will increase the separation and mask reactive sites on the steel and consequently decrease the paint film adhesion. So, surface preparation is very important step before paint application.

Chapter 7 - Paint and coatings are almost as old as human itself. Over 35,000 years ago, when man was living in caves, he decorated his cave walls with drawings using natural materials such as clays, chalks and animal fats. Now, in the 21st century, the paint and coating global market value is as high as $86 billion. It is clear that drive provided by the emergence of new technologies, the diversity of coatings and their impact on the society will be dramatically increased in coming years, an impact even much greater than what has been observed over the last 100 years. While the first production of paint goes back to the 19th century, a lot of development such as introducing synthetic binders, coil coatings, powder coatings, high solid paints and industrial water based coatings have been seen since then. In fact, paint and coatings industry is growing day by day around the globe. Today, coatings are used to protect metals and buildings from corrosion, in addition to their traditional role for decoration. For highlighting the importance of protective coating, it should be noted that annual global cost of corrosion is about $300 billion which accounts for almost 4% of the worlds GNP. Therefore, all major paint and coating companies are investing huge amounts on their research and development sector to formulate products compatible and suitable for today's aggressive environment. The current trends and challenges in paint and coatings technology are briefly reviewed in this chapter.

Chapter 8 - New regulations for hazardous air pollutants drive operations to compliant coatings. Powder coatings are the most popular choice. Also the sizes of furnaces for powder coatings have increased in the past. Today large parts up to 7 m length and 5 tons weight can be treated. These recent developments open the possibility to manufacture building parts made of steel and protected by intumescent coatings like columns and pipes continuously: in a first step the specimens are cast or moulded. In a second step the hot parts are powder coated. Intumescent additives may be added to many classical formulations of powder coatings but the amount of additives is limited and cross liking binders allow low or no intumescence. Therefore special interest deserve polyurethanes comprising organophosphorus polyesters which exert intrinsic intumescence without further additives. Another limitation has to be seen in the requirement that the curing temperature has to be lower than the temperature of intumescence. UV curing intumescent powder coatings avoid this limitation and combine the advantages of UV curing and powder coating. As powder coating is not restricted to steel a large field of activities is opened.

Chapter 9 - Over the last decades silver has been engineered into nanoparticles, structures from 1 to 100 nm in size. At present day silver nanoparticles are widely used as antibacterial/antifungal agents in a diverse range of consumer products. This paper deals with the authors' research in the field of preparation and studying antimicrobial properties of silver nanoparticle-embedded paints. The silver nanoparticles have been synthesized using the technique based on using cellulose fibers as reductant. For conducting microbiological tests the commercially available water paint was mixed with 200 ppm silver nanoparticles solutions in various ratios of 20:1, 50:1 and 100:1 in order to impart antimicrobial properties to the paint. To evaluate the antibacterial and fungicidal properties of Ag nanoparticles in paints the authors have used Escherichia coli, Salmonella typhimurium, Aspergillus niger, Staphylococcus aureus, Pseudomonas aeruginosa, Candida albicans cultures. The tests conducted have demonstrated that synthesized silver nanoparticles added to water paints show a pronounced antibacterial/antifungal effect, despite the fact that they tend to be agglomerated into clusters. It has been shown that larger concentrations of silver nanoparticles have a greater antibacterial/antifungal efficacy in Ag-nanoparticle-embedded paints.

Chapter 10 - Scanning electron microscopy (SEM) combined with energydispersive spectrometry (EDX) and X-ray diffraction (XRD) nondestructive analytical techniques are well established in materials characterization and, recently, in archaeometric investigations, since their simultaneous application allows mineralogical, chemical and topographic analysis. There is still a lack of detailed information regarding their suitability for studying samples such as thin paint layers on pottery or micro-granular phases, where standard procedures are not always appropriate.

This chapter focuses on the use of X-ray and SEM techniques for the mineralogical and chemical characterization of pigments, and surface treatments and paintings in pottery from Ambato and Portezuelo styles of the Aguada Culture (Catamarca, Argentina, ca. 600-1000 AC). An image-treatment software was developed to solve the difficulties for paint discrimination, which implements a new methodology to process backscattered electron images. This software brings to evidence small mean atomic number contrasts among paints and the paste in ceramics with a minor detail loss. Quantitative mineral compositions were obtained by Rietveld refinement of XRD patterns. In Tricolor Ambato pottery, reddish paint resembles the paste due to the presence of hematite and Fe-clays; sometimes the white paint contains Pb-rich instead of Ca-rich phases. Black paint has scarce Mn-minerals but often, like in the case of Black Incised type sherds, no particular phase is identified as a color source, suggesting possible organic pigments ("carbon black") or resulting from the firing technique.

The polychrome paints in sherds of Aguada Portezuelo style were made over a white Ca-rich base and they contain Fe-Mn (black), Fe- Mn-Ca (burgundy) and Fe-Ca (reddish). The white ones correspond to gehlenite, a firing product (possibly above 900-1000_C); but calcite and CaO also occur (above 900_C).

White and reddish pigments found at Piedras Blancas (Aguada Ambato) were also characterized. Due to their scarcity a new methodology was developed complementing XRD Rietveld refinements for mineral quantification with quantitative elemental analysis by SEM-EDX spectra, which proved to be consistent. Aspecial sample holder for fewmilligrams was developed; surface charge accumulation effects were considered by determining the Duane-Hunt limit to assess the effective incident energy, which remarkably improved the sets of concentrations obtained.

The mineralogical and chemical differences found between Ambato and Portezuelo styles suggest that they are two distinctive entities not only on their designs but also on the materials chosen and the technology used.

The results obtained by all means were consistent. The methodologies and characterization tools proposed here are suitable and can be recommended for routine analyses of different materials.

Chapter 11 - In this chapter, a 3D design and machining system based on a 3-axis NC machine tool with a rotary unit is introduced to efficiently produce the artistic design of wooden paint rollers. The paint rollers are used to execute a relief wall just after painting. A simple post-processor is first proposed for the NC machine tool to transform a base tool path called cutter location data (CL data) to NC data, mapping the y-directional pick feed to the rotational angle of the rotary unit. Also, the post-processor has a novel function that elaborately adjusts feed rate values according to the curvature of each design so as not to chip the carved surface. The suitable feed rate values are generated by using a simple fuzzy reasoning method while checking edges and curvatures in a relief design. The postprocessor allows the 3-axis NC machine tool with a rotary unit to easily carve an artistic relief design on a cylindrical wooden workpiece without an undesirable edge chipping. Experimental results show that wooden paint rollers with an artistic relief design can be successfully machined without any chipping. When a wooden paint roller is popular with the design, its mass production is further required. To cope with the mass production, rubber type paint rollers can be efficiently formed by using an aluminum mold with the same design of the wooden paint roller.

In: Paints: Types, Components and Applications
Editor: Stephanie M. Sarrica

ISBN: 978-1-61761-813-0
© 2011 Nova Science Publishers, Inc.

Chapter 1

ANALYTICAL METHODS IN PAINT EVALUATIONS: FOSTERING INFORMATION

Fabíola Manhas Verbi Pereira[*1], *Edenir Rodrigues Pereira-Filho*[2] *and Maria Izabel Maretti Silveira Bueno*[3]

[1] Embrapa Instrumentação Agropecuária, São Carlos, SP, Brazil,
[2] Universidade Federal de São Carlos - UFSCar, Departamento de Química,
Grupo de Análise Instrumental Aplicada, São Carlos, SP, Brazil,
[3] Universidade Estadual de Campinas - UNICAMP,
Instituto de Química, Departamento de Química Analítica,
Campinas, SP, Brazil

ABSTRACT

This chapter shows the potentialities of some prominent analytical techniques for painting evaluations considering performance and formulation. Other topics mentioned are the uses of paint analyses for forensic purposes and heritage objects. The analytical techniques discussed in this chapter are described in the literature from 2004 up to 2010. Spectroanalytical techniques like Atomic Absorption Spectrometry (AAS), Infrared Spectroscopy (IR), Mass Spectrometry (MS), Gas Chromatography and Liquid Chromatography, Electroanalytical methods and X-ray Fluorescence (XRF) are some examples of techniques that are discussed. In addition new trends are highlighted, such as, Chemometrics and image applications. In order to amplify the information about paints, an exhaustive bibliographic searching is done in the main data bases, looking for the many aspects of studied methods and their applications. The foremost intention of this chapter is to show the relevance of paints-related studies focused on (i) forensic analysis, (ii) heritage importance, (iii) performance of the products and (iv) formulation. The principal expected audiences for this publication are graduate students and professionals dedicated to paints quality control or even manufacturing. This chapter can foster the field of paints by means of fresh scientific information.

---

[*] Corresponding author: fmverbi@uol.com.br

Keywords: paints; analytical techniques; quality control; heritage

INTRODUCTION

This chapter is a compilation of many analytical methods and their contributions to the analysis of paint products appearing in the literature in the last six years (2004-2010). Most of published works are dedicated to Fourier Transform Infrared Spectroscopy (FTIR) in organic composition studies [1] and X-Ray Fluorescence (XRF) used to identify and quantify inorganic species [2, 3]. Due to the non-destructive property of XRF, its most general use in paints field is applied to the investigation of archeological and art objects [4]. Other contributing techniques are Raman Spectroscopy (RS) [5], chromatography [6] and atomic spectrometry [7]. Laser techniques are utilized in arts and paints for qualitative/quantitative analysis [8] and Nuclear Magnetic Resonance Spectroscopy (NMR) are employed to study the oxidative and hydrolysis processes associated with paints [9].

The American Standard Test and Methods (ASTM) are to be mentioned for establishing quality control and performance evaluations paint films and related products. These tests are specially designed to evaluate the durability and performance of consumer or trade products, under defined weathering conditions and working as parameter for the developing of other methods [10].

The topics of this chapter are organized in four different sections: forensic analysis, works of art, evaluations of performance of the paints and related products and; at last, formulation of products.

The writing team believes that this chapter can foster the paints field from the described information as follows.

1. FORENSIC ANALYSIS

Bentlin et al. [7] propose an analytical method for serigraphy, acrylic and tattoo paints analysis using Direct Sampling and Graphite Furnace Atomic Absorption Spectrometry (DS-GFAAS) for determination of six elements (Cd, Pb, Cr, Ni, Co and Cu). The samples are digested with concentrated acids and palladium is used as chemical modifier for Cd, Pb and Cu, while $Mg(NO_3)_2$ is used in Co determinations. In the case of Ni the graphite platform is covered with carbon powder. The proposed method is compared with ICP-MS (Inductively Coupled Plasma Mass Spectrometry) and GFAAS.

Smith et al. [11] use LA-ICP-MS (Laser Ablation ICP-MS) to propose a method for authentication of paintings. The authors observe that using trace element profiles of the backing substrate and binder is a good alternative to identify paint authenticity. The method collect debris directly generated by LA-ICP-MS and minimizes the amount of damage when scrapings are used.

Brominated Flame Retardants (BFRs) used in paint formulations are a present concern due to their potential risk in animal and human health. In this case, Vazques et al. [12] propose a method combining flow injection and ICP-MS for bromine determination in polymers and paintings. The samples are digested in a microwave oven and the limit of detection (LOD) is 4.2 mg/kg. This method is also applied in a screening system to control bromine in different commercial samples. In another paper from the same group, Vazquez et al. [13] employ Radiofrequency - Glow Discharge - Optical Emission Spectrometry (RF-GD-OES) and the obtained LOD is 0.044%.

Color identification is an effective parameter for forensic investigation. In this case, Trzcinska et al. [14] propose a microspectrometer to compare colors, avoiding subjective visual inspections. Fragments of red, blue, brown and green car paints are measured in reflectance mode and the reflectance curve obtained can be used to represent the color. Using FTIR and RS, Bell et al. [15] distinguish white color paints. This approach is possible due to the different organic resins used in paints preparation. The authors conclude that RS gives superior results than FTIR.

In another paper Bell et al. [16] use RS to identify lilac paints. The best wavelength excitation is 785 nm for reduced fluorescence and increased discrimination by moving off-resonance from the coloring agents in the paint.

Skenderovska et al. [17] also use FTIR and RS to investigate automotive topcoats. The authors present the results for four successfully solved cases from the police investigation in hit-and-run accidents. Deconinck et al. [18] also study hit-and-run accidents but they employ LA-ICP-MS using the signal profiles to identify the samples involved.

Szafarska et al. [19] propose a method using ATR-FTIR (Attenuated Total Reflectance FTIR). The authors subtract and normalize IR spectra and extract mathematically the pure paint spectrum from the spectrum of paint coat on different bases. The method is successfully tested in five different paints sprayed on plastic foil and cotton.

Mazzeo et al. [20] use ATR-FTIR to characterize pigment-binder interaction in reconstructured paint films. The authors observe the formation of metal soaps in oil and egg yolk tempera paint.

Payne et al. [21] use images obtained in visible and near-IR spectral region for forensic analysis of paints, tapes and adhesives, inks and firearm propellants. The proposed method presents advantages when compared to traditional techniques because sample preparation is reduced and the evidences for further investigations are preserved.

Govaert and Bernard [22] combine three techniques to discriminate red spray paints: Optical Microscopy (OM), FTIR and XRF. An information database is built and FTIR spectra are classified according to binder type, filler and pigment composition.

Milczarek and Zieba-Palus [23] propose a method to examine spray paints. This type of sample is normally analyzed by IR but the inorganic components can interfere in the data interpretation because bands of plaster components can overlay signals from the paint. In this case, the authors proposed the use of Py-GC-MS (Pyrolysis - Gas Chromatography - Mass Spectrometry) and the influence of the inorganic constituents are negligible. The method is tested in 15 spray paints and its samples being effectively differentiated.

Burns and Doolan [6, 24] also use Py-GC-MS to discriminate automotive clear coat paints. The authors demonstrate the potential of this analytical technique for discrimination of automotive paints. This information can be used to improve a database for forensic proposes.

The same technique is also applied for forensic purposes by Zieba-Palus et al. [25] to identify polymer binders in 60 automobile paint samples using infrared spectra and Py-GC-MS. The last technique is valuable and complementary to the FTIR spectroscopy in the investigation of car paint samples according to the authors´ results.

Bowman et al. [26] compare cyclic voltammetry and Rutherford Backscattering Spectrometry (RBS) for Pb determination in paints collected at various sites in the historic campus at A&M University (AAMU). The authors observe that most of the investigated samples had Pb content in agreement with EPA USA (Environmental Protection Agency of United States of America).

Micro-IR Spectrometry (μ-IRS) is evaluated in the identification of paints spray coated on various surfaces. In order to test the paints, Zieba-Palus [27] verified different surfaces such as, metal, glass, foil, fabric and a wall made of brick, plaster board and mineral plaster. The results show that the identification of the paint using this technique depends on thickness of the paint coat and on type of the base.

Zieba-Palus and Borusiewicz [28] utilize the same technique and RS for examination of multilayer fragments of paints, for forensic purposes. The authors observe that Raman spectra in the visible range (633 nm) provide suitable information only for the pigments, whereas μ-IRS is useful to characterize the polymer. The authors confirm the presence of identified pigments by μ-XRF technique.

2. WORKS OF ART

The study of works of art is a dificult task, due to the applied analytical methods have to perform a non-destructive analysis and, in some special cases, needs good spatial resolution for mapping the samples surfaces. Methods with spatially analytical resolution have significantly enhanced the possibilities to study heritage objects considering they cause minimal or not any damage to the studied material. In this context, a few analytical techniques operating within the requested spatial resolution are applicable for the investigation of paints components of works of art objects. In particular, there is an increasing need for non-destructive analytical techniques [2, 3].

Otherwise, to conserve or restore painted artwork, it is necessary to understand the nature of gradual chemical and physical degradation that occurs within the several components of the system. According to Dawson, there is a range of analytical techniques (mostly spectrographic), which have often been specifically developed to aid the identification of the wide range of inorganic colors and organic binders (or their degradation products) that may be present in each paint layer of an old masterpiece. Although an art gallery or museum strives to conserve the artifacts it displays, over time there will be varying degrees of mechanical or chemical damage [29].

Red lake pigments and dyes used in works of art are characterized by Microspectrofluorimetry by Claro et al. [30]. Emission and excitation spectra are obtained with high spatial resolution (8-30 μm) in cross-sections from paintings by Vincent van Gogh and Lucien Pissarro and from millenary Andean textiles. The fluorophores are identified by comparing their spectra with those from historic reconstructions assembled in a database. In the paints, purpurin and eosin lakes are detected. In the Paracas and Nasca textiles, dated from B.C. 200 to A.D. 1476, purpurin and pseudopurpurin are the red dyes used. Carminic acid is detected in textiles dated close to the Inca Empire, A.D. 1000-1476. The results obtained with this new technique are confirmed and are in agreement with those obtained with conventional methods requiring microsampling, such as HPLC-DAD-MS (High Performance Liquid

Chromatography - Diode Array Detection - Mass Spectrometry) and SEM-EDS (Scanning Electronic Microscopy - Energy Dispersive X-Ray Spectrometry).

Painted glass magic lantern plates from the Museo Nazionale del Cinema, Torino (Italy), are studied [31] using Mid-IR Fiber-Optic Reflectance Spectroscopy (FORS), a non-invasive technique, to test its potential for the identification of the types of binding media used in the paints. The authors identified gum, oil and resin media on the plates and the amounts of these media varied from place to place.

In the conservation literature, little information with respect to artists' alkyd paints are reported. Recently, analytical methods have been developed to identify the components in these polymers, rates of cross-linking and mechanical properties of these paints. Ploeger et al. [32] show the results of artists' alkyd paints using THM-GC-MS (Thermally Assisted Hydrolysis and Methylation - Gas Chromatography - Mass Spectrometry) and FTIR - ATR. Four brands of artists' alkyd paints containing alkyd resin have been characterized: one containing a phthalic anhydride and pentaerythritol based alkyd resin, two containing isophthalic acid and pentaerythritol based alkyd resins, and the final one containing both phthalic anhydride and isophthalic acid and pentaerythritol based resins among the colors studied.

In order to characterize the original medieval painting technique and the subsequent editions of the painting panel "S. Francesco d'Assisi" (Museo diocesano "Mons. A. Marena", Bitonto, Italy), some analytical techniques are evaluated, as follows [33]. OM, SEM-EDS, μ-RS and Py-GC-MS are applied on various samples taken from significant parts of the painting. In situ μ-Raman analyses are also performed. The most interesting information are that the results confirm that the painting belongs to the 13th century Italian painting tradition. Furthermore, combination of various analytical techniques reveals that the 13th century original background, which now appears dark gray, is realized by applying a tin foil covered by a mecca layer composed of siccative oil and heated Pinaceae resin. Thus, originally the background should have had a gold-like appearance. The most important manipulation of the painting of S. Francesco probably dates back to the 16th century but shows a quite traditional technique.

Chinese ink is based on a mixture of soot and animal glue, has been used in East Asia for centuries as the sole black paint of choice. This combination create a distinctive dispersion system giving Chinese ink its unique properties among paints and inks. To reveal subtle differences in particle size and aggregation among inks of different soot origin, Swider et al. [34] utilize Photon Correlation Spectroscopy (PCS) size measurements and SEM imaging. Another property like surface chemistry of the particles is examined using Laser Doppler Electrophoresis (LDE) for determination of the isoelectric point (IEP). The authors verified that the IEPs of different inks are not distinct, but reflectes the presence of the collagen-based glue on the particles' surface. The information was supported by the fact that the IEP and size drops significantly when inks are treated with collagenase and when soot and carbon blacks alone are measured, pointing to the important role of animal glue in this dispersion system.

Duran-Benito et al. [35] analyze the color, chemical composition, and mineralogical phases of the superpositioned layers of materials in the architectural construction elements of the pond found in the middle of the "Patio de las Doncellas" Mudejar style Palace, during archaeological research on the "Reales Alcazares" palace in Sevilla, southern Spain. This analysis is executed using Visible Spectroscopy (Vis Spectroscopy), FTIR, OM, SEM coupled with an ED-X-Ray analyzer, and XRD. The results show that in the 14th century, the pond was covered with a lime mortar finished with a geometric drawing of Mudejar bow manufacture. The pigments used for the paints were made using albero (sedimentary material) that has a very light yellow color and iron oxide and vermilion or cinnabar for the red color. In the 15th century this lime mortar was peaked, and another drawing was applied over it. The pigment used for the black color in the geometric drawing of waves over this last mortar was carbon obtained from burned animal bone. The red color that appears in the surface of the mortar was made with iron oxide and vermilion or cinnabar. These results are important to prove that the two interventions made regarding the pond and the time at which they were made and the characterization of pigments has facilitated the choice of materials and colors used to carry out the restoration of this historical building.

The potential of X-ray diffraction (XRD) complemented by RS analyses of synthetic organic pigments in powder samples, layered paint systems, and commercial artists' paints bound in acrylic, alkyd, and oil media are presented by Brostoff et al. [36]. In the described study is used a stratified model paint systems that mimic the layering structure typically found in modern paintings to evaluate the effect of the μ-XRD experimental parameters. The advantages of XRD is demonstrated for the specific identification of synthetic organic pigment mixtures and fillers in acrylic and alkyd bound artists' paints. The limitation of XRD is an identification of these pigments in oil bound paints. Then, detailed crystallographic information provided by XRD is shown to be complementary to molecular information provided by Raman analysis. The authors observed that the combination use of these techniques improve more information of compound identification than would be possible using one technique alone.

A method using Py-GC-MS in works of art is presented by Peris-Vicente et al. [37]. The proposed analytical method show some advantages such as, does not require previous treatment of the sample and only a small sample quantity in the microgram range can be used. The authors introduce two case studies where direct Py-GC-MS and thermally assisted hydrolysis and methylation GC-MS are applied on art objects: first, a modem gluing material of a medieval reverse glass painting, and the second example, the binding medium of a painting by Georg Baselitz ("Senta", 1992-1993) from the Sammlung Moderne Kunst at the Pinakothek der Moderne, Munich.

Another analytical technique is efficiently utilized in the study of painted artworks [38]. The binding medium from two original oil paintings, dated from the early 20th and the late 17th century, is studied via high-resolution 1D and 2D NMR. The suitable result is the establishing of the advanced state of hydrolysis and oxidation of the oil paint. The method propose speed, simplicity, and nondestructive nature.

Asensio et al. [39] report the ATR-FTIR analysis of samples of polymeric materials used in the conservation of artworks. These samples were examined directly in the solid material without preparation.

Non-destructive and non-invasive μ-Raman fiber optic and μ-XRF analyses are performed to study wallpaper from the beginning of the 19th century [40]. The authors showed that both techniques are complimentary. The analyzed artwork is considered one of the most beautiful wallpapers ever manufactured according to the catalogues and books; it is known as Chasse de Compiegne, manufactured by Jacquemart, Paris, in 1812. During the analysis, an unexpected pigment is detected by both analytical techniques: lead-tin yellow type II. This pigment was used until *ca.* 1750, when other yellow pigments started to be used, thus it is very difficult to find it in paintings afterwards. Together with this pigment, red lead,

Prussian blue, brochantite, yellow iron oxide, calcium carbonate, vermilion, carbon black of animal origin (bone black), lead white, and raw and burnt sienna were also determined by combining the analytical information provided by both techniques.

Stained glass windows incorporating dark blue and purple enamel paint layers are irradiated by μ-XRF for detection of elements content in their composition [41]. The samples of enamels glass paints are from Northwestern Europe of period between 16 - early 20th centuries. These enamels coatings are basically paints layers vitrified after firing processes. These paints were also analyzed by electron probe microanalysis and Transmission Electron Microscopy (TEM) to understand the causes of the degradation. The data reveled the inorganic composition on the enamels such as, Cu, Au, Fe, and Pb.

A proteomics approach is used for the identification of protein binders in historical paints: the proteins are digested enzymatically into peptides using trypsin before being separated and detected by High Performance Liquid Chromatography - Electrospray Ionization - Tandem Mass Spectrometry (HPLC-ESI-MS-MS) [42]. Mascot (Matrix Science) is used to analyze the resulting data and for protein identification. The authors used amino acid sequences due to the fact they retain much more information about the proteins. The best extraction strategy is selected based on the number of peptides that are identified in the protein content of paint replicas using different methods. Finally this method is applied to historical paint microsamples on the anonymous early 15th century panel painting Crucifixion with St Catherine and St Barbara (Calvary of the Tanners), the St Catherine Altarpiece by Joes Beyaert (*ca.* 1479) and two painting by Pieter Brueghel the Younger (1617-1628).

Bearing in mind that every analytical method holds particular limitations, two complementary spectroscopic techniques, namely Confocal μ-RS and μ-XRF, are joined in one instrument [43]. The combined μ-XRF and μ-RS device were joined in one mobile setup, called PRAXIS, which allows micrometric and in situ analysis. μ-XRF allows also to collect elemental and spatially-resolved information in a non-destructive way on major and minor constituents of a variety of materials. Confocal RM was able to offer information from single pigment grains. However, in some cases the presence of a strong fluorescence background limits the applicability.

Some samples from late Aztec period (15th century) of raw blue pigments coming from an archaeological rescue mission in downtown Mexico City are characterized using different techniques [44]. The striking characteristic of these samples is that they seem to be raw pigments prior to any use in artworks, and it is possible to collect a few μg of pigment after manual grain selection under a microscopy monitoring. All pigments are made of indigo, an organic colorant locally known as *xiuhquilitl*. The results of characterization using high resolution powder diffraction recorded at the European Synchrotron Radiation Facility (BM25A, SpLine beamline) complemented with other techniques are presented. All of them give consistent results on the composition.

D'agata et al. [45] prepare iron-gall ink according to ancient recipes and its composition is evaluated by AP-MALDI-MS (Atmospheric Pressure Matrix-assisted Laser Desorption/Ionization – Mass Spectrometry). The latter is demonstrated to be a very valuable tool for the study of the organic components of ancient works of art as it combines the advantages of the vacuum MALDI-MS analytical approach to the possibility of analyzing samples in air.

Portable equipments are useful for paint analysis, mainly in works of art. The advantage is the non-destructiveness property of the XRF. Ida and Kawai [46] assemble a portable XRF with an X-ray generator that is driven by a 9 V dry electric battery. The authors detect some elements in paints with same color but related to different pigments in their chemical composition.

FORS is used to characterize pigment mixtures in paints used in art [47]. Measurements are non invasive, without any contact with the sample. The experimental device is portable; therefore measurements can be performed in situ. The protocol is validated using modern gouache paints: 10 pure gouaches are used as references and 27 binary mixtures of these pure gouaches are studied. Reflectance spectra are processed using the Kubelka-Munk theory in order to get scattering and absorption coefficients of the references. Assuming a linear dependence of these optical properties with the pigment volume concentration (PVC) of the components of paint layers, the protocol enables qualitative and quantitative interpretation of the reflectance spectra measured on binary mixtures of references. Indeed, for most cases, numerical processing of FORS-measurements performed on a mixture leads to the identification of its components.

Angelini et al. [48] showed the potentialities for chemical characterization via XRF portable instrument of the red decorative pigment and of the stone surface of the Capestrano Warrior (Archaeological Museum of Chieti, Italy), which is considered the most important Italic stone statue found in Italy. The authors also demonstrated the use of a portable Electrochemical Impedance Spectroscopy (EIS) measurement system for the characterization of the protective effectiveness of the paints on the railing of Palazzo Reale in Torino, Italy.

A portable XRF spectrometer is applied in situ for the non-destructive elemental mapping of the pigment components of the 15th century mural painting and frescos of the Little Christopher chamber in the Main Town Hall of Gdansk, Poland [49]. For better data interpretation, the authors applied Principal Component Analysis (PCA) to associate the most intense lines of the elements Ca, Cu, Fe, Pb, and Hg in the XRF spectra with the palette of colors: white, brown, green, blue, red, yellow, and black observed in the painting. This allows limiting the number of extractions of the micro-samples for the complementary Raman measurements, thus assuring the practically non-destructive character of the entire analysis.

Gautier et al. [50] report the multi-analytical investigation of ready-mixed house paints used by artists such as Pablo Picasso (1881-1973). The pigment composition of paint swatches on four historic paint sample cards from the Art Institute of Chicago reference collection is characterized by thorough screening using FTIR and XRF spectroscopes. Spectroscopic investigations highlighted the high concentration of zinc-based whites.

Noninvasive handheld X-ray fluorescence is used for in situ measurements in restoration probes with the aim to distinguish original paints and repaints [51], during the restoration of the Altarpiece of Matejovce, Northern Slovakia, a rare example of the so-called Cracow school of painting of European 15th century. The layer stratigraphy is then precisely described on cross-sections of microsamples by light and electron microscopes. The most important feature is that Master of Matejovce altarpiece used Zn-dolomites containing iron red, which could come from oxidation zones of carbon ate-hosted Zn-Pb deposits near Cracow, Poland, and azurite with admixtures of Cu and Zn arsenates possibly from the Slovak Cu deposit at L'ubietova.

The analyses of pigments originating from well dated ancient boat models found in Egyptian graves are used for characterization and for dating tasks of unknown objects [52].

Cotton buds were applied for sampling these valuable artifacts for a subsequent Total Reflection-XRF (TXRF) analysis. The results of the analyses performed on micro-amounts of paints (< 1 µg) show that some artifacts are misclassified and belong to other epochs. Some others are retouched with modern colors.

Adriaens [53] gives an overview of research in or associated with the pan-European network COST Action G8, which aims at achieving a better preservation and conservation of their cultural heritage by increasing the knowledge of art and archaeological objects through advanced chemical and physical analyses. The paper aimed on the use of various analytical techniques for the examination of cultural heritage materials and includes research examples on painted works of art, ceramics, glasses, glazes and metals.

Thermally Assisted Hydrolysis and Methylation - Gas Chromatography -Mass Spectrometry (THM-GC-MS) in conjunction with FTIR are successfully used for the identification of different binding media, including synthetic resins and natural binders, employed by Picasso on Still Life (1914), Nude Woman in a Red Armchair (1932) and Weeping Woman (1937), works owned by the Tate, London [54].

Throughout the 20th century a number of binding media, including synthetic resins, have been employed in paints. Cappitelli [55] shows as THM-GC-MS is successfully used in conjunction with FTIR for the study of binding media in samples from Yellon, Islands by Jackson Pollock (Tate Collection, T00436) and Break Point by Fiona Banner (Tate Collection, T07501).

Thermal analyses, specifically, Differential Scanning Calorimetry (DSC) and Thermogravimetric Analysis (TGA), are used to study the changes in properties of artists' alkyd paint films with age [56]. Both techniques showed changes in the thermal properties as the films increased in age. The cross-linking rate during accelerated photo-ageing is observed with DSC by monitoring the exothermic reaction after the decomposition of peroxide groups, which form during auto-oxidation of the alkyd resin.

Extensive PIXE (Particle-Induced X-ray Emission) investigations [57] have been performed at the LABEC accelerator laboratory in Florence on the painting "Ritratto Trivulzio" by Antonello, da Messina, one of the great Italian masters of 15th century and a pioneer in modern oil painting. The combination of advanced variants of PIXE, such as differential and scanning-mode analysis, provides, in a totally non-invasive and non-destructive way, a relevant contribution to the characterization of paint layers, in terms of composition and structure. Szokefalvi-Nagy et al. [58] present illustrative applications of XRF devices with radioisotope excitation. The detection of the presence of Ti in white spots of a painting provided scientific basis to decide that the painting in question is a fake. The difficulties caused by the simultaneous presence of Ti and Ba (a very frequent component in white paints) are also discussed.

A project aiming to characterize the wide range of pigments found in the paints of Sam Francis (American, 1923-1994) prompted work to investigate more fully the sensitivity of a Direct Temperature - Resolved Mass Spectrometry (DTMS) system by comparing various ionization conditions, including electron impact (EI) at 70 and 16 eV, and chemical ionization (CI) with iso-butane in both positive and negative-ion modes [59]. Negative-ion CI conditions showed the best results for detecting the majority of synthetic organic pigments tested.

Cholesterol constitutes approximately 5% of the lipid fraction of eggs. The compound is therefore abundant in fresh egg tempera paints. The fate of cholesterol upon light ageing of egg tempera paint binding medium is investigated by DTMS and tandem mass spectrometry

(DT-MS-MS) [60]. Cholesterol oxidation products (COPs) such as 5,6-epoxycholestan-3-ol and 3-hydroxycholest-5-en-7-one are positively identified in light-aged egg binding medium. Cholesterol oxidation products are better markers for egg tempera than the cholesterol molecule itself. Cholesterol and COPs are discovered in paints on German baroque altar pieces from the 16th and 18th centuries and in a 20th century glaze on a Mark Rothko Seagram Mural painting at Tate by DTMS fingerprinting analysis of paint microsamples.

Naples yellow-containing oil paints aged under natural and artificial conditions are investigated as model systems to evaluate the potential of Secondary Ion Mass Spectrometry (SIMS) when used in combination with other MS and spectroscopic analytical methods [61]. Compared to SIMS, DTMS offers information into how the various constituents are incorporated into the paint film.

Poly(ethylene glycol) (PEG) compounds in artists' acrylic emulsion paint products from different paint manufacturers are characterized with a newly developed mass spectrometric method which combines data from MALDI-MS and Nano-ESI-MS [62, 63]. MALDI-MS is used for the determination of the molar mass distribution (MMD) and calculation of the molar mass averages (M-w and M-n), the polydispersity index (D) and the relative amount of a specific distribution if multiple PEGs are present. Nano-ESI-MS is used for the end-group analysis. Three different classes of polymers is found being PEG, polypropylene glycol (PPG) and a block copolymer of polyethylene glycol/polypropylene glycol (PEG/PPG) with molar mass averages ranging from 400 to 4200 Da (Dalton).

In the last decade, μ-RS is recognized as one of the powerful non-destructive analytical techniques for physical-chemical characterization of cultural heritage artifacts. Activities regarding the use of μ-RS in characterization of archaeological objects are recently carried out in Republic of Macedonia [64]. This technique was used to estimate the firing temperature of some Byzantine glazed ceramic finds in Republic of Macedonia, as well as to obtain information on their manufacturing technology.

Voltammetry of microparticles, an electrochemical methodology based on the record of the voltammetric response of sparingly soluble solids mechanically transferred to the surface of inert electrodes in contact with suitable electrolytes, is able to provide significant analytical information in the fields of conservation and restoration of cultural objects. Using this methodology, identification, speciation, and relative and absolute quantification of analytes from works of art samples can be achieved [65].

One-hundred seventy organic pigments are analyzed with FTIR spectroscopy and 125 pigments of particular relevance to artist's paints are characterized with RS so far [66]. The pigment collection encompasses pigment classes and subgroups. Besides peak tables and spectra patterns, flow charts based on color, pigment class, group and individual color index are presented to help identification of unknowns and mixed paint samples.

RS is a suitable technique for the in situ non-destructive identification of synthetic organic pigments in the presence of the complex binding media characteristic of synthetic resin paints or color lithographic inks. The Raman spectra of 21 yellow synthetic organic pigments are presented [67]. Since modern artists frequently mixed paint developed for other applications, in addition to colorants developed as artists' paints, other synthetic organic pigments are included in the spectral database.

Raman and IR Microscopy are used to characterize the black pigments on prehistoric Southwest American black-on-white pottery [68]. Conclusive spectroscopic evidence for the use of carbon-based paints on these sherds has been provided using the Raman technique.

Maghaemite (gamma-Fe_2O_3) and magnetite (Fe_3O_4), found alternatively or mixed with a carbonaceous pigment, are also identified on some sherds. The presumption that lipids are used as binders/vehicles in pictograph paints from southwestern Texas is tested using GC-MS [69]. A one-step transesterification/derivatization procedure is used to convert bound and unbound fatty acids to fatty acid methyl esters for the analysis. Approximately 30 organic compounds are detected in the natural rock coating that encapsulates the paints, but there are no compounds unique to the paints.

Laser-based analytical techniques find applications in the field of cultural heritage, giving information about the chemical composition of materials, at the atomic or molecular level [70], when associated to other techniques, as Raman or Fluorescence Spectroscopy. Also, they can produce new schemes of analysis, including, for instance, non-linear or remote sensing spectroscopy as well as laser ablative sampling and excitation.

The Confocal Microfluorescence Spectroscopy is successful tested by Claro et al. [71] for the characterization of selected red lake pigments and paints based on alizarin, purpurin and eosin. These pigments are used by artists since ancient times and impressionist painters. The technique is proved useful for characterization of paints with some characteristics such as, fastness and high spatial resolution being reproducible.

Leo et al. [72] propose a method of identification of proteinaceous components in paintings using proteomic strategies. The method of identification of proteins in binders carries out a minimal invasive sample preparation. The authors use a microwave to enhance the enzymatic digestion yield, followed by the analysis of the peptide mixtures by nanoLC-MS-MS with ESI. According to Leo et al. the method allowed the identification of milk proteins in a sample from Cimabue and Giotto paintings referent to 13th century Italian masters located in the Basilica of St. Francis in Assisi, Italy.

3. EVALUATIONS OF PERFORMANCE OF THE PAINTS AND RELATED PRODUCTS

X-ray fluorescence is usually employed for inorganic determinations. Verbi *et al.* [73] present a novelty in detecting organic modifications in paints with ED-XRF and PCA. The authors apply three different paints (latex, corrective fluid and varnish) onto wood substrates. The coated substrates are treated with infrared radiation and solar light at room temperature. The selection of specific region of the ED-XRF spectra, from 18 to 24 keV, is useful to classification and separation of the paints according to treatment. This region includes Compton and Rayleigh scatterings from Rh tube-EDXRF system.

Properties related to quality control of paints and varnishes can be calibrated using ED-XRF and the Chemometrics method Partial Least Squares Regression (PLS). The main idea of Pereira and Bueno study [74] is to substitute some of the standard tests for evaluating parameters of paints performance. Most of these standard tests are recommended by ASTM. Firstly, ED-XRF spectra are obtained for all samples. ASTM methods are also applied and the values are considered as reference. Reference values and spectra are then used as dependent and independent variables, respectively on the matrix data. Varnish density is the best calibrated property with absolute errors up 0.0014 and 0.055 g cm^{-3}. For paints, Stormer viscosity is possible to be calibrated with absolute errors up 0.4 to 8 Krebs Units (KU). The authors also predict C and H contents (% w/w) on the paint products. The main advantage is to measure properties using only the ED-XRF spectra, permitting to save time and reducing residue generation.

Primers are important adherence agents of topcoat paintings, also improving resistance to corrosion processes. Four primers of distinct formulations are applied on steel plates submitted to accelerated laboratory and outdoor exposure tests for distinct intervals [75]. Plate images are obtained using a commercial scanner after the exposure tests. The images are converted into a matrix; histograms of gray scale color are generated for each image [73]. Finally, the histograms are submitted to Chemometrics tools to identify the best primer performances.

The performance of 27 commercial products, including 17 varnishes and 10 paints are evaluated using ED-XRF spectra and digital images from a conventional scanner and Chemometrics tools [76]. The flat surfaces coated by varnishes are wood substrates of 75 x 150 mm (width x height) and 5 mm thick. The paints are applied on metallic plates of the same dimensions, but 1 mm thick. Using ED-XRF spectra the authors investigate chemical modifications in the different exposure tests, following ASTM recommendations. These data are corroborated by mathematical and statistical treatments of digital images. Chemometrics tools are applied to both data and a ranking of sample performances is presented, without subjective interpretations.

Spyros and Anglos [9] verify that small quantities of acrylates in paintings can be detected by NMR. As example, the molar fraction of methyl methacrylate units in an acrylic copolymer is 0.29 and the molar fraction of poly(ethylene glycol) in an aged acrylic medium is 0.08. The values are in good concordance with another acrylic medium used as reference. The copolymer composition of acrylic media is determined and the presence of additives is also identified. The main advantage of the method is the reduced sample size needed for NMR analyses.

Investigations with samples of cultural heritage are focused by Marengo et al. [77]. The authors describe a method for monitoring degradation processes in paints using FTIR and PCA. Samples are also submitted to distinct degradation tests. Two inorganic pigments, ultra-marine blue ($Na_nAl_6Si_6O_{24}S_n$) and red ochre (Fe_2O_3) are blended with linseed oil and spread on canvas. Each canvas is submitted to accelerate ageing in the presence of typical degradation agents, such as UV radiation and acidic solution. FT-Raman periodic analyses are evaluated to identify alterations on sample surfaces. By applying PCA, an effective separation in two groups is verified (before and after the expositions). The authors use this information to construct multivariate control charts, as example, CUSUM (Cumulative Sums) and SMART (Scores Monitoring and Residuals Tracking). The process evaluates the conservation state of pigments adhered to canvas, being able to indicate degradation processes before the definitive damage. In a previous study, Marengo et al. [5] utilize two analytical techniques NIR and FTRS allied to multivariate control charts to evaluate the conservation state of pigmented and wooden surfaces in face to different accelerated ageing processes.

Laser - Induced Breakdown Spectroscopy (LIBS) is a versatile technique that scans samples in different positions and detects organic and inorganic species from a wide spectral range. Most of the analyses can be performed without pre-treatment of the sample, being a fast and low cost analytical alternative. For paints, it is possible to use small sample amounts, being available as portable equipments. Kim et al. [8] show a method based on LIBS system for analyses of paints and coatings, considering the characterization of metal substrates, pretreatment layers, different paints and formulation constituents. Commercial softwares with LIBS spectra of many types of paints are ready for use. The goal of Kim et al. is to distinguish paints of the same color and/or with similar composition. Each LIBS spectrum represents a fingerprint of a product. The authors open new possibilities of using LIBS for quality control and quality assurance.

Bethencourt et al. [78] propose an accelerated method to study degradation processes of two water-based paints applied to carbon steel surfaces. The method is based on the application of cycles, combining EIS spectra, cathodic polarization steps, and recording the corrosion evolution in function of time. The results indicate that the proposed method does not introduce alterations in the degradation mechanism. Additional contributions are a considerable reduction of the time required for tests and adequate sensitivity to detect changes in the protective properties of the samples.

Lee and Pyun [79] evaluate the corrosion resistance of the surface-coated galvanized steel with various resin coating layers using AC impedance spectroscopy. The measurements are performed with a specimen exposed previously to a salt-sprayed corrosive environment and the results enabled the authors to rank the corrosion resistances.

Bressy et al. [80] apply EIS to evaluate (i) water sorption and (ii) solubility of chemically active paints. The samples are erodible acrylic-based coatings containing biocidal tertiary amines. The authors verify that for higher hydrophobic character, higher resistances to erosion are obtained.

Floyd et al. [81] show that the combination of the two DC electrochemical techniques can be used to evaluate the relative corrosion contribution (or inhibition) of various constituents in water-based coatings. It is observed that this technique is useful to ranking finished liquid paints for their subsequent corrosion resistance. Also, the batch consistency of liquid paints is possible to be checked.

Uchida et al. [82] investigate the curing and degradation thin films of oil paints exposed to pollutant gases (as NO_2 and SO_2) using NIR-excited FTRS.

The performance of the different paints (polyurethane, acrylic water-based and acrylic solvent-based enamel) coating carbon steel and Al panels are ranked by Shalaby et al. [83]. The panels are exposed at 85 sites in the arid desert atmosphere of Kuwait during five years. The evaluated parameters are gloss measurements, rust grade and rust rating.

Fraticova et al. [84] verify that chemiluminescence is a powerful feature to evaluate the oxidation stability of various polyurethane systems. In order to evaluate degradation of poly(ester-urethanes) and poly(acrylic-urethanes), the samples are exposed to several doses of artificial weathering. Deya et al. [85] test different formulations of alkyd, epoxy and water-borne paints to study the efficiency of calcium tripolyphosphate and zinc tripolyphosphate as anticorrosive pigments for paints in aggressive environments. The paints are submitted to accelerated tests, such as exposure to salt spray and humidity in closed chambers. EIS is used to assess the protective performance of the coatings. Calcium and zinc tripolyphosphates are considered efficient to protect the painted steel panels against corrosion.

Erhardt et al. [86] study hydrolysis and soap formations in sonic naturally aged drying oil and paint films. The extractability of these materials in organic solvents and changes over time in the physical properties are evaluated. Long-term physical and mechanical changes caused by aging show minor impacts in comparison to those produced by over cleaning or excessive exposure to heat.

4. FORMULATION OF PRODUCTS

Antifouling paints are investigated by Fay et al. [87] to optimize the formulation of the products. The main parameters for the investigation are biocide amounts and film surface erosion. SEM and EDX are important for detection of elements in the paints and also to elucidate erosion processes occurring in film surfaces.

Debnath and Vaidya [88] report a XRD-based method to evaluate 22 commercial pigments used in decorative, automotive and refinish paints. The main objective is to generate a data bank for quality control of this type of product. The authors emphasize XRD importance for establishing crystal phases and chemical identities of crystalline pigments and extenders. The method is useful for fast identification of individual components in a multi-component unknown pigment sample. The limitation is for complex pigment systems, with mixed phases. In this last case, it is necessary to use EDX to obtain complete information.

Hazards products in car repairs procedures are evaluated by Boutin et al. [89]. Special attention is given to isocyanate for understanding the mechanisms involved in thermal degradation. Samples are produced by means of heating in a laboratory furnace under inert and oxidative conditions. Isocyanates in the gaseous effluents are derivatized using specific reagents. Then, the derivatives are analyzed by High Performance Liquid Chromatography hyphenated with Tandem Mass Spectrometry (HPLC-MS-MS). Hazard compounds are detected, such as, isocyanic acid, aliphatic isocyanates, alkenyl isocyanates and their structural isomers.

The composition of carboxylated styrene butadiene rubber is investigated using central composite experimental design [90]. Some constituents of latex are talc, titanium dioxide (modified Rutile) and additional hindered phenolic stabilizer. With this study, it is possible to verify the effects produced by the formulation items and to optimize the formulation. The thermal performance of cool colored acrylic paints containing infrared reflective pigments are compared to conventional colored acrylic paints of similar colors applied on sheets of corrugated fiber cement roofing, by Uemoto et al. [91]. The studied properties are evaluated under the ASTM procedures denoted as ASTM D 2244-89 and ASTM E 90396. The sample exposure to infrared radiation is correlated with the measurements of surface temperatures of the specimens with thermocouples connected to a data logging system. The results show that the cool colored paint formulations produce significantly higher NIR reflectance than conventional paints of similar colors. The study concludes that cool paints enhance thermal comfort inside buildings, considering cost minimizations with air conditioning.

Micciche et al. [92] describe the identification of metal clusters in drier solutions using ESI-MS.

An impact of a specific coalescent agent of paints named Texanol is evaluated by Gallagher et al. [93] using naive (unfamiliar with paint constituents) and experienced (familiar with paint constituents) subjects. Odor properties of paints with and without this product are considered by the subjects. VOC (Volatile Organic Compound) emissions from neat paint and paint applied to gypsum wallboard are collected via solid-phase microextraction and analyzed by gas chromatography-mass spectrometry and gas chromatography/olfactometry. The authors consider the technique association as an effective method for analyzing the structure of paint volatiles and their sensory properties that hope solving many odorous indoor air problems.

Zhang et al. [94] apply a combination of techniques to evaluate the curing agent content on the performance of epoxy coatings on mild steels such as, EIS, DSC and FTIR.

The experimental procedure to prepare aluminum basic benzoate ($AlC_{14}H_{11}O_{15}$) for use in anticorrosive paints is described by Blustein et al. [95]. The authors utilize many techniques to determine anticorrosive properties of aluminum basic benzoate. The results show that basic aluminum benzoate is adequate to formulate solvent and water-borne epoxy anticorrosive paints with improved anticorrosive performance when it is used in combination with zinc oxide (ZnO). Following observations from the authors, solvent alkyd paints can also be prepared but with lower performance.

Shtykova et al. [96] study the interaction between an antifouling agent and binder in a model marine paint system by means of NMR and FTIR techniques. NMR diffusometry is used to quantify the strength of binder-antifouling agent interaction and FTIR spectroscopy to study the mechanism of binding.

The influence of aging in oil paints is studied using FTIR. The authors verify that pigments can significantly alter the infrared spectra of drying oil paints [1].

Eighteen samples of solid and metallic paints are taken from new and repainted cars for tests using VIS microspectrometry. The CIELAB (Commission Internationale d'Eclairage, LAB) color units are measured and used as parameter to evaluation distinct samples. LAB is represented by La*b* and describes the color through three axes in dimension space where, L* indicates a measure of the lightness (a positive L* value indicates lightness, whereas a negative L* value is darker); a* is a measure of the color on the red to green axis (positive a* is nearest to red and negative a* is nearest to green) and b* represents a measure of the color from the yellow to blue axis (positive b* is closest to yellow and negative b* is closest to blue). The authors propose a complementary analytical method for car paint traces evaluations [97].

Escola et al. [98] present a method to determine the time evolution of epoxy conversion to epoxy/amine and epoxy/amide formulations in paints using NIR. According to the author's information the cure reaction involves a simple addition mechanism between the oxirane ring and the amine/amide hydrogen functional groups. The addition reactions can be observed from the NIR absorption band at around 4530 cm^{-1}, relative to the oxirane rings concentration.

Formulation tests are studied by Carlson et al. [99] in pressure sensitive paints (PSP). The particularity of these paints is that they can be used to measure changes in air pressure, being useful in the design of aircraft and other vehicles. Highly luminescent divalent osmium complexes are added into paints. The complexes are considered attractive luminescent dyes for PSP.

A review discussing the previous editions and future printings of Principles of Color Technology is reported by Berns and Mohammadi [100].

Eterradossi et al. [101] test the tool based on Multidimensional Scaling and Procrustes Analysis known as Appearance Maps. This statistical tool is useful for sensory evaluations of automotive paints. The observations from paint experts and sensory assessors are compared with information taken from the maps. Good correlation is seen between both data sets.

CONCLUSION

In this chapter, valuable new methods are described that can be very useful to professionals leading with paints. This subject needs more contributions to help improvements in industrial processes and quality control. A positive outlook is related to innovating information from analytical methods associated with mathematical and statistics methods and also images.

ACRONYMS

AAS	Atomic Absorption Spectrometry
AP-MALDI-MS	Atmospheric pressure Matrix-assisted Laser Desorption/ionization – Mass Spectrometry
ASTM	American Standard Test And Methods
ATR	Attenuated Total Reflectance
Brominated Flame Retardants	BFRs
CI	Chemical Ionization
CIELAB	*Commission Internationale D'eclairage*, LAB
COPs	Cholesterol Oxidation Products
CUSUM	Cumulative Sums
DSC	Differential Scanning Calorimetry
DS-GFAAS	Direct Sampling and Graphite Furnace Atomic Absorption Spectrometry
DTMS	Direct Temperature - Resolved Mass Spectrometry
EI	Electron Impact
EIS	Electrochemical Impedance Spectroscopy
EPA USA	Environmental Protection Agency of United States of America
ESI-MS	Electrospray Ionization - MS
FORS	Fiber-Optic Reflectance Spectroscopy
FTIR	Fourier Transform Infrared Spectroscopy
HPLC-DAD-MS	High Performance Liquid Chromatography - Diode Array Detection - Mass Spectrometry
HPLC-ESI-MS-MS	High Performance Liquid Chromatography - Electrospray Ionization - Tandem Mass Spectrometry
HPLC MS-MS	High Performance Liquid Chromatography hyphenated with Tandem Mass Spectrometry
ICP-MS	Inductively Coupled Plasma Mass Spectrometry
IEP	Isoelectric Point
IR	Infrared Spectroscopy
LA-ICP-MS	Laser Ablation ICP-MS
LDE	Laser Doppler Electrophoresis
LIBS	Laser - Induced Breakdown Spectroscopy
LOD	Limit of Detection
MALDI-MS	Matrix Assisted Laser Desorption/Ionization MS
MMD	Molar Mass Distribution

MS	Mass Spectrometry
μ-IRS	Micro-IR Spectrometry
NIR	Near Infrared Spectroscopy
NMR	Nuclear Magnetic Resonance Spectroscopy
OM	Optical Microscopy
PCA	Principal Component Analysis
PCS	Photon Correlation Spectroscopy
PEG	Poly(Ethylene Glycol)
PIXE	Particle-Induced X-Ray Emission
PLS	Partial Least Squares
PPG	Polypropylene Glycol
PSC	Photon Correlation Spectroscopy
PSP	Pressure Sensitive Paints
PVC	Pigment Volume Concentration
Py-GC MS	Pyrolysis - Gas Chromatography - Mass Spectrometry
RBS	Rutherford Backscattering Spectrometry
RF GD-OES	Radiofrequency - Glow Discharge - Optical Emission Spectrometry
RM	Raman Microscopy
RS	Raman Spectroscopy
SEM-EDS	Scanning Electronic Microscopy-Energy Dispersive X-Ray Spectrometry
SIMS	Secondary Ion Mass Spectrometry
SMART	Scores Monitoring and Residuals Tracking
SR	Synchrotron Radiation
TEM	Transmission Electron Microscopy
TGA	Thermogravimetric Analysis
THM-GC-MS	Hydrolysis and Methylation - Gas Chromatography - Mass Spectrometry
TXRF	Total Reflection-XRF
VIS	Visible
VOC	Volatile Organic Compound
XRD	X-Ray Diffraction
XPS	X-ray Photoelectron Spectroscopy
XRF	X-Ray Fluorescence

REFERENCES

[1] van der Weerd, J; van Loon, A; Boon, J. *J. Stud. Conserv.,* 2005, 50, 3-22.

[2] Jenkis, R. *X-ray fluorescence spectrometry.* Wiley-Interscience: New York, NY, 1999.

[3] Tsuji, K; Injuk, J; Van Grieken, R. Eds., *X-Ray Spectrometry: Recent Technological Advances.* John Wiley & Sons: Chichester, WS, 2004.

[4] Potts, PJ; Ellis, AT; Kregsamer, P; Streli, C; Vanhoof, C; West, M; Wobrauschek, P. *J. Anal. At. Spectrom.,* 2006, 21, 1076-1107.

[5] Marengo, E; Robotti, E; Liparota, MC; Gennaro, MC. *Talanta,* 2004, 63, 987-1002.

[6] Burns, DT; Doolan, KP. *Anal. Chim. Acta,* 2005, 539, 145-155.

[7]   Bentlin, FRS; Pozebon, D; Mello, PA; Flores, EMM. *Anal. Chim. Acta.*, 2007, 602, 23-31.

[8]   Kim, T; Nguyen, BT; Minassian, V; Lin, CT. *J. Coat. Technol. Res.*, 2007, 4, 241-253.

[9]   Spyros, A; Anglos, D. *Appl. Phys. A-Mater.*, 2006, 83, 705-708.

[10]   *Annual Book of ASTM Standards*, vol. 06-01 through 06-04. ASTM International: West Conshohocken (PA) USA, 2010.

[11]   Smith, K; Horton, K; Watling, RJ; Scoullar, N. *Talanta*, 2005, 67, 402-413.

[12]   Vazquez, AS; Martín, A; Costa-Fernandez, JM; Encimar, JR; Bordel, N; Pereiro, R; Sanz-Medel, A. *Anal. Chim. Acta.*, 2008, 623, 140-145.

[13]   Vazquez, AS; Martín, A; Costa-Fernandez, JM; Encimar, JR; Bordel, N; Pereiro, R; Sanz-Medel, A. *Anal. Bioanal. Chem.*, 2007, 389, 683-690.

[14]   Trzcinska, B; Zieba-Palus, J; Koscielniak, P. *J. Mol. Struct.*, 2009, 924, 393-399.

[15]   Bell, SEJ; Fido, LA; Speers, SJ; Armstrong, WJ; Spratt, S. *Appl. Spectrosc.*, 2005, 59, 1340-1346.

[16]   Bell, SEJ; Fido, LA; Speers, SJ; Armstrong, WJ; Spratt, S. *Appl. Spectrosc.*, 2005, 59, 100-108.

[17]   Skenderovska, M; Minceva-Sukarova, B; Andreeva, L. *Maced. J. Chem. Chem. Eng.*, 2008, 27, 9-17.

[18]   Deconinck, I; Latkoczy, C; Gunther, D; Govaert, F; Vanhaecke, F. *J. Anal. At. Spectrom.*, 2006, 21, 279-287.

[19]   Szafarska, M; Wozniakiewicz, M; Pilch, M; Zieba-Palus, J; Koscielniak, P. *J. Mol. Struct.*, 2009, 924, 504-513.

[20]   Mazzeo, R; Prati, S; Quaranta, M; Joseph, E; Kendix, E; Galeotti, M. *Anal. Bioanal. Chem.*, 2008, 392, 65-76.

[21]   Payne, G; Wallace, C; Reedy, B; Lennard, C; Schuler, R; Exline, D; Roux, C. *Talanta*, 67, 334-344.

[22]   Govaert, F; Bernard, M. *Forensic Sci. Int,*. 2004, 140, 61-70.

[23]   Milczarek, JM; Zieba-Palus, J. *J. Anal. Appl. Pyrolysis.*, 2009, 86, 252-259.

[24]   Burns, DT; Doolan, KP. *Anal. Chim. Acta.*, 2005, 539, 157-164.

[25]   Zieba-Palus, J; Milczarek, JM; Koscielniak, P. *Chem. Anal.*, 2008, 53.

[26]   Bowman, L; Spencer, D; Muntele, C; Muntele, I; Ila, D. *Nucl. Instrum. Meth. B*, 2007, 261, 557-560.

[27]   Zieba-Palus, J. *J. Mol. Struct.*, 2005, 744, 229-234.

[28]   Zieba-Palus, J; Borusiewicz, R. *J. Mol. Struct.*, 2006, 792, 286-292. 109-121.

[29]   Dawson, TL. *Color Technol.*, 2007, 123, 281-192.

[30]   Claro, A; Melo, MJ; de Melo, JSS; van den Berg, KJ; Burnstock, A; Montague, M; Newman, R. *J. Cult. Heritage.*, 2010, 11, 27-34.

[31]   Ploeger, R; Scalarone, D; Chaintore, O. *J. Cult. Heritage.*, 2010, 11, 35-41.

[32]   Ploeger, R; Scalarone, D; Chaintore, O. *J. Cult. Heritage.*, 2008, 9, 412-419.

[33]   van der Werf, ID; Gnisci, R; Marano, D; De Benedetto, GE; Laviano, R; Pellerano, D; Vona, F; Pellegrino, F; Andriani, E; Catalano, IM; Pellerano, AF; Sabbatini, L. *J. Cult. Heritage.*, 2008, 9, 162-171.

[34]   Swider, JR; Hackley, VA; Winter, J. *J. Cult. Heritage.*, 2003, 4, 175-186.

[35]   Duran-Benito, A; Herrera-Quintero, LK; Robador-Gonzalez, MD; Perez-Rodriguez, JL. *Color Res. Appl.*, 2007, 32, 489-495.

[36]   Brostoff, LB; Centeno, SA; Ropret, P; Bythrow, P; Pottier, F. *Anal. Chem.*, 2009, 81,

6096-6106.

[37]   Peris-Vicente, J; Baumer, U; Stege, H; Lutzenberger, K; Adelantado, J. V. G. *Anal. Chem.*, 2009, 81, 3180-3187.

[38]   Spyros, A; Anglos, D. *Anal. Chem.*, 2009, 76, 4929-4936.

[39]   Asensio, RC; Moya, MS; de la Roja, JM; Gomez, M. *Anal. Bioanal. Chem.*, 2009, 395, 2081-2096.

[40]   Castro, K; Perez-Alonso, M; Rodriguez-Laso, MD; Etxebarria, N; Madariaga, JM. *Anal. Bioanal. Chem.*, 2007, 387, 847-860.

[41]   Schalm, O; Van der Linden, V; Frederickx, P; Luyten, S; Van der Snickt, G; Caen, J; Schryvers, D; Janssens, K; Cornelis, E; Van Dyck, D; Schreiner, M. *Spectrochim. Acta Part B*, 2009, 64, 812-820.

[42]   Fremout, W; Dhaenens, M; Saverwyns, S; Sanyvova, J; Vandenabeele, P; Deforce, D; Moens, L. *Anal. Chim. Acta.*, 2010, 658, 156-162.

[43]   Van Der Snickt, G; De Nolf, W; Vekemans, B; Janssens, K. *Appl. Phys. A: Mater. Sci. Process.*, 2008, 92, 59-68.

[44]   Del Rio, MS; Gutierrez-Leon, A; Castro, GR; Rubio-Zuazo, J; Solis, C; Sanchez-Hernandez, R; Robles-Camacho, J; Rojas-Gaytan, J. *Appl. Phys. A: Mater. Sci. Process.*, 2008, 90, 55-60.

[45]   D'agata, R; Grasso, G; Parlato, S; Simone, S; Spoto, G. *Appl. Phys. A: Mater. Sci. Process.*, 2007, 89, 91-95.

[46]   Ida, H; Kawai, J. *X-ray Spectrom.*, 2005, 34, 225-229.

[47]   Dupuis, G; Menu, M. *Appl. Phys. A: Mater. Sci. Process.*, 2006, 83, 469-474.

[48]   Angelini, E; Grassini, S; Corbellini, S; Ingo, GM; De Caro, T; Plescia, P; Riccucci, C; Bianco, A; Agostini, S. *Appl. Phys. A: Mater. Sci. Process.*, 2006, 83, 643-649.

[49]   Sawczak, M; Kaminska, A; Rabczuk, G; Ferretti, M; Jendrzejewski, R; Sliwinski, G. *Appl. Surf. Sci.*, 2009, 255, 5542-5545.

[50]   Gautier, G; Bezur, A; Muir, K; Casadio, F; Fiedler, I. *Appl. Spectrosc.*, 2009, 63, 597-603.

[51]   Hradil, D; Hradilova, J; Bezdicka, P; Svarcova, S. *X-ray Spectrom.*, 2008, 37, 376-382.

[52]   Huhnerfuss, K; von Bohlen, A; Kurth, D. *Spectrochim. Acta Part B*, 2006, 61, 1224-1228.

[53]   Adriaens, A. *Spectrochim. Acta Part B*, 2006, 60, 1503-1516.

[54]   Cappitelli, F; Koussiaki, F. *J. Anal. Appl. Pyrolysis.*, 2006, 75, 200-204.

[55]   Cappitelli, F. *J. Anal. Appl. Pyrolysis.*, 2004, 71, 405-415.

[56]   Ploeger, R; Scalarone, D; Chiantore, O. *Polym. Degrad. Stab.*, 2009, 94, 2036-2041.

[57]   Grassi, N. Nucl. Instrum. *Methods Phys. Res., Sect.*, B 2009, 267, 825-831.

[58]   Szokefalvi-Nagy, Z; Demeter, I; Kocsonya, A; Kovacs, I. *Nucl. Instrum. Methods Phys. Res., Sect. B*, 2004, 226, 53-59.

[59]   Menke, CA; Rivenc, R; Learner, T. *Int. J. Mass Spectrom.*, 2009, 284, 2-11.

[60]   van den Brink, OF; Ferreira, ESB; van der Horst, J; Boon, JJ. *Int. J. Mass Spectrom.*, 2009, 284, 12-21.

[61]   Keune, K; Hoogland, F; Boon, JJ; Peggie, D; Higgitt, C. *Int. J. Mass Spectrom.*, 2009, 284, 22-34.

[62]   Hoogland, FG; Boon, JJ. *Int. J. Mass Spectrom.*, 2009, 284, 66-71.

[63]   Hoogland, FG; Boon, JJ. *Int. J. Mass Spectrom.*, 2009, 284, 72-80.

[64]   Minceva-Sukarova, B; Grupce, O; Tanevska, V; Robeva-Cukovska, L; Mamucevska-

Miljkovic, S. *Maced. J. Chem. Chem. Eng.*, 2007, 26, 103-110.

[65] Domenech-Carbo, A. *J. Solid State Electrochem.*, 2010, 14, 363-379.

[66] Scherrer, NC; Stefan, Z; Françoise, D; Annette, F; Renate, K. *Spectrochim. Acta Part A*, 2009, 73, 505-524.

[67] Ropret, P; Centeno, SA; Bukovec, P. *Spectrochim. Acta Part A*, 2008, 69, 486-497.

[68] van der Weerd, J; Smith, GD; Firth, S; Clark, RJH. *J. Archaeol. Sci.*, 2004, 31, 1429-1437.

[69] Spades, S; Russ, J. *Archaeometry*, 2005, 47, 115-126.

[70] Anglos, D; Georgiou, S; Fotakis, C. *J. Nano Res.*, 2009, 8, 47-60.

[71] Claro, A; Melo, MJ; Schafer, S; de Melo, JSS; Pina, F; van den Berg, K. J; Burnstock, A. *Talanta*, 2008, 74, 922-929.

[72] Leo, G; Cartechini, L; Pucci, P; Sgamellotti, A; Marino, G; Birolo, L. *Anal. Bioanal. Chem.*, 2009, 395, 2269-2280.

[73] Verbi, FM; Pereira-Filho, ER; Bueno, MIMS. *Microchim. Acta*, 2005, 150, 131-136.

[74] Pereira, FMV; Bueno, MIMS. *Chemom. Intell. Lab. Syst.*, 2008, 92, 131-137.

[75] Pereira, FMV; Bueno, MIMS. *Anal. Chim. Acta.*, 2007, 588, 184-191.

[76] Pereira, FMV; Bueno, MIMS. *J. Coat. Technol. Res.*, 2009, 6, 445-455.

[77] Marengo, E; Liparota, MC; Robotti, E; Bobba, M; Gennaro, MC. *Anal. Bioanal. Chem.*, 2005, 381, 884-895.

[78] Bethencourt, M; Botana, FJ; Cano, MJ; Osuna, RM; Marcos, A. *Prog. Org. Coat.*, 2004, 49, 275-281.

[79] Lee, SJ; Pyun, SI. J. *Solid State Electrochem.*, 2007, 11, 829-839.

[80] Bressy, C; Hugues, C; Margaillan, A. *Prog. Org. Coat.*, 2009, 64, 89-97.

[81] Floyd, FL; Tatti, S; Provder, T. *J. Coat. Technol. Res.*, 2007, 4, 111-129.

[82] Uchida, T; Yoshida, A; Takahashi, K; Higuchi, S. *Bunseki Kagaku*, 2004, 53, 715-722.

[83] Shalaby, HM; Al-Sabti, F; Riad, WT; Al-Hashash, H. *Mater. Performance*, 2006, 45, 32-37.

[84] Fraticova, M; Simon, P; Schwarzer, P; Wilde, HW. *Polym. Degrad. Stab.*, 2006, 91, 94-100.

[85] Deya, M; Di Sarli, AR; del Amo, B; Romagnoli, R. *Ind. Eng. Chem. Res.*, 2008, 47, 7038-7047.

[86] Erhardt, D; Tumosa, CS; Mecklenburg, MF; *Stud. Conserv.*, 2005, 50, 143-150.

[87] Fay, F; Linossier, I; Langlois, V; Haras, D; Vallee-Rehel, K. *Prog. Org. Coat.*, 2005, 54, 216-223.

[88] Debnath, NC; Vaidya, SA. *Prog. Org. Coat.*, 2006, 56, 159-168.

[89] Boutin, M; Lesage, J; Ostiguy, C; Pauluhn, J; Bertrand, M. *J. Anal. Appl. Pyrolysis.*, 2004, 71, 791-802.

[90] Jubete, E; Liauw, CM; Allen, NS. *Polym. Degrad. Stab.*, 2007, 92, 2033-2041.

[91] Uemoto, KL; Sato, NMN; John, VM. *Energ. Buildings*, 2010, 42, 17-22.

[92] Micciche, F; van Straten, MA; Ming, WH; Oostveen, E; van Haveren, J; van der Linde, R; Reedijk, J. *Int. J. Mass Spectrom.*, 2005, 246, 80-83.

[93] Gallagher, M; Dalton, P; Sitvarin, L; Preti, G. *Environ. Sci. Technol.*, 2008, 42, 243-248.

[94] Zhang, JT; Hu, JM; Zhang, JQ; Cao, CN. *Corrosion*, 2005, 61, 872-879.

[95] Blustein, G; Romagnoli, R; Jaen, JA; Di Sarli, AR; del Amo, B. *Corrosion*, 2007, 63, 899-915.

[96] Shtykova, LS; Ostrovskii, D; Handa, P; Holmberg, K. Nyden, M. *Prog. Org. Coat.*, 2004, 51, 125-133.

[97] Zieba-Palus, J; Trzcinska, B; Koscielniak, P. *Anal. Lett.*, 2010, 43, 436-445.

[98] Escola, MA; Moina, CA; Gomez, ACN; Ybarra, GO. *Polym. Test.*, 2005, 24, 572-575.

[99] Carlson, B; Bullock, JP; Hance, TM; Phelan, GD. *Anal. Chem.*, 2009, 81, 262-267.

[100] Berns, RS; Mohammadi, M. *Color Res. Appl.*, 2007, 32, 201-207.

[101] Eterradossi, O; Perquis, S; Mikec, V. *Color Res. Appl.*, 2009, 34, 68-74.

In: Paints: Types, Components and Applications
Editor: Stephanie M. Sarrica

ISBN: 978-1-61761-813-0
© 2011 Nova Science Publishers, Inc.

Chapter 2

PART I: EXAMINATION OF UNTREATED AND TREATED OIL PAINT SURFACES BY 3D-MEASUREMENT TECHNOLOGY AT THE UNIVERSALMUSEUM JOANNEUM, GRAZ, AUSTRIA

Paul-Bernhard Eipper
Alte Galerie am Landesmuseum Joanneum,
Raubergasse, Graz, Austria

ABSTRACT

Grime and dirt are hazards to oil-paint-surfaces. To remove these impurities, paintings are usually cleaned dry or wet with surfactants in aqueous medium. Historic paint material (oil-wax colours produced by Schoenfeld Lukas, Düsseldorf) used by the Rhenish painter August Deusser (1870-1942) were obtained and studied. To examine the effects of different cleaning methods, numerous paint surfaces were treated dry with a latex sponge and wet with (individually) water; demineralized water; saliva; methyl cellulose; carboxymethyl cellulose; Marlipal®1618/25; and SDS. The surfaces of the treated paints were examined by 3D-measuring technology based on micro mirrors. This new, transportable technology provides measurements in seconds during the cleaning process and produces measurable images which show changes on the surface and craquelé.

Some aqueous cleaning systems can provoke craquelés with different width up to five times as much as dry cleaning methods on oil paint surfaces. However, dry methods are not sufficient to completely clean the surfaces. Therefore modification of aqueous cleaning methods are necessary and include using mild non-ionic surfactants, thickening of the solutions used, reduction of contact humidity, and increasing temperature and pH.

INTRODUCTION

Most of the hundreds of surfactant solutions in use worldwide by conservators (e.g. Orvus WA, Surfynol 61, Triton-X 100, Synperonic N, etc.) as well as other substances (e.g. potassium oleate, ammonium citrate, desoxycholic acid, saponin, etc.) [1-12] have been shown to be harmful for treating art objects [13-16]. Oil- and oil resin paints should not be eroded, leached out, weakened, or discolored by the use of improper or unsafe materials. To take into account differing paint binding materials, conservators want and need the potential to modify a cleaning solution. Because of the scarce financial resources available for such investigation in Europe, research at a sophisticated level is not possible for many of the numerous institutions, not even to mention private conservators.

The wet cleaning of originally unvarnished paint surfaces is a frequently performed activity in conservation work. By the term "wet cleaning", only the removal of water soluble surface impurities is implied here (grime, particles, etc. which have accumulated over time) and not the removal of additional layers, such as overpainting, varnish etc. Not infrequently museums in all parts of the world request that such paintings (e.g. impressionist, expressionist, abstract works of art) be cleaned, and that the work should be performed in a timely and above all cost effective manner. Conservators and curators, even those at renowned institutions of art, are often not absolutely certain which materials/methods are safe and non damaging for surface cleaning these paintings. It has been reported in the literature that the use of certain detergents can be harmful and provoke serious damage to paint film. [17, 14, 15]. In collaboration with the Medical University of Hanover and the Max Planck Institute, Stuttgart, for example, we found that aqueous cleaning of surfaces always provokes microtears in the surface of oil paints [18, 19]. After finding reaction differences between newer and older oil paints and oil-resin colors, a cleaning solution was developed and examined.

Criteria for surface active substances include carbon chain length, critical micelle concentration (CMC), maximum foaming temperature (MFT), ethylene oxide number (EO/mol), hydrophobicity index (HI), hydrophilic lipophilic balance (HLB), and Zein Number. It must be taken into account that the selected surfactant should not tend to discolor the oil paint or catalyse the decomposition of the painting material.

A combination of 2% methyl cellulose and 0,2 - 0,3% of linear, nonionic, pH-neutral, minimally hygroscopic, non-degradable, non-discoloring fatty alcohol ethoxylate Marlipal®1618/25 (C-chain length between C 16 and C 18), produced by Sasol (formerly Hüls, then CONDEA, Marl, Germany), in aqueous solution (100 ml water) was developed and found to be exceptionally good for cleaning oil paint surfaces. In the years following, we have attempted to demonstrate the safety and usefulness of this solution using SEM [18,19], ESEM [22], Laser profilometry [20, 21], and 3D-Stripe projection [22] to detect any undesirable changes on the paint surface.

It is well known that there are differences in the formulation of various oil paints used during the span of art history. Nevertheless, in the last years we have increasingly found more similarities between many oil and oil-resin paints [21, 22]. A flexible solution for treating a given painted surface would enable conservators to make safe and confident decisions for handling works of art and thus prevent harm to treated paint surfaces. Expected results would show the practicability of formulating special modifications of the test solution, for use in wet cleaning various oil and oil-resin paint surfaces. Through rigorous evidence-based investigation, the safety of such a system and its adaptability for use by conservators in the field as an effective solution for treating an particular painting may be demonstrated.

APPARATUS

This new, transportable technology, called MicroCad, was presented to the public for the first time in 2001 at Washington D.C., U.S.A. and Düsseldorf, Germany. The 3D-Stripe projection technique based on micro mirrors was developed at the GFMesstechnik, Berlin-Teltow, Germany.

The micro mirror projektion can be differentiated from previous such stripe projections primarily through the use of digital projection. This measurement system produces and projectes on the measured area different grey values, which may be regulated by the use of micromirrors. This permits local reflections from various areas on the surface of the object to be balanced, in contrast to the use of traditional laser profilometry which causes artefacts.

The micro mirror technology is based on a development by Texas Instruments, Dallas. 1024 x 768 micro mirrors of 16 x 16 micrometers each, are arranged in arrays (13 mm x 8 mm). Each of these micro mirror can be regulated in brightness and/or colour by a computer. In that way various colours and/or light intensities can be produced. The advantage of this technology is that the computer is calculates a pattern and this is then projected and finally an image is recorded with a CCD-camera. The difference between the recorded pattern which sent by the computer via the projector and that which was recorded by the camera and sent to the computer, gives the so called measurement effect. An additional important advantage of the micro mirrors is that one has a great deal of available light, for measuring the paint surface. These type of micro mirror projectors were developed for the use in Power Point-presentations projectors. In 1996/97 the idea for using the micro mirror projectors for measurement technology was developed by GFM.

The difference between the GFM-MicroCad-technology compared to usual stripe projection may be summarized as follows:

1. With micro mirror projection, one can project digitally, that means, that through specific control of the micro mirrors, measurement points of defined grey values between 0 and 255 Bits can be produced and projected. Thus differing local reflexions on surfaces can be equalized.
2. Through the use of micro mirrors for light projection, one observes (when compared to conventional projection systems such as LCD´s) up to 90% of the light intensity from the lamp directed on the object to be measured. Thus one is the position to carry out optical measurement techniques in normally lighted rooms without the need to darken the room.
3. Through a special development of GFM and Texas Instruments a so called high speed projection is feasibile; that is, by using conventional lighting with structured white light, the video pulse time of the projectors/camera used are able to produce a 3D-profile in 68 milliseconds. In contrast to the 64 hour measuring time needed by laser- profilometry this is of great advantage.

The MicroCad projects stripe patterns on the surface using microscopic and macroscopic digitale stripe projections reflected from micro mirrors (refered to as digital micro mirror devices).

The light used, produced by a cold halogen source of 270 Watt turned the highest intensity level (about 20.000 Lux) is conveyed through the projector to the surface to be measured for 300 and 1000 milliseconds.

Following measurement of the surface with light projection only, stripe patterns are the projected on the surface of the object to be measured. These intersecting patterns are photographed by a CCD-camera of 1300 x 1024 Pixel and in seconds displayed on the computer screen. The measured surface is shown in the form of intersecting patterns of the projected lines on the object. The distance between measurement points is 1,5 micrometers pro pixel in the lateral (x-, y-) resolution and 0,3 micrometers in the vertical (z-) resolution with 1000 points/millimeter (ca 500 x enlargement). The photo technique delivers colour and black/white scaled (for orientation on the measuring field) camera photos, a photorealistic representation (which is very similar to SEM) and scaled colour micro topographic 3D-pictures. The last two, because based on 3D-point clouds, produce pictures with are both objectively measurable. Currently, the smallest measurement field is 2 x 1,5 mm^2.

The outstanding advantage of the MicroCad for conservators is that this device allows immediate inplace examination and measurement in seconds during the cleaning process and producing measurable images which can show surface and craquelé changes before, during, and after treatment. This mobile measuring technique allows one to vary the surfactants solutions used during the cleaning process. The exact, scaled colour micro topographic images, give exact information concerning possible surface changes which can also be useful in legal court cases. The MicroCad can also be used for measuring whole picture surfaces, for determing the edges of overpainted areas, and comparing different brush strokes of questionable attributions. In the future no damaging paint sampling will need to be performed and no surface replicas will have to be produced.

EXPERIMENTAL

To evaluate the impact of cleaning methods (wet/dry) as well as various types of wet cleaning agents on the surfaces of paint samples, the MicroCad technology was applied. Measurements at ten to fifteen centimeter distance above the samples were carried out before and after treatments as described below.

MATERIALS

Paint Samples

Historic painting material ("Lukas"- oil-wax colours from the firm of Schoenfeld, Düsseldorf) from a painting by the Rhenish artist August Deusser (1870-1942) were taken [23].

To examine the effects of different cleaning methods the paint surfaces were first treated dry with the latex sponge ("wishab", "akapad") [24], and then wet (seperately) with

1. water
2. demineralized water
3. saliva
4. 2 g methyl cellulose in 100 ml demineralized water
5. 2 g carboxymethyl cellulose in 100 ml demineralized water
6. the non-ionic Marlipal®1618/25
7. the combination 2 g methyl cellulose and 0,2 g Marlipal®1618/25 in 100 ml demineralized water
8. the anionic SDS
9. the combination methyl cellulose and 0,2 g SDS in 100 ml demineralized water.

Before and after treatment, the chosen surfaces were examined using the MicroCad.

Basic Properties of the Wet Cleaning Agents Used in the Study

Demineralized water

Demineralized water has little or no cleaning effect and cannot dissolve oily or fatty substances. As is already known [25], treatment of surfaces with demineralized water can, however, produce changes in the surfaces of oil paintings. Since the publication of these studies, we have been aware that water and especially demineralized water interacts unspecifically with dirt and oil paints and can thereby harm the surface of paintings. In practical work, this means that it is not advisable to use demineralized water. It is better to use, depending on the pH-value of the dirt on the painted surface, water which includes carbon, calcium and magnesium ions. Water with ferrum, copper, and manganese ions cannot be recommended.

Reducing the amount of water, for example, by producing pastes is advisable. In our present study demineralized water has been used in order to compare it with non-demineralized water as well as to compare it with the effectiveness of surfactants.

Saliva

The use of saliva has demonstrated generally good results when used for the cleaning of paintings. The unspecific cleaning effect of saliva must be considered. Saliva contains a number of ingredients, e.g. enzymes, digesting lipids, and carbohydrates, which can decompose binding media. Saliva solubilizes polar components from varnishes and binding media [26 - 29].

Methyl cellulose (MC)

The usual Na-methylcellulose is made of wood or cotton cellulose and caustic soda solution. MC soluble in cold water comes in several types, some of which are alkaline. Normally, MC is used for wallpaper adhesive. MC is a bacteriocide and fungicide. In the cleaning of surfaces of paintings, MC is used to lower the surface tension of water, or to thicken surfactant solutions.

Carboxymethyl Cellulose (CMC)

CMC is, like MC, a derivate of cellulose. CMC can form complexes, that is, CMC can carry dirt in solution, even in absence of electrolytes; that is the reason that CMC is used in detergents. In the cleaning of paintings, CMC is used to lower the surface tension of water or to thicken surfactant solutions although it is not as effective for thickening surfactant solutions as MC.

Marlipal®16181/25 Powder

It is still problematic finding an ideal surfactant for conservators. According to information obtained from chemists working at Sasol, Germany's largest producer of surfactants, from the huge range of non-ionic surfactants only one can be used for cleaning paint surfaces. Marlipal®1618/25-powder, a sebum-fattyalcohol surfactant (Alkyl-polyethylene-glycolether) has a water-like character. Its specific chemical details are as follows:

- chain length of C 16 - C 18
- EO/mol number 25
- HLB number approximately 16
- reaches critical micelle concentration at a concentration of 0.2% at pH 6.5-7
- is 90% biologically degradable
- very low adipose effect
- HI number approximates 1; therefore it is hygroscopic only to a small extent. To examine this quality, 1 cm^3 concentrated granulate of Marlipal®1618/25 powder (100%) was stored at 100% relative humidity. After 14 days the concentrated granulate was swollen and sticky, but only on its surface. In contrast to this, 1 cm^3 concentrated granulate of SDS powder (see below) was wet after only six hours. For this examination solutions of 0.2% concentrations were used, because Marlipal®1618/25 powder reaches the critical mycelle concentration at this point. If surfactant concentrations exceed the critical mycelle concentration, the cleaning effect will not be more effective [22].

Sodium dodecyl sulphate (SDS)

SDS ($NaC_{12} H_{25} O_4 S$, a monoester of sulphuric acid with dodecanol-1) is an anionic surfactant.

- the HLB number of the surfactant is approx. 35
- SDS reaches its critical micelle concentration at 0,2%
- its pH ranges between 6.5 - 7 (0.5 g/100ml)
- its Zein number is 640 (see appendix)

SDS can be considered an aggressive cleaning agent for paintings. Since SDS is a semi-ester of sulphuric acid, the ester binding may not be stable. It is known that catalytic agents, such as heavy metals found in paints, may hydrolyse SDS over time. The product is highly reactive and very hygroscopic sulphuric acid, provided the paint layer and the ground layer of the support does not contain enough cations to buffer the sulphuric acid.

SDS interacts with the hydrophobic side chains of proteins (molecular weights >5000 Dalton) and through its hydrophilic sites it will solubilize formerly insoluble proteins.[30 - 32] This must be taken into account. It is possible to choose SDS for the surface cleaning of paint films and paint layers which are made of linseed oil, walnut oil, poppyseed oil, soya oil or hempseed oil.

For this examination solutions of 0,2% concentrations were used, because SDS reaches the critical micelle density at this point. If surfactant concentrations exceed the critical micelle concentration, the cleaning effect will not be more effective.

EXPERIMENTAL: CLEANING TESTS

In accordance with our experiences with using computer assisted laser profilometry [20, 21, 33-38] areas were cleaned with the above mentioned solutions. The greyish dirt layer consisted of loose dirt and grime. Sampling was done according to the treatment listed below.

- dry cleaning using latex-sponge ("wishab", "akapad") (duration 20 sec)
- wet cleaning using saliva on cotton swabs (duration 20 sec)
- wet cleaning using tap water on a microporous sponge (duration 20 sec)
- wet cleaning using demineralized water on a microporous sponge (duration 20 sec)
- wet cleaning using methyl cellulose (2g/100ml demineralized water) on a microporous sponge (duration 20 sec)
- wet cleaning using carboxy methyl cellulose (2g/100ml demineralized water) on a microporous sponge (duration 20 sec)
- wet cleaning using Marlipal® 1618/25 (0,2g/100ml demineralized water) on a microporous sponge (duration 20 sec)
- wet cleaning using Marlipal® 1618/25 (0,2g/100ml demineralized water) and methyl cellulose (2 g/100ml demineralized water) on a microporous sponge (duration 20 sec)
- wet cleaning using SDS (0,2 g/100ml demineralized water) on a microporous sponge (duration 20 sec)
- wet cleaning using SDS (0,2%) and methyl cellulose (2 g/100ml demineralized water) on a microporous sponge (duration 20 sec)

RESULTS

Dry cleaning of oil paints with latex sponges did not provoke new craquelé in the oil paint. However, dry cleaned surfaces did not appear clean enough. Unfortunately aqueous cleaning harms the surface of oil paints.

The wet treated surfaces of the oil paints after cleaning showed more homogeneous profiles than the dry cleaned oil paints. We discovered, similar to previous studies [18, 19, 22] that all wet cleaned surfaces demonstrated huge ranges in the width of the craquelé after wet cleaning (The craquuelé was 10 – 20 micrometers wide before treatment, and 20 - 50 micrometers wide after wet treatment).

If the craquelés before cleaning measured between 10 and 20 micro meters wide, then the following wet cleaning and time to dry all of the samples measured between 20 and 50 micro meters wide.

We found that demineralized water provoked very wide craquelé in the paint surfaces. Non demineralized water which includes carbonate, calcium, and magnesium ions is recommended over plain demineralized water.

We found that runny, unthickened solutions provoke a wider craquelé in the surface than thickened cellulose ether pastes.

Surprisingly, we found that combinations of Marlipal®1618/25 (0,2g/100ml demineralized water) and methyl cellulose (2 g/100ml demineralized water) used on microporous sponge are the most sensitive cleaning agents for oil paints. This paste reacts more sensitively than saliva. A rinse cleaning with demineralised water does not have to be carried out; water which includes carbon, calcium and magnesium ions is sufficient.

Cellulose ethers (MC or CMC) offer a careful, safe surface cleaning. Cellulose ethers tend to accumulate on surfaces of resinous oil paints [22]. Therefore a second cleaning with water as above is necessary.

It has been observed throughout the cleaning process, the sponge should follow the direction of the brushstrokes of the paint. Tissue paper should not be used for drying the surfaces. Cotton swabs provokes scratches on the paint surface. Microporous sponges made of polyvinyl acetate (e.g. "blitzfix", "wondersponge") should be used.

If the combination Marlipal®1618/25 (0,2g/100ml water) and methyl cellulose (2 g/100ml water) used on microporous sponge does not produce sufficiently clean surfaces, it is possible to prolong the time of contact, increase the temperature of the solution, or change the pH-value of the solution (max. 6, max. 8) depending on the pH of the surface impurities to increase efficiency. This would be better than using a stronger, anionic surfactant.

CONCLUSIONS

It should be pointed out that shiny, rough, and dark surfaces of oil-wax paints were problematic for the MicroCad because their high reflectivity caused interruptions during the measurements. Smooth, matte, and/or non-shiny paint surfaces are well suited for these examinations.

OUTLOOK

Some important questions still remain unanswered concerning the use of Marlipal®1618/25 solution. For example: which components of oil paints are affected or dissolved by the solution in contrast to combinations of anionic sodium dodecyl sulfate (an ingredient in many detergent combinations, e. g. Orvus WA)? Which residues and what amounts of residues can be detected following the surface treatment? What interactions between this cleaning solution and oil paint are possible and unavoidable? What are the effects of a second cleaning on the surface? Can treatment to remove remaining residues be recommended, or does this affect the surface once again? Are there any and if so, what are the synergistic effects between the nonionic surfactant Marlipal®1618/25and methyl cellulose? What effects cause these interactions?

Why is the cleaning effect of an combination methyl cellulose/Marlipal®1618/25 better than using the single components alone? How can the use of modifications/additions, such as increase in temperature, increasing or decreasing pH (are there ideal buffer solutions?), time of surface contact, addition of small amounts of solvents, etc. optimize the results of cleaning?

Can these modifications provoke damage to the treated surfaces and can this damage be prevented/minimized? Especially important and timely is the question: how harmful is demineralized water used in combination with this solution? Which mineralized water and at what pH is best for cleaning soiled surfaces of varying acidity?

First, materials for the cleaning solution as well as the oil paint surfaces to be cleaned will be examined with diagnostic tools such as EDX, XRD, and FT-IR. Before and after treating various oil paint and oil-resin color surfaces with differing surfactant treatments, the surfaces will be carefully examined with available instruments. Remnant residues of prior cleaning agents can thus be detected. The treated surfaces can be examined with ESEM or 3D-Stripe Projection, the latter better because it requires no samples to be taken and it is easy to examine the same area before and after treatment. Effects provoked by various modifications (solution, viscosity, pH, time of contact, and temperature changes, etc.) will be evaluated for their effects to the surface.

APPENDIX: GLOSSARY OF TECHNICAL TERMS

Chain length: The higher the chain length of a surfactant, the better this surfactant can be used for the cleaning of painting surfaces. Surfactants with a chain length of, for example, C 12 are aggressive to painting materials. The recommended chain length of C 16 - C 18 for surfactants protects the surface from damage because these surfactants are mild. Please note that the chain length is not the only criterion for the selection of a suitable surfactant

Critical Micelle Concentration (CMC): at a certain critical concentration a surfactant will spontaneously form small agglomerates of itself in solutions, called micelles. This tendency is also crucial for the detergent-like activity of a surfactant. At the CMC several characteristics of the solution change.

Ethylene oxide number (EO/mol): High numbers mean greater similarity to water, i.e. these surfactants do not tend to leach the lipids of oil colours. Surfactants with high ethylene oxide numbers are ideal for paint films with an oil binding media (of, say, < 10 EO/mol).

Hydrophobicity Index (HI): Low HI-numbers (<1) mean a low hygroscopical surfactant. Remnants of a hygroscopical surfactant on the surface of a painting after a surface cleaning can cause damages. These remnants may expand and shrink by a change of temperature and relative humidity. They can also work as catalysts for degrading processes of the painting materials.

Hydrophilic Lipophilic Balance Number (HLB): A numerical scale (0-40), on which HLB 40 indicates an exceptionally strong emulsifying capacity for oil and water systems. This system was originally developed to quantify the hydrophobic or hydrophilic character of non-ionic surfactant solutions. Meanwhile it is also used for anionic surfactants, which causes sometimes different research numbers for the same surfactant. An approximate HLB Number of 15 is suitable for a surfactant used for oil painting surface cleaning.

Zein-Test: Corn contains the protein zein. Zein cannot be dissolved in water and alcohol. It can be solubilized in 80% alcohol and 20% water. In the Zein-test the ability of solubilizing zein in a solution of an anionic surfactant is tested. The resulting number stands in relation to the irritation of the human skin caused by a solution of a surfactant. The highest number is 755 mg N 2 /100 ml (i.e. highest adipose effect). Numbers above 400 imply a very irritating substance. This indicates, applied on oil paintings, that the ideal surfactant for painting surface cleaning must have a number below 200 mg N_2 /100 ml (i.e. very low adipose effect).

3D-STRIPE PROJECTIONS

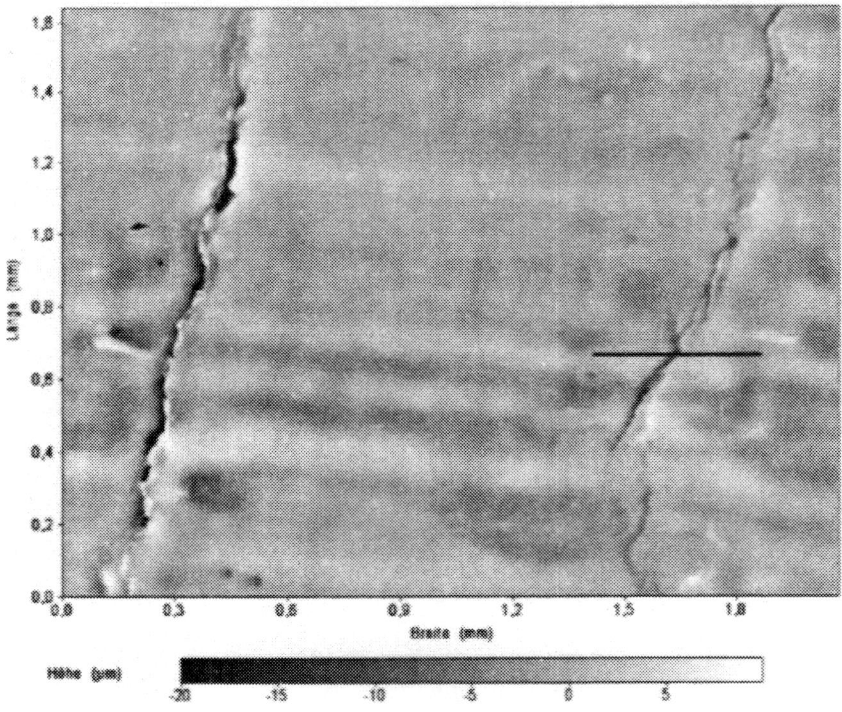

Figure 1a. Before dry cleaning with "wishab" (latex)-sponge.

Figure 1 Continued

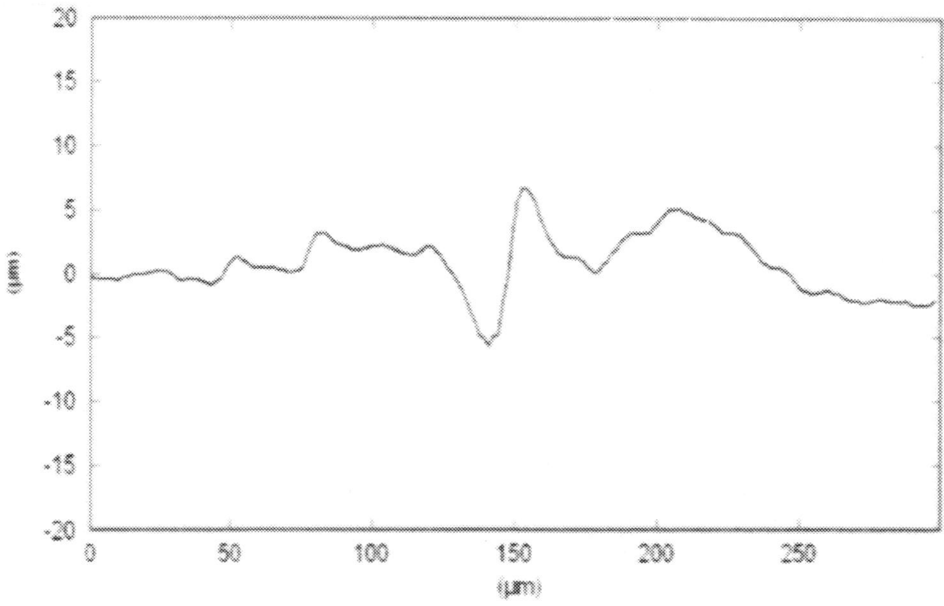

Figure 1b. Measurement (blue) before dry cleaning with "wishab" (latex)-sponge.

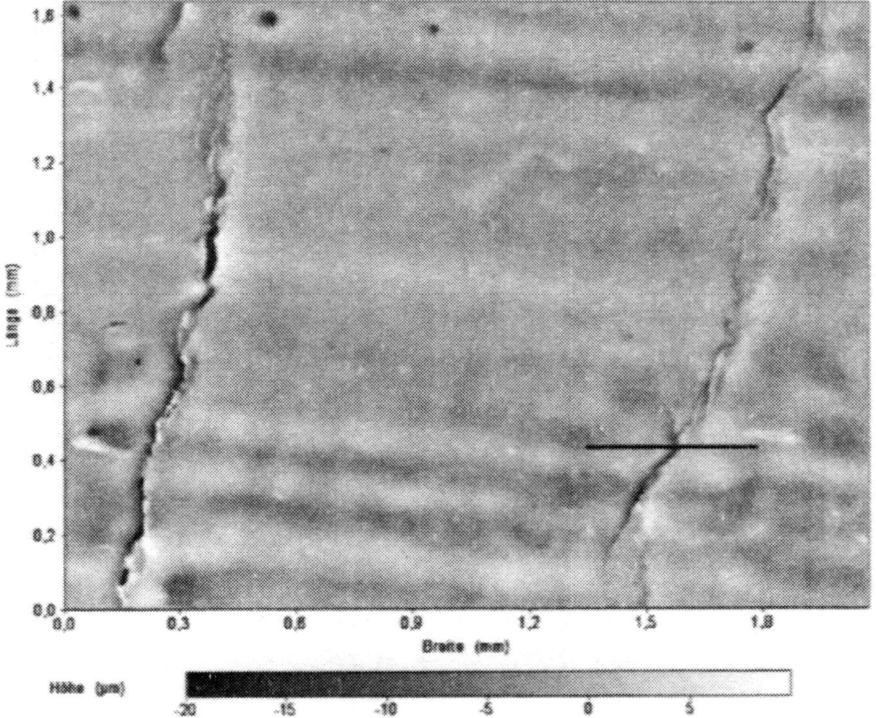

Figure 1c. After dry cleaning with "wishab" (latex)-sponge.

Figure 1 Continued

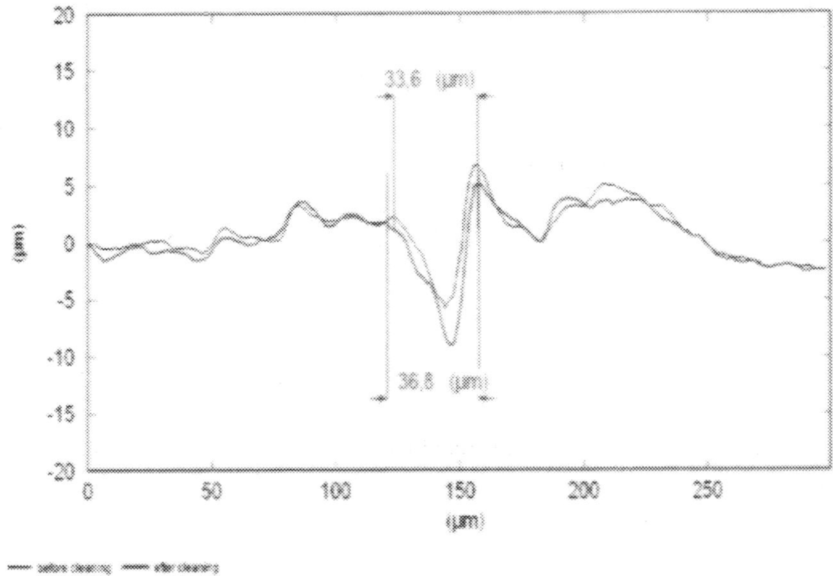

Figure 1d. Measurement (black) after dry cleaning with "wishab" (latex)-sponge.

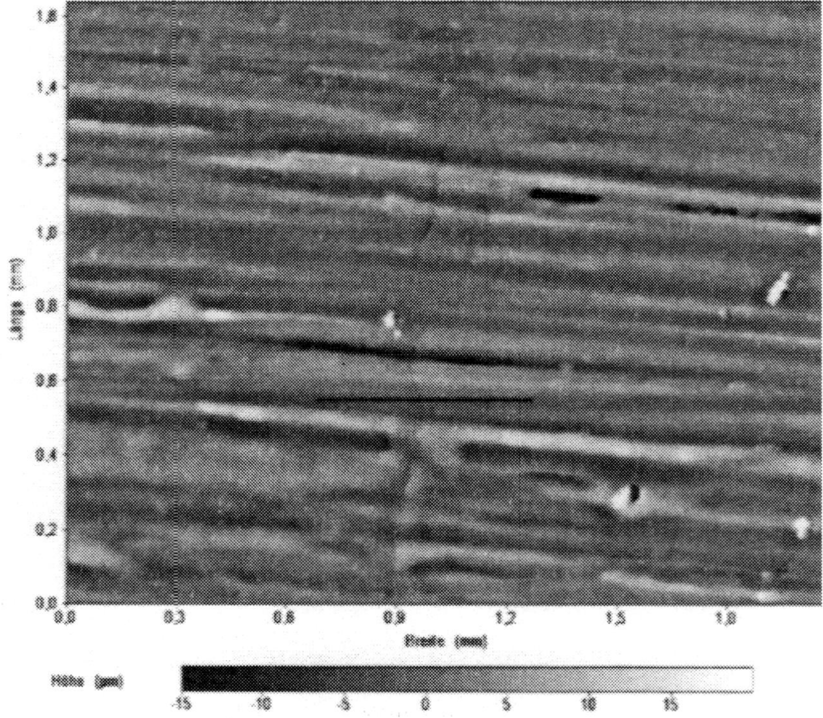

Figure 2a. Before wet cleaning with saliva on cotton swabs.

Figure 2 Continued

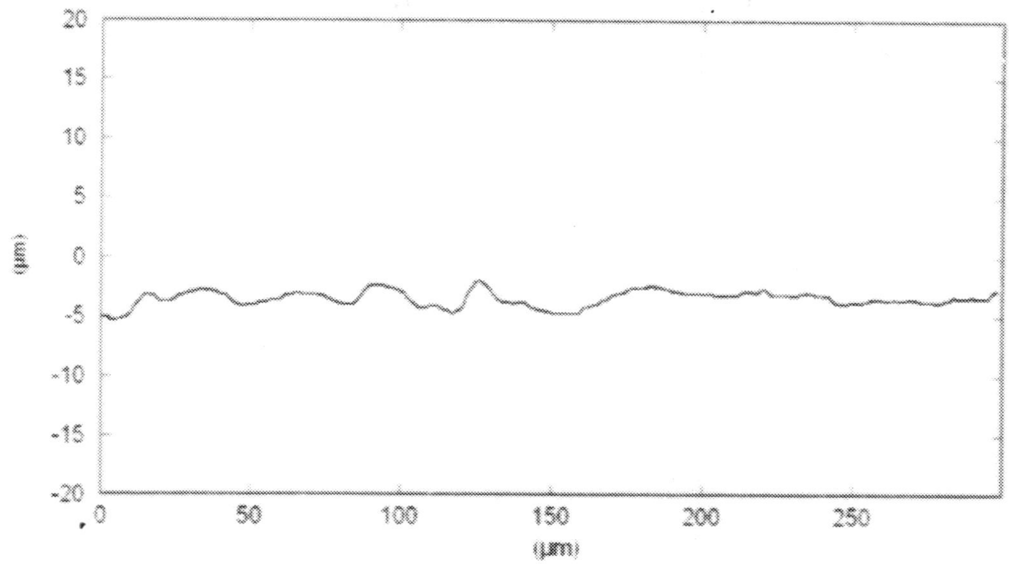

Figure 2b. Measurement (blue) before wet cleaning with saliva on cotton swabs.

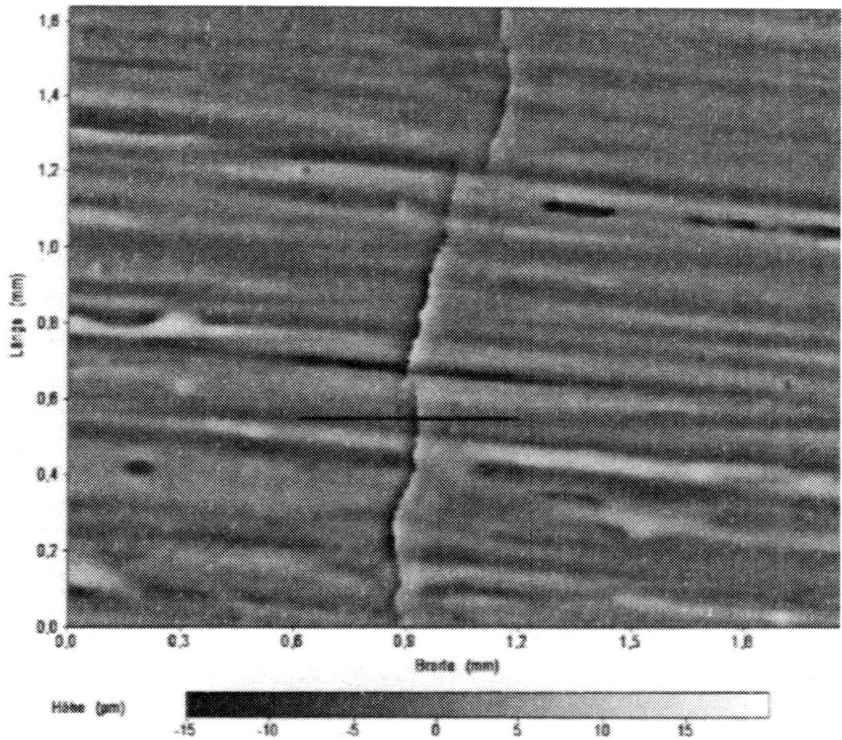

Figure 2c. After wet cleaning with saliva on cotton swabs.

Figure 2 Continued

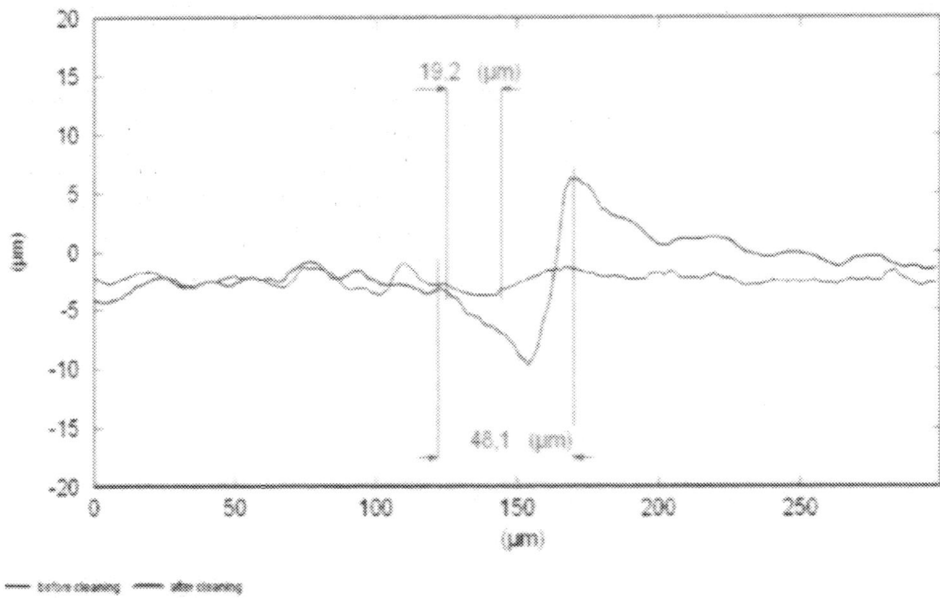

Figure 2d. Measurement (black) after wet cleaning with saliva on cotton swabs.

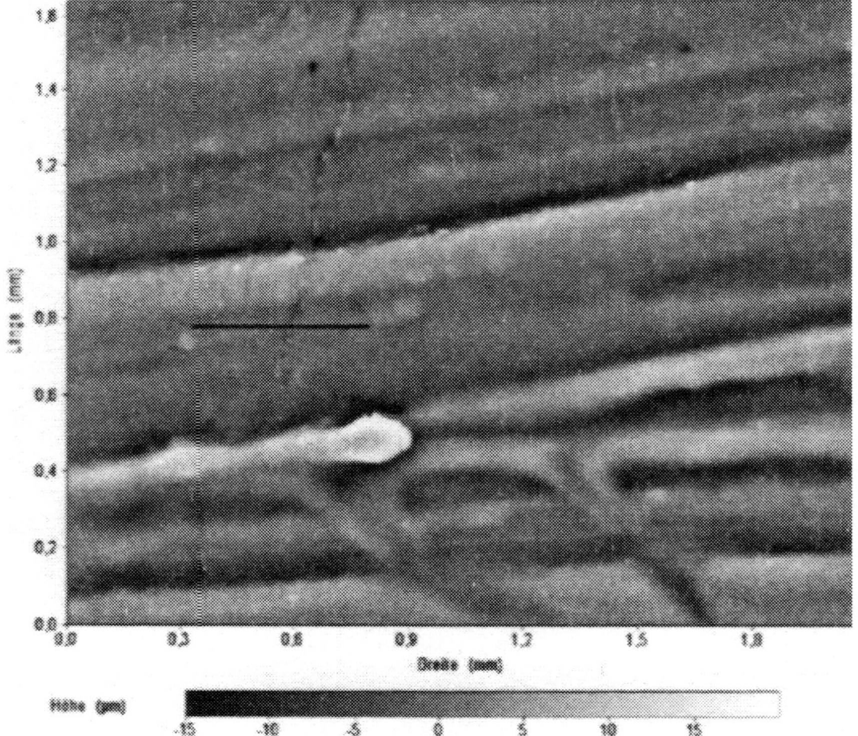

Figure 3a. Before wet cleaning with tab water on microporous sponge.

Figure 3 Continued

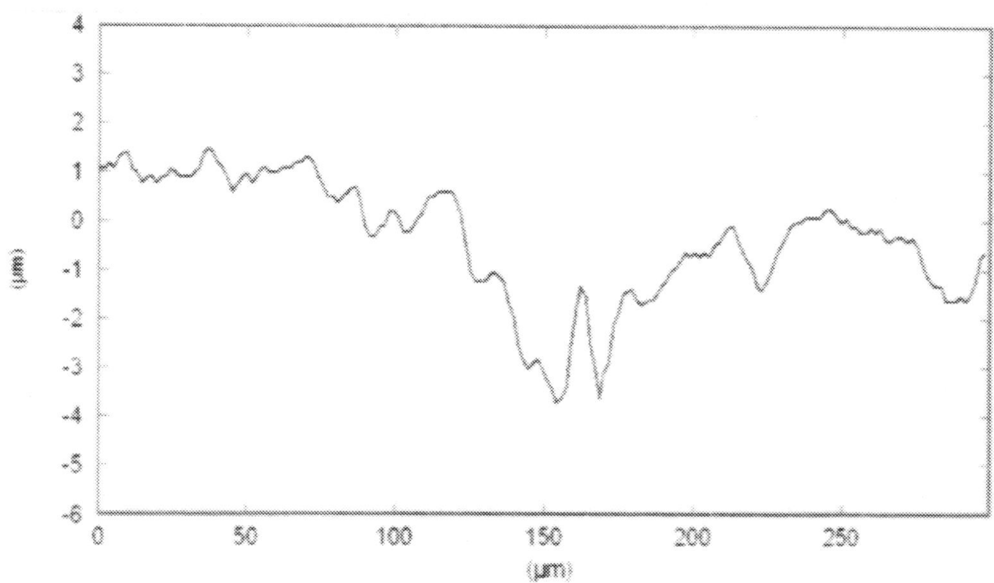

Figure 3b. Measurement (blue) before wet cleaning with tab water on microporous sponge.

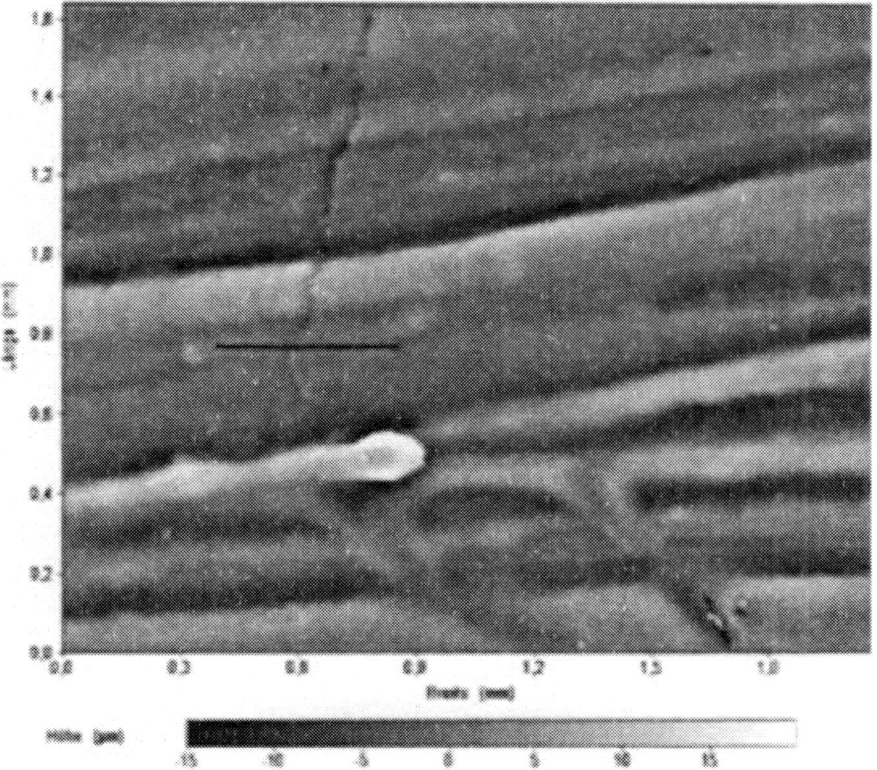

Figure 3c. After wet cleaning with tab water on microporous sponge.

Figure 3 Continued

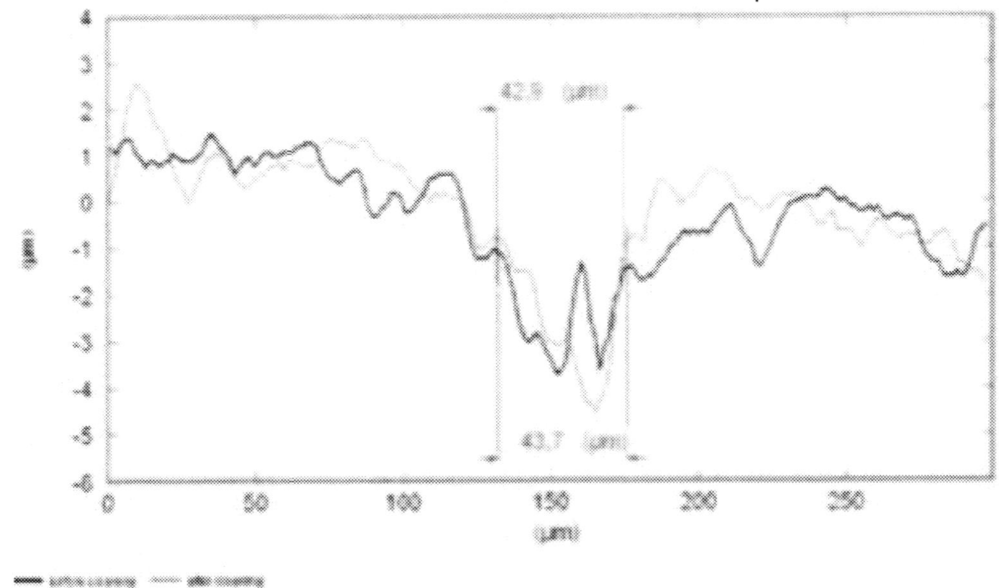

Figure 3d. Measurement (black) after wet cleaning with tab water on microporous sponge.

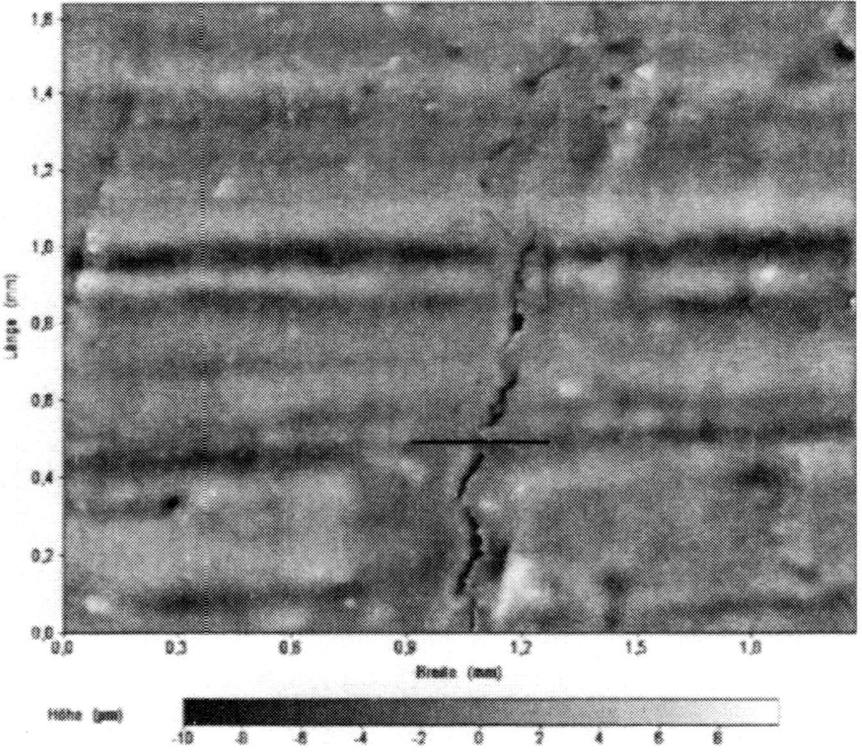

Figure 4a. Before wet cleaning with demineralized water.

Figure 4 Continued

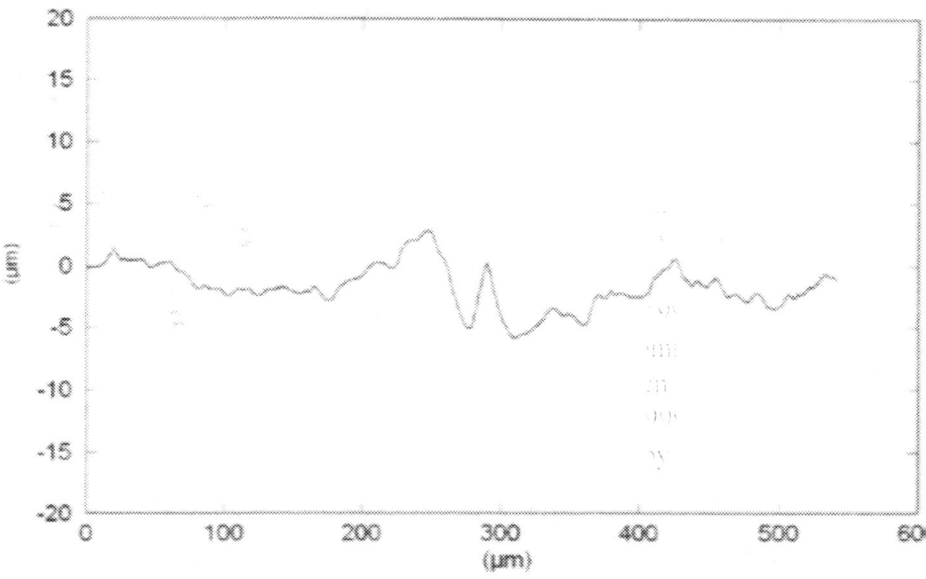

Figure 4b. Measurement (blue) before wet cleaning with demineralized water.

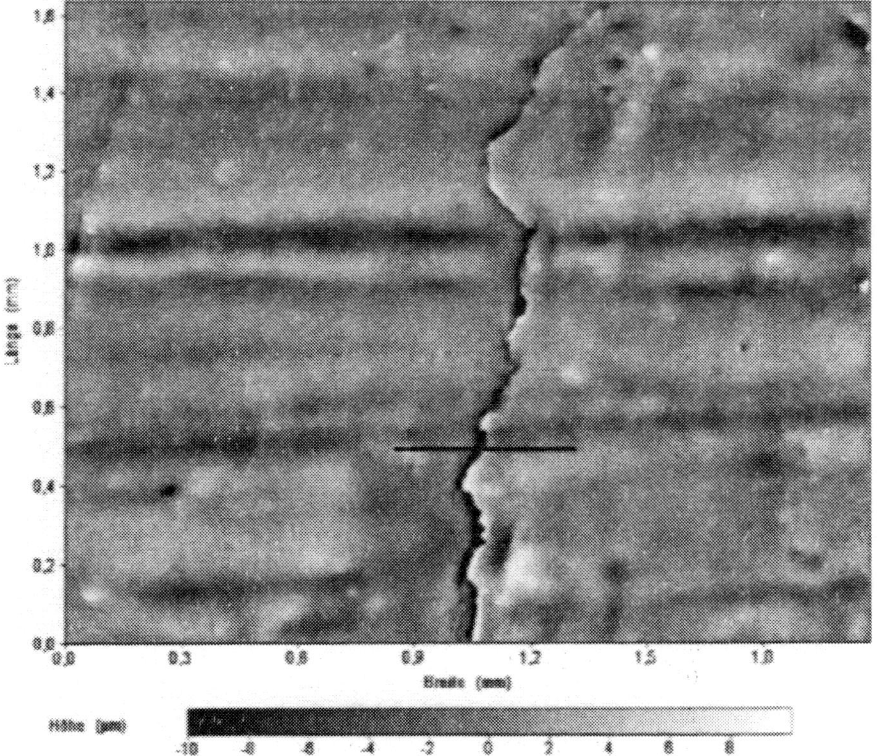

Figure 4c. After cleaning with demineralized water.

Figure 4 Continued

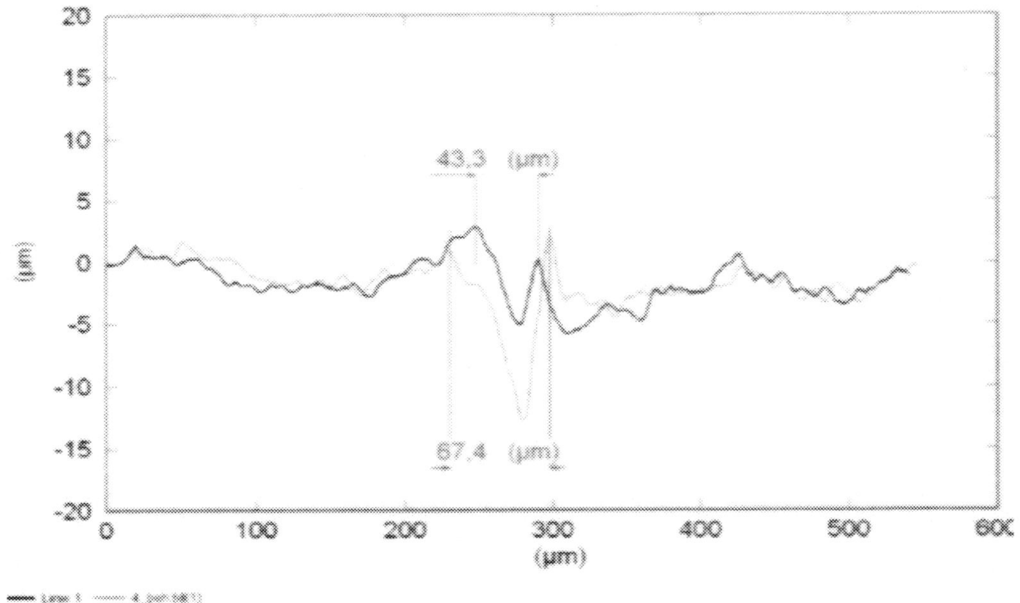

Figure 4d. Measurement (black) after wet cleaning with demineralized water.

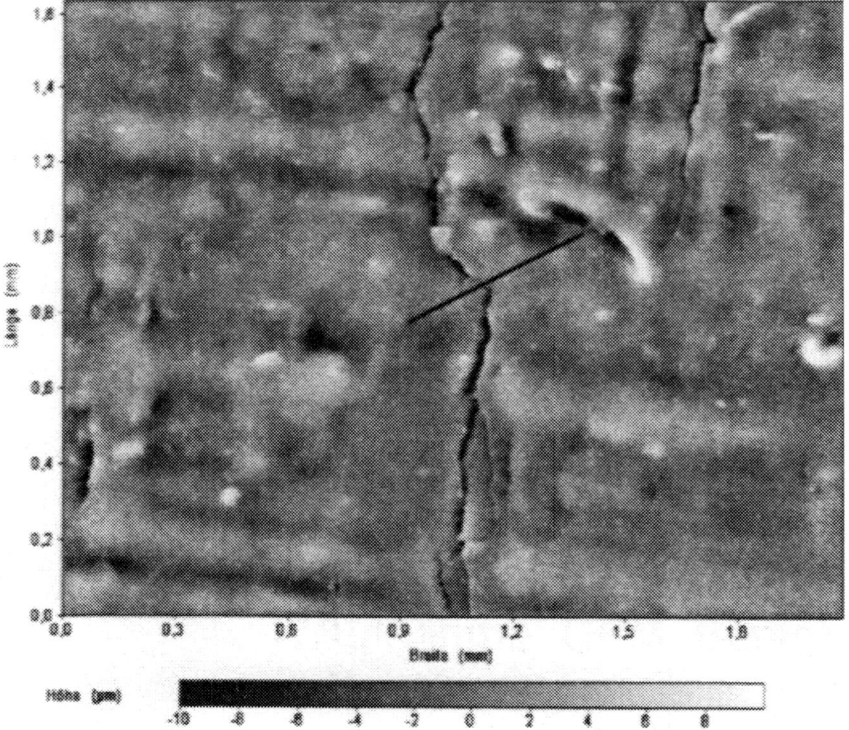

Figure 5a. Before wet cleaning with methyl cellulose.

Figure 5 Continued

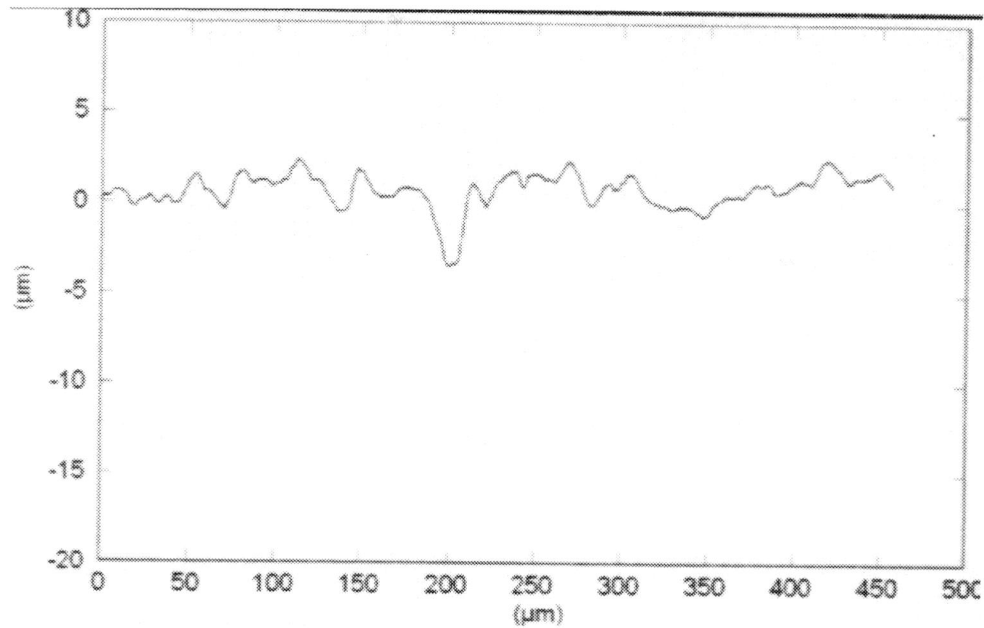

Figure 5b. Measurement (blue) before wet cleaning with methyl cellulose.

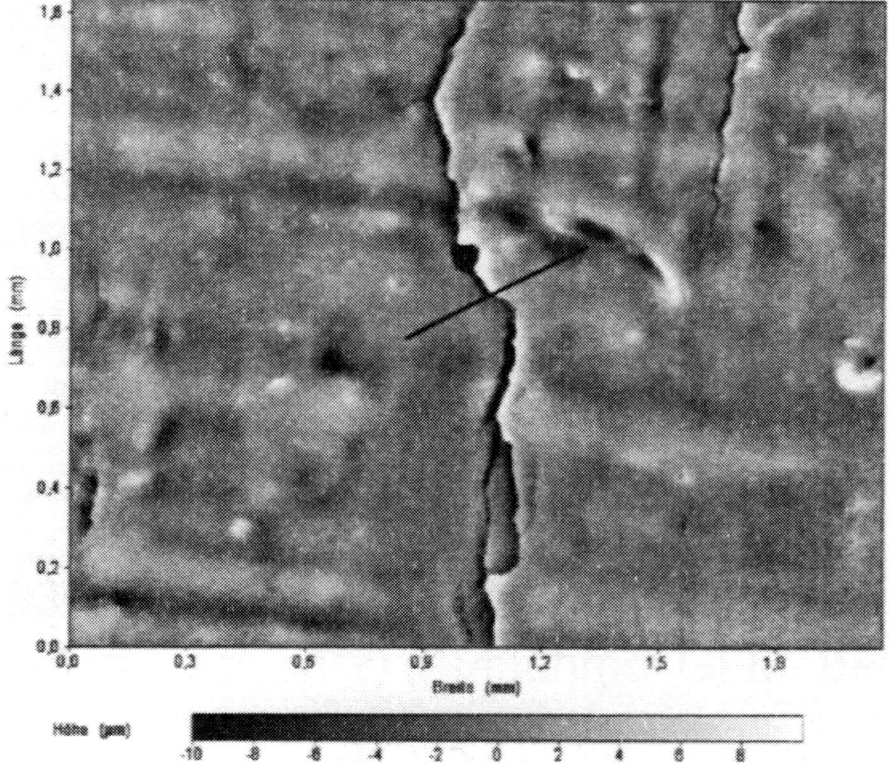

Figure 5c. After wet cleaning with methyl cellulose.

Figure 5 Continued

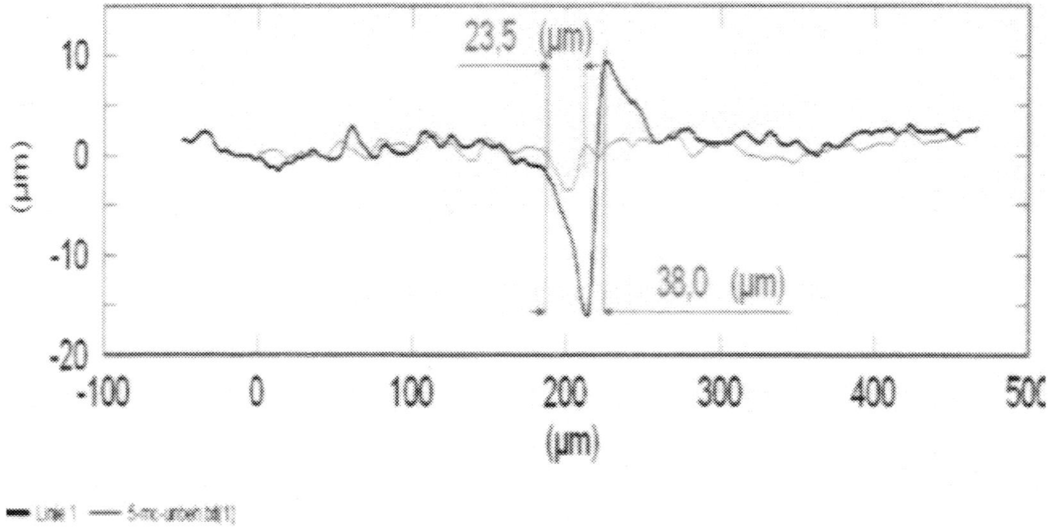

Figure 5d. Measurement (black) after wet cleaning with methyl cellulose.

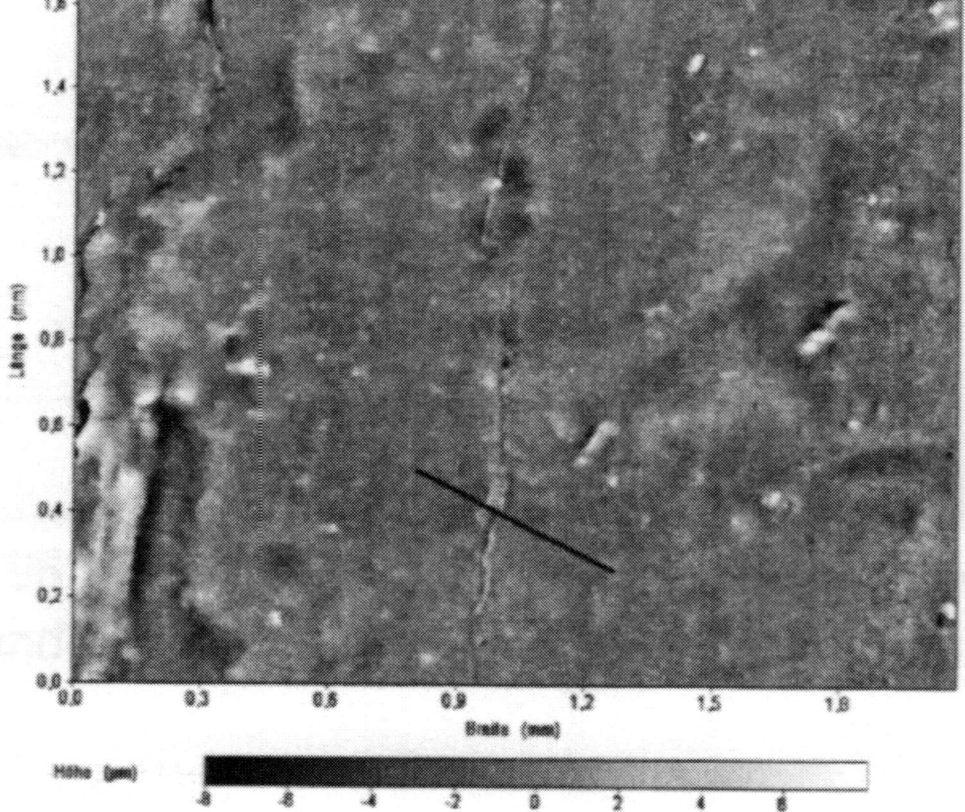

Figure 6a. Before wet cleaning with carboxy methyl cellulose.

Figure 6 Continued

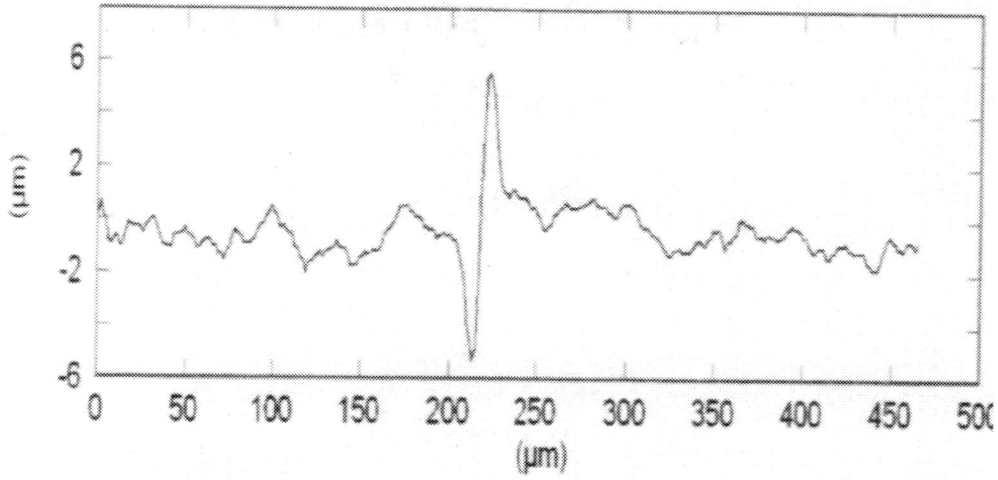

Figure 6b. Measurement (blue) before wet cleaning with carboxy methyl cellulose.

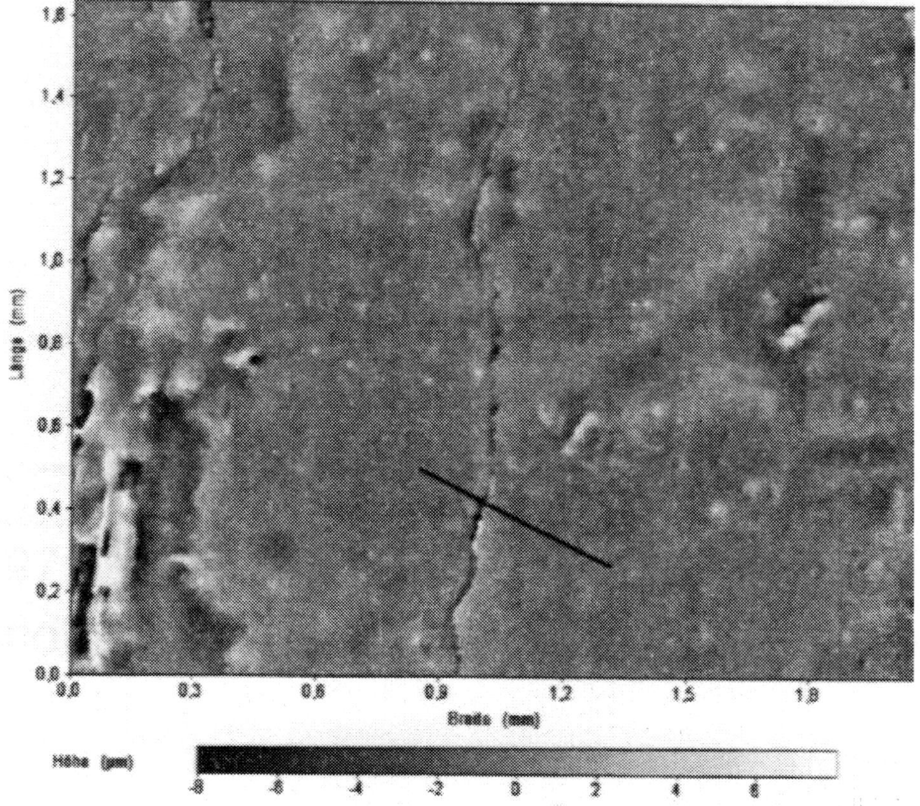

Figure 6c. Before wet cleaning with carboxy methyl cellulose.

Figure 6 Continued

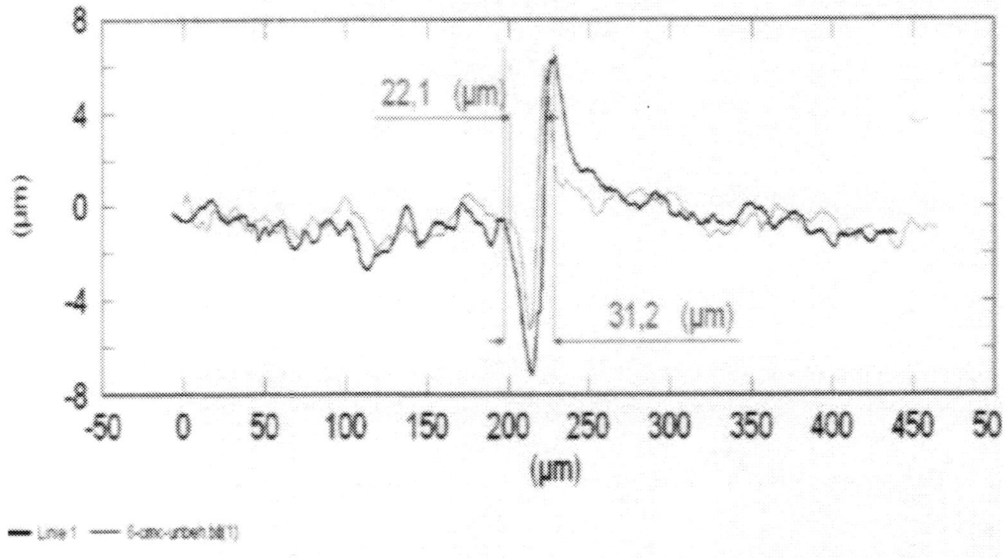

Figure 6d. Measurement (black) after wet cleaning with carboxy methyl cellulose.

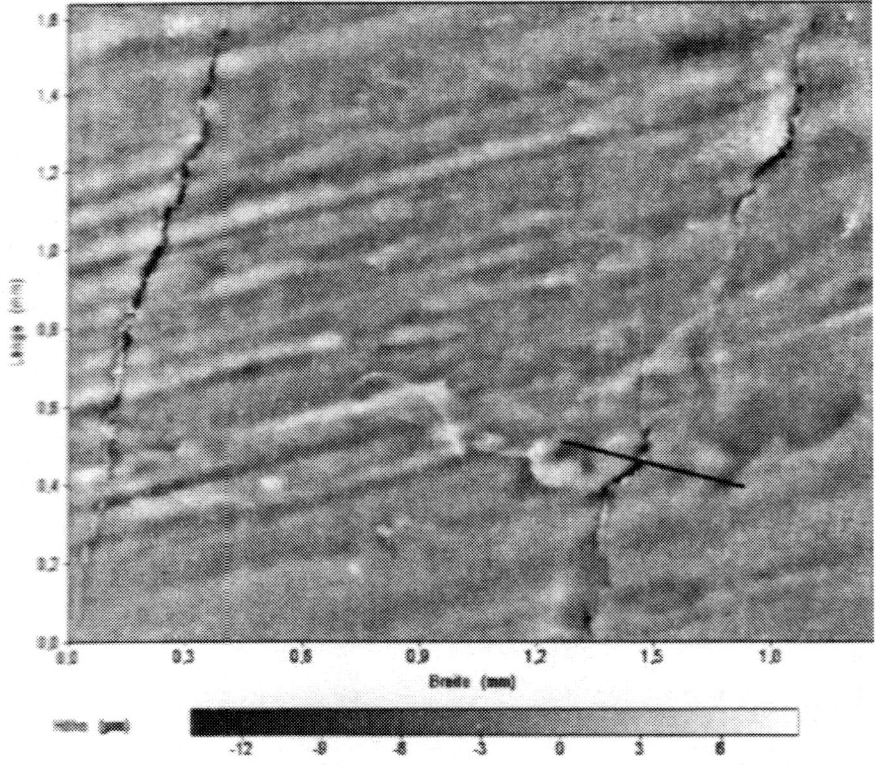

Figure 7a. Before wet cleaning with Marlipal® 1618/25 on microporous sponge.

Figure 7 Continued

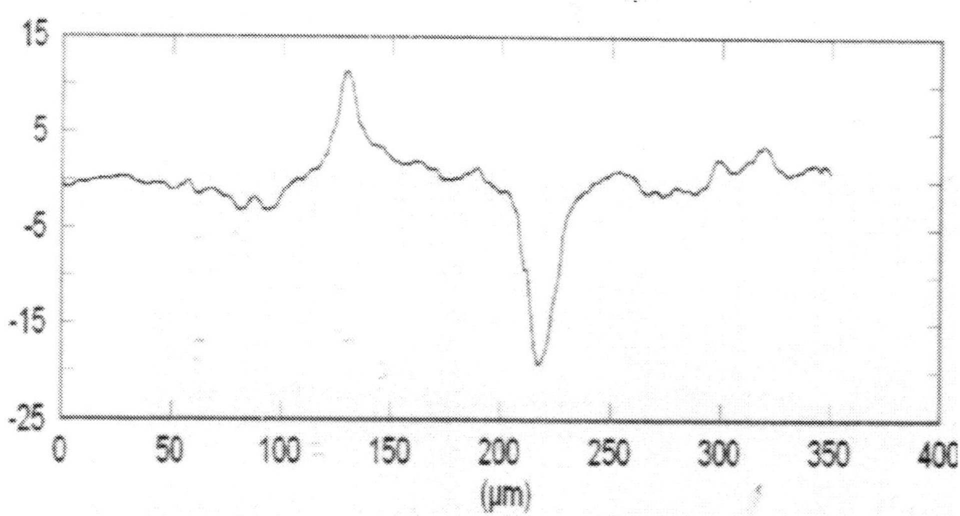

Figure 7b. Measurement (blue) before wet cleaning with Marlipal® 1618/25 on microporous sponge.

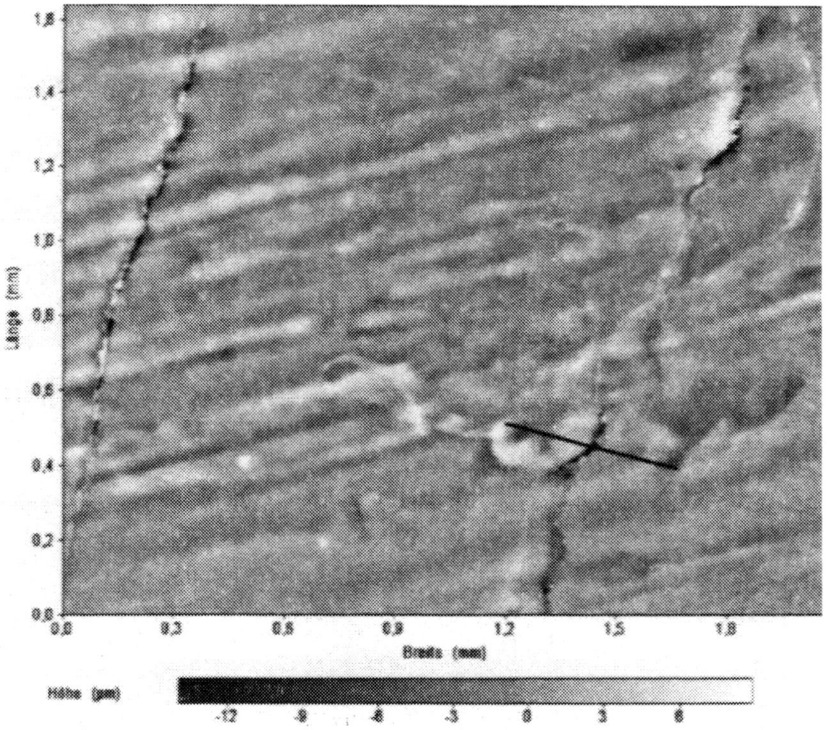

Figure 7c. After wet cleaning with Marlipal® 1618/25 on microporous sponge.

Figure 7 Continued

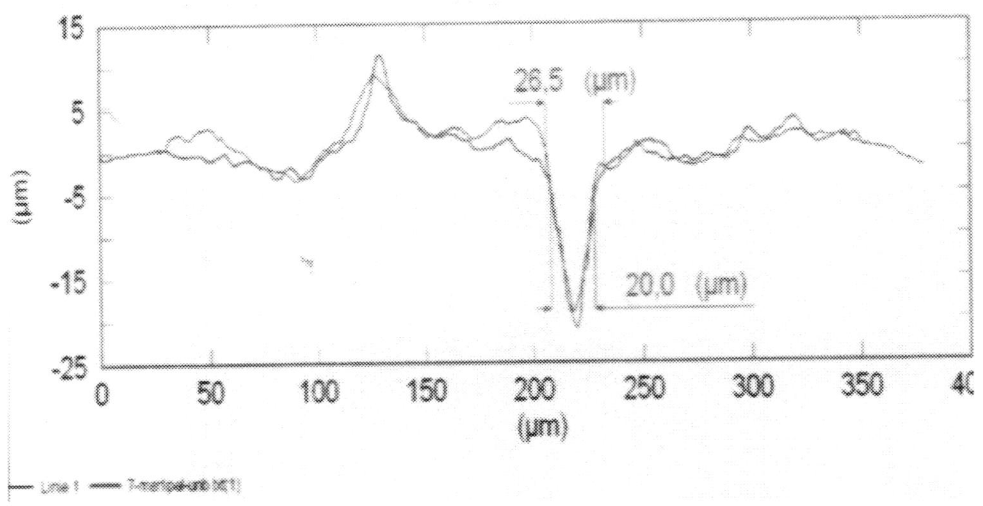

Figure 7d. Measurement (black) after wet cleaning with Marlipal[®] 1618/25 on microporous sponge.

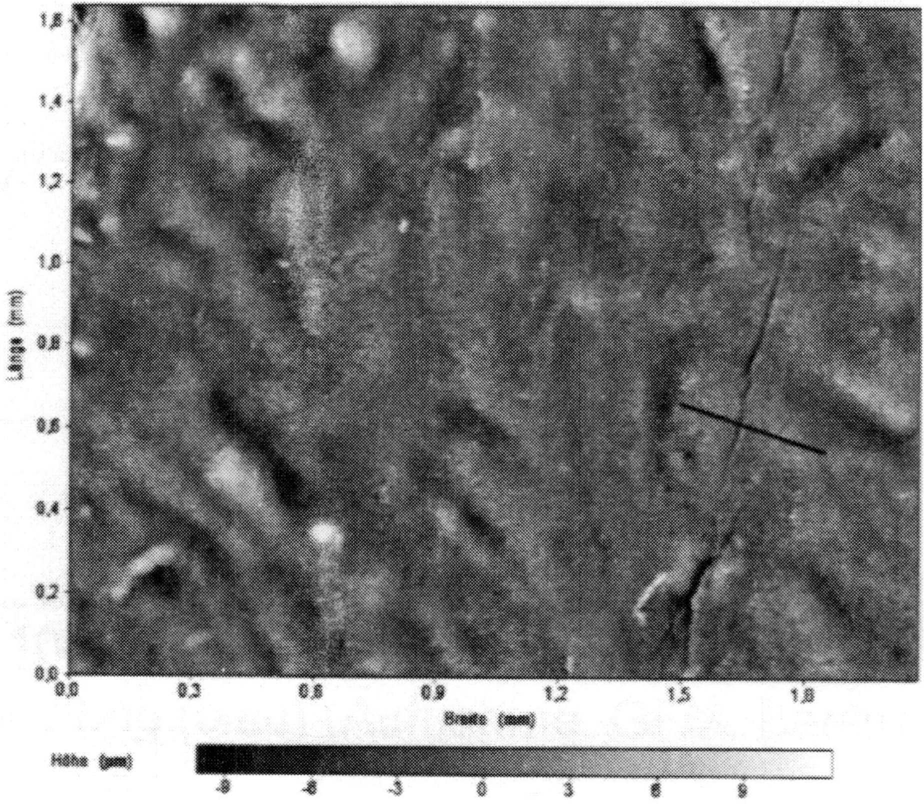

Figure 8a. Before wet cleaning with Marlipal[®] 1618/25 and methyl cellulose on microporous sponge.

Figure 8 Continued

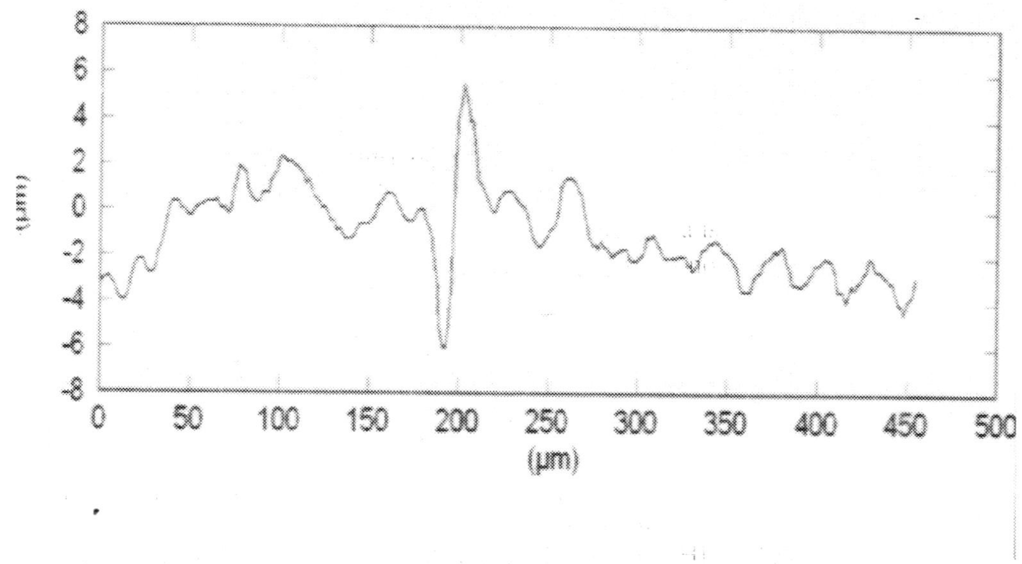

Figure 8b. Measurement (blue) before wet cleaning with Marlipal[®] 1618/25 and methyl cellulose on microporous sponge.

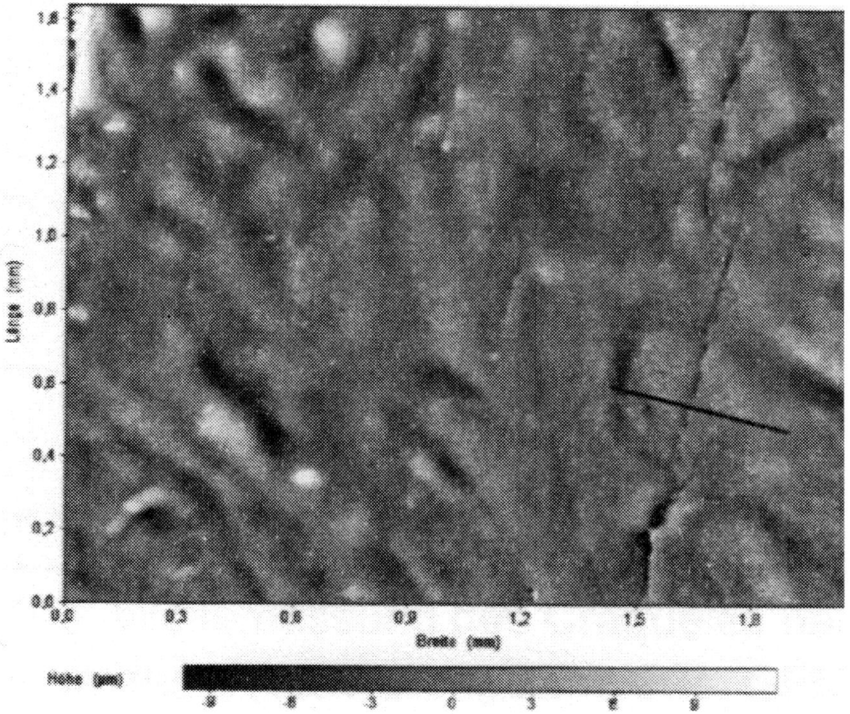

Figure 8c. After wet cleaning with Marlipal[®] 1618/25 and methyl cellulose on microporous sponge.

Figure 8 Continued

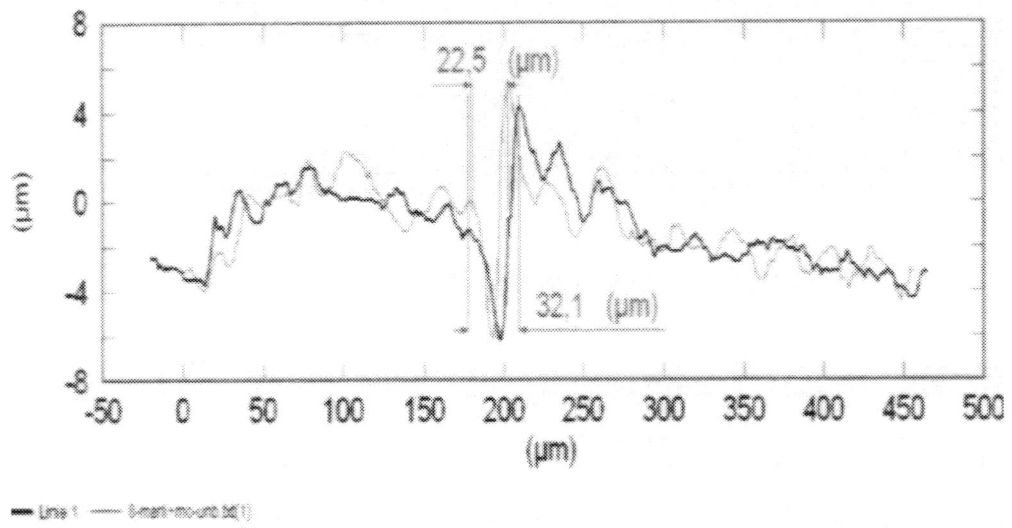

Figure 8d. Measurement (black) before wet cleaning with Marlipal® 1618/25 and methyl cellulose on microporous sponge.

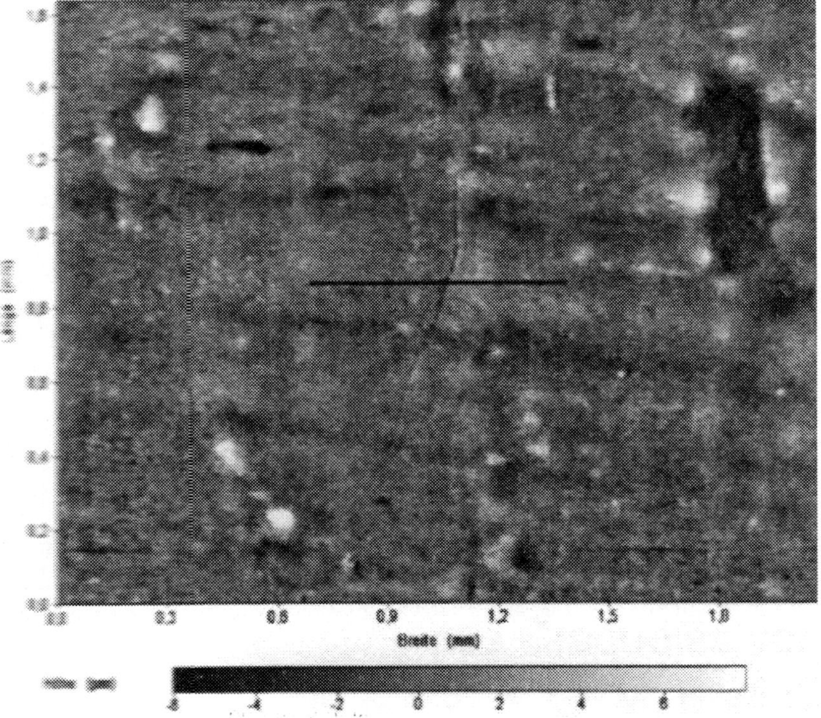

Figure 9a. Before wet cleaning with SDS on microporous sponge.

Figure 9 Continued

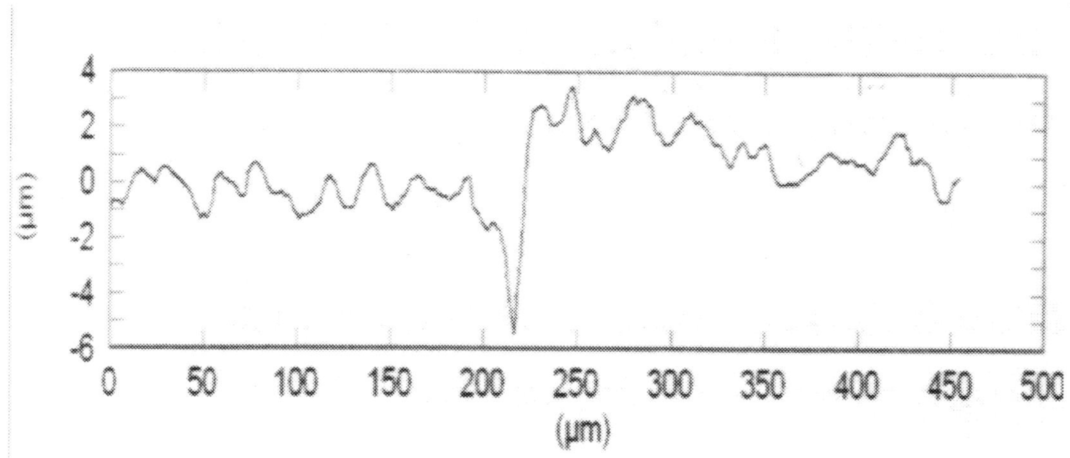

Figure 9b. Measurement (blue) before wet cleaning with SDS on microporous sponge.

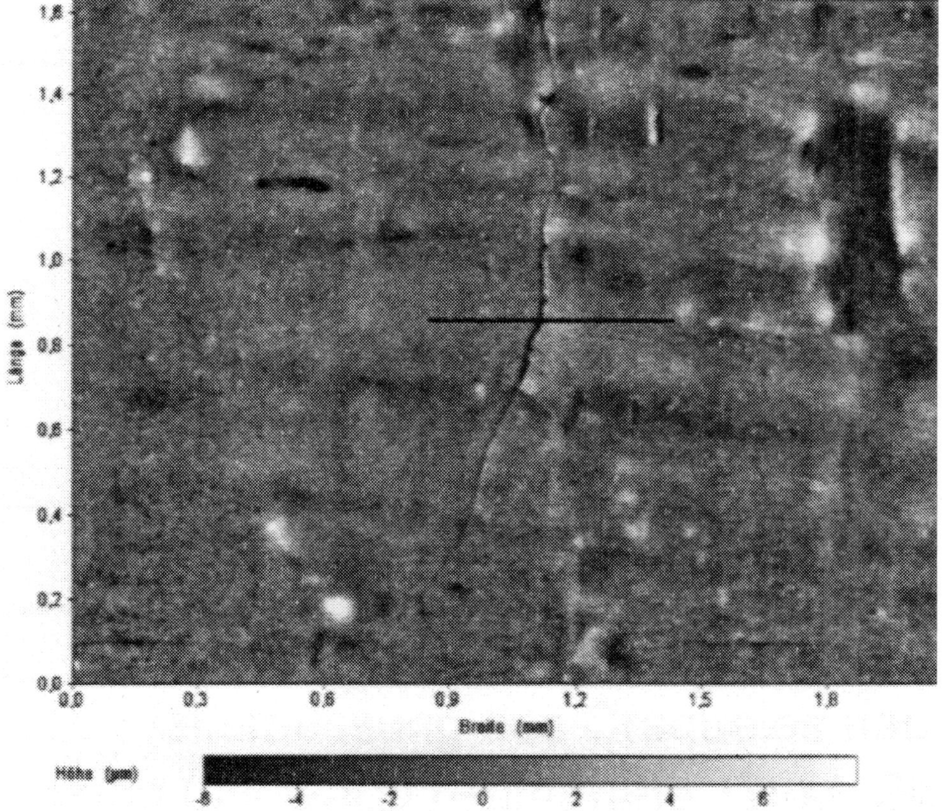

Figure 9c. After wet cleaning with SDS on microporous sponge.

Figure 9 Continued

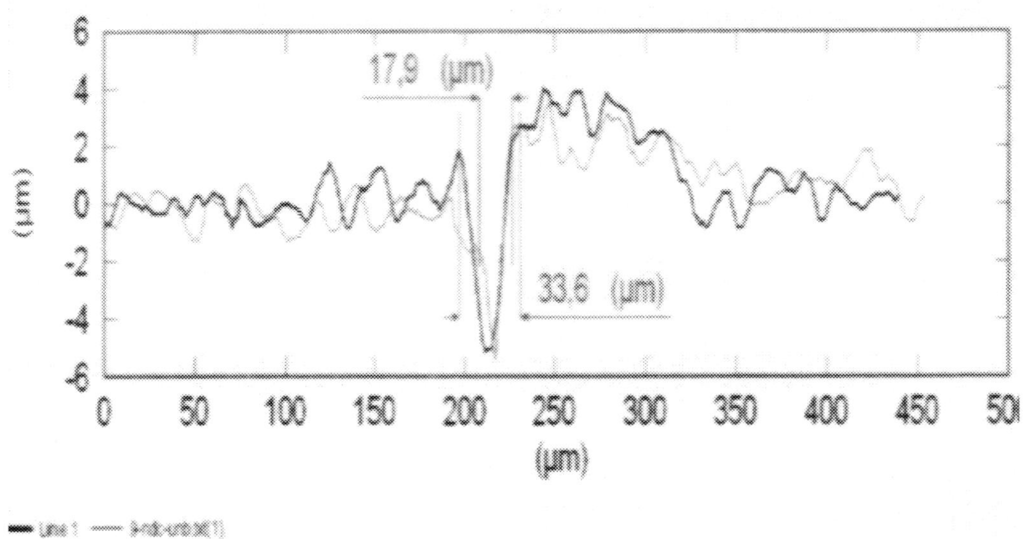

Figure 9d. Measurement (black) before wet cleaning with SDS on microporous sponge.

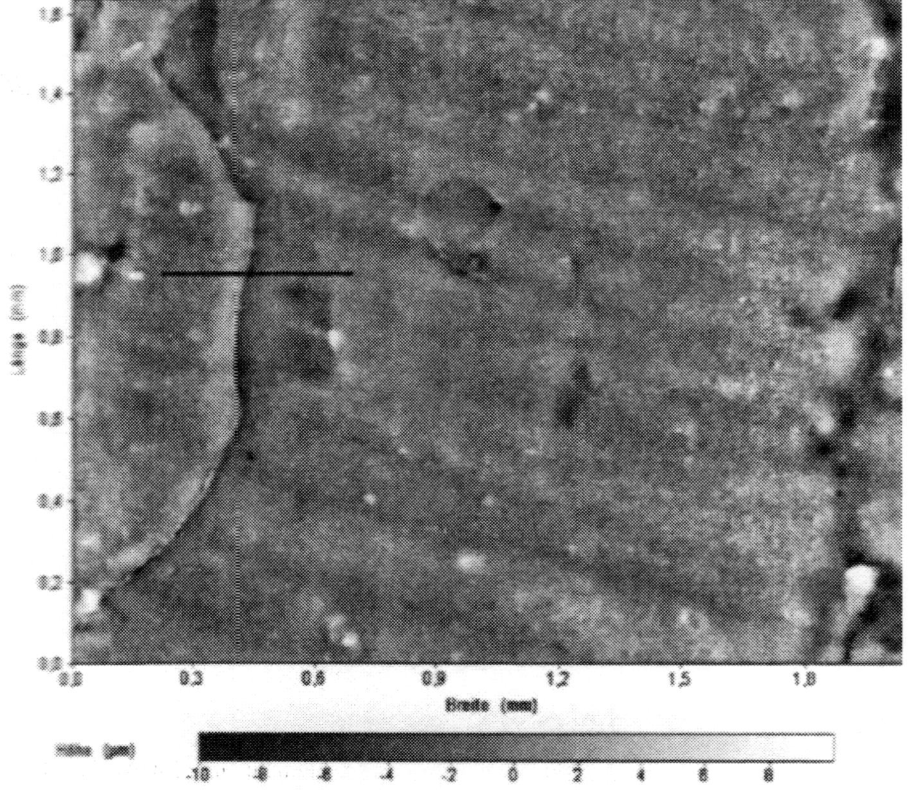

Figure 10a. Before wet cleaning with SDS and methyl cellulose on microporous sponge.

Figure 10 Continued

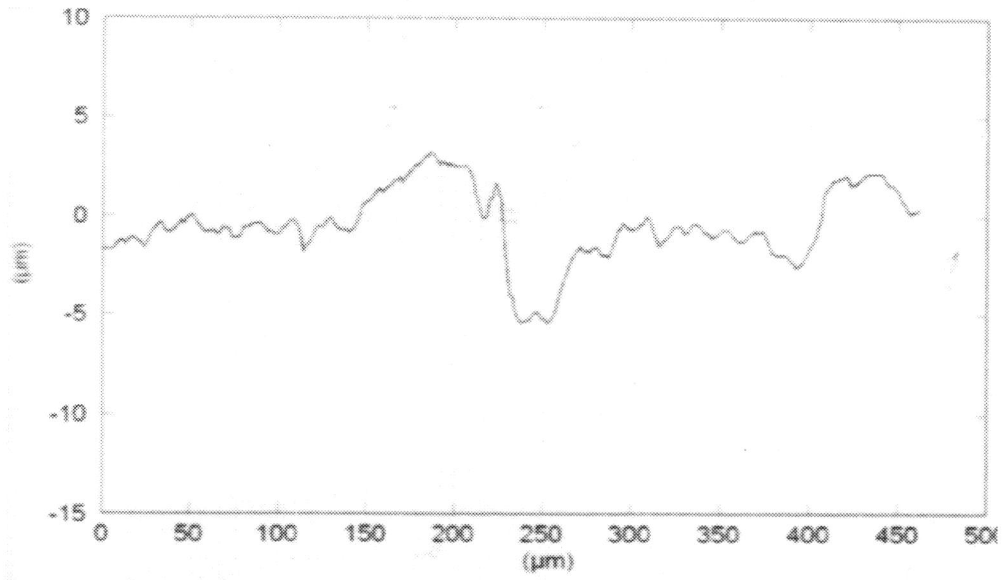

Figure 10b. Measurement (blue) before wet cleaning with SDS and methyl cellulose on microporous sponge.

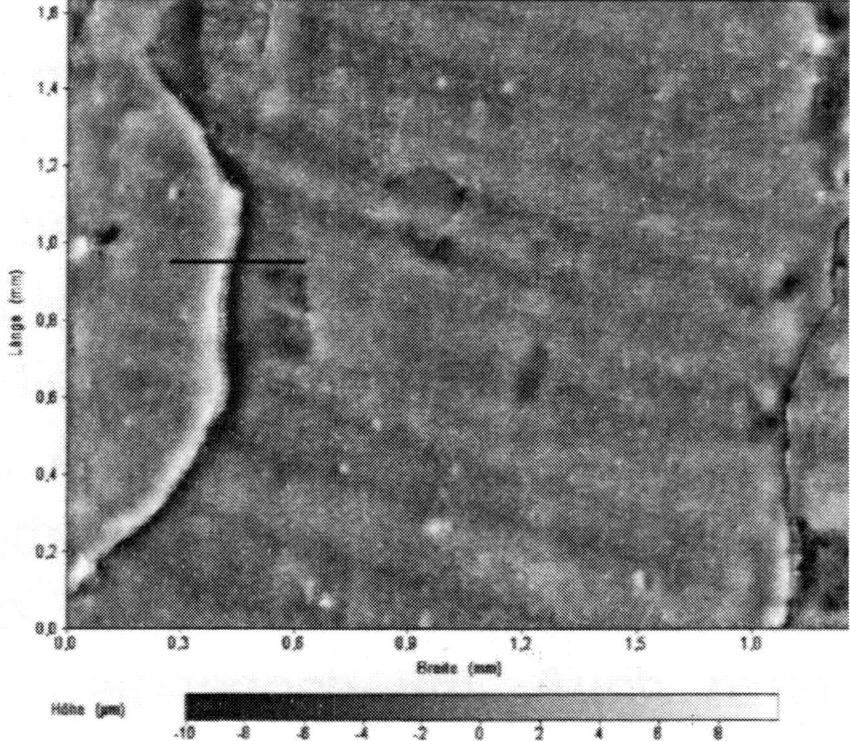

Figure 10c. After wet cleaning with SDS and methyl cellulose on microporous sponge.

Figure 10 Continued

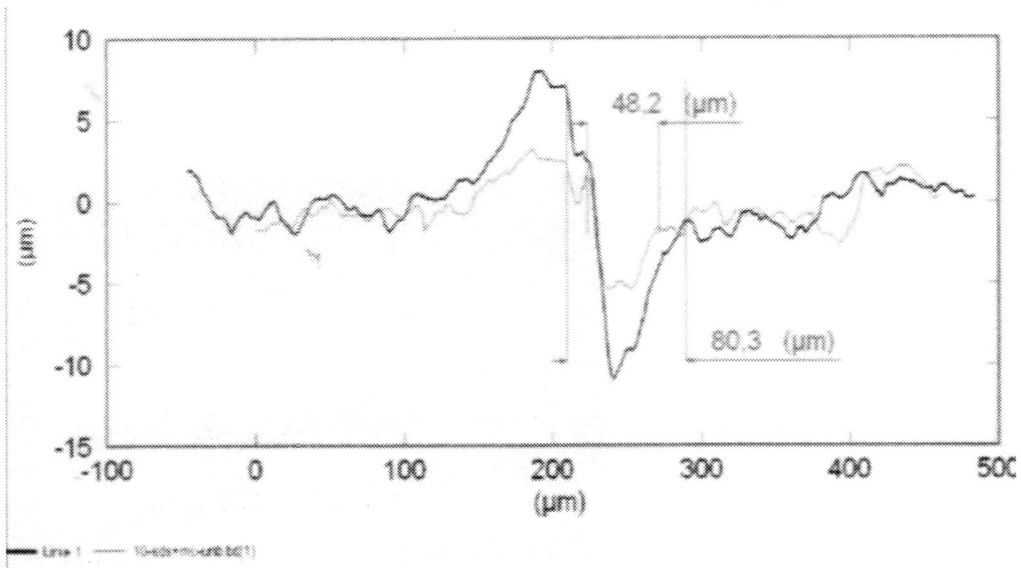

Figure 10d. Measurement (black) after wet cleaning with SDS and methyl cellulose on microporous sponge.

ACKNOWLEDGMENTS

The author would like to thank Dr. Gottfried Frankowski, Dipl.-Ing. (FH) Silvio Zepke (GFMesstechnik, D-Berlin-Teltow), Prof. Dr. Julia Welzel (Klinikum, D-Augsburg), Prof. Dr. Gebhard Reiss (the University of Witten-Herdecke, D), Dipl.-Ing. Gert Opielka (the Max Planck Institute for Metal Research, D-Stuttgart), Dr. Claus-Dierk Hager (chemical firm Sasol, D-Marl) and Hugo Ammann (August Deusser Museum, CH-Bad Zurzach).

ABOUT THE AUTHOR

Dr. Dipl.-Rest. PAUL-BERNHARD EIPPER received his Diploma as a Conservator in 1992 from the University of applied Sciences, Cologne, Germany. From 1990- 2005 he has been Head of Conservation at the August Deusser Museum, Zurzach, Switzerland. From 2000-2009 he has been publicly appointed expert by the Chamber of Commerce, he serves as an expert in questions concerning art restoration under legal consideration in Germany. Since 2006 he has been Head of Conservation at the Universalmuseum Joanneum, Graz, Austria. In 2009 he received his PhD from the Faculty of Medicine at the University of Witten-Herdecke, D). He is the author of numerous books and articles. Author's address: Universalmuseum Joanneum, Weinzöttlstrasse 16, 8045 Graz, Austria, paulbernhardeipper@museum joanneum.at

REFERENCES

[1] Erhardt, D. & Bischoff, J. J. (1992). "Resin soap components", Todd, V. et al.: Preprints to Restoration '92 Conservation. *Training, Materials and Techniques: Latest Developments*, held at the RAI International Exhibition and Congress Centre, Amsterdam, UK Institute of Conservation, London, 77-79.

[2] Ford, B. & Byrne, A. (1991). "The lipid stripping potential of resin soap gels used for cleaning oil paintings", Australian Institute for the *Conservation of Cultural Materials Bulletin, 17*, 51-60.

[3] Gilsa, B. (1991). Gemäldereinigung mit Enzymen, Harzseifen und Emulsionen", *Zeitschrift für Kunsttechnologie und Konservierung, 1*, 48-58.

[4] Hedley, G., Odlyha, M., Burnstock, A., Tillinghast, J. & Husband, C. (1990). "A study of the mechanical and surface properties of oil paint films treated with organic solvents and water", IIC Summaries of the Posters at the Brussels Congress Cleaning, *Retouching and Coatings*, 98-105.

[5] Khandekar, N. (2000). A survey of the conservation literature relating to the development of aqueous gel cleaning on painted and varnished surfaces", *Reviews in Conservation, 1*, 10-20.

[6] Mittal, K. L. & Bothorel, P. (eds), (1986). "Surfactants in Solution: Proceedings of two international symposia on surfactants in solution", *Plenum Publishing Corporation*, 1-562.

[7] Urbaneja, M. A., Arrondo, J. R. L., Alonso, A. & Goñi, F. M. (1986). "The Interaction of Triton X-100 with Multilammellar Phosphaticylcholine Liposomes", *Mittal & Bothorel*, 759-771.

[8] Watherston, M. M. (1976). *"The cleaning of colourfield paintings. The great decade of american abstract modernist art 1960-1970"*, The Museum of Fine Arts, Houston, Texas.

[9] Wolbers, R. C. (1990). *"Notes for workshop on new methods in the cleaning of paintings"*, Getty Conservation Institute, 1-158.

[10] Wolbers, R. C. *"Slides for workshop on new methods in the cleaning of paintings. Laboratory exercises"*, University of Delaware, Newark, DE, USA without year, without pages.

[11] Wolbers, R. C. (2000). *"Cleaning painted surfaces. Aqueous methods"*, Archetype Publications, London, 1-198.

[12] Eipper, P. B. (2003). Oberflächenreinigung mit Tensiden", *Museum aktuell, 93*, 4012-4019.

[13] Ashani, Y. & Catravas, G. N. (1980). "Highly reactive impurities in Triton X-100 and Brij 35: partial characterization and removal", *Anal. Biochem., 109*, 55-62.

[14] Eipper, P. B. (1993). Die Reinigung von Gemäldeoberflächen mit Tensiden. Der Einsatz von modifizierten Polyvinylacetaten zur Konservierung von textilen Bildträgern", *Paul Haupt Verlag, Bern*, 1-134.

[15] Eipper, P. B. (1997). Vier Künstlerfarbenhersteller zwischen 1900 und 1970. Die Reinigung von Gemäldeoberflächen mit wässrigen Systemen", *Paul Haupt Verlag, Bern*, 1-115.

[16] Mansmann, K. (1998). Oberflächenreinigung mit Ammoniumcitraten", *Zeitschrift für*

Kunsttechnologie und Konservierung, 2, 220-237.

[17] Hackney, S., Townsend, J. & Eastaugh, N. (Ed), (1990). *"Dirt and Pictures Separated"*, The U.K. Institute for Conservation, London.

[18] Eipper, P. B. & Reiss, G. (1995). Vergleichende Versuche zur Reinigung von Gemäldeoberflächen mit Tensiden", *Mitteilungsblatt 49, Museumsverband Niedersachsen und Bremen* e.V., Hannover, 51-60.

[19] Eipper, P. B. & Reiss, G. (1996). "Comparison of cleaning agents applied to surfaces of paintings", *The Picture Restorer, 10*, 5-10.

[20] Eipper, P. B. & Welzel, J. (1997). Die Überprüfung von mit Tensiden gereinigten Ölfarboberflächen durch die computergestützte Laserprofilometrie", *Mitteilungsblatt 54* Museumsverband Niedersachsen und Bremen e.V., *Hannover*, 69-80.

[21] Eipper, P. B. & Welzel, J. (1999)., "Studies Towards an Examination of Surfactant-cleaned Oil Paint Surfaces by Computer-aided Laser Profilometry", *The Picture Restorer, 15*, 5-13.

[22] Eipper, P. B., Frankowski, G., Opielka, H. & Welzel, J. (2004). "Ölfarbenoberflächenreinigung. Die Überprüfung gereinigter Ölfarbenoberflächen durch das Raster-Elektronen-Mikroskop, das Niederdruck-Raster-Elektronen-Mikroskop, die Laser-Profilometrie und die 3D-Messung im Streifenprojektionsverfahren", Verlag Dr. Christian Müller-Straten, *München*, 1-250.

[23] Eipper, P. B. (1995). Malmittel und Maltechnik August Deussers. Zwei Beiträge zu grundlegenden Fragen der Restaurierung Düsseldorfer Malerei der Jahrhundertwende*", Wienand Verlag, Köln, 1*-24.

[24] Eipper, P. B. (2003). Trocken-Schwämme zur Oberflächenreinigung", *Museum aktuell, 91*, 3898-3902.

[25] Hedley, G., Odlyha, M., Burnstock, A., Tillinghast, J. & Husband, C. (1990)., "A study of the mechanical and surface properties of oil paint films treated with organic solvents and water", op. cit., 98-105; Michalski, S., "A physical model of varnish removal from oil paint", op. cit., 85-92; Phenix, A. & Burnstock, A., "The deposition of dirt: a review of the literature, with scanning electron microscope studies of dirt on selected paintings", *"Dirt und Pictures Separated"*, UKIC, London, 11-18.

[26] Romao, P. M. S., Alarcao, A. M. & Viana, C. A. N. (1990). '"Human saliva as a cleaning agent for dirty substances", *Studies in Conservation, 35*, 153-155.

[27] Lehmann, R. R. (1992). "Ökologie der Mundhöhle", *Dental Magazin, 1*, 102-106.

[28] Ramsay-O´ Hoski, B. (1976). "An investigation into the composition and properties of saliva in relation to surface-cleaning of oil paintings", National Gallery of Canada, Ottawa.

[29] Girling, J. (1992). "The use of human saliva in conservation", MA Thesis. *Institute of Archaeology.*, University College, London.

[30] Nielsen, T. B. & Reynolds, J. A. (1978). Measurements of molecular weights by gel electrophoresis", *Methods in Enzymology, 48*, 4-12.

[31] Stegemann, H. (1979). "SDS-gel-electrophoresisinpoly-acrylamide, merits and limits", Righetti, Ossand, Vanderhoff (eds.) *Electrokinetic Separation Methods,* 313-336.

[32] Stegemann, H. & Pietsch, G. (1983). "Methods for Characterization of Seed Proteins of Cereals and Legumes", *Seed Proteins, Biochemistry, Genetics, Nutritive Value,* 45-75.

[33] Saur, R., Schramm, U., Steinhoff, R. & Wolff, H. H. (1991). Strukturanalyse der Hautoberfläche durch computergestützte Laser-Profilometrie", *Der Hautarzt, 42*, 499-

506.

[34] Rohr, M. & Schrader, K. (1994). Surfactant-induced Skin Roughness. Quantitative Analysis of the Structure of the Skin Via Automated Non-touch Laser Scanning", *Euro Cosmetics*, 24-28.

[35] Rohr, M. & Schrader, K. (1995). Hautphysiologische In-vivo-Untersuchungen verschiedener Tenside" *Skin Care Forum, 12*, 12-16.

[36] Efsen, J., Handsen, H. N., Christiansen, S. & Keiding, J. (1995). "Laser profilometry", Serup, J. and Jemec, G. B. E. (eds.), *Handbook of Non-Invasive Methods* und the Skin CRC Press Boca Raton, Ann Arbor, London, Tokyo 97-105.

[37] Müller, U. (1995). "Roughness Measured by Profilometry: Mechanical, Optical and Laser", In: E., Berardesca, P., Elsner, K. P. Wilhelm, & H. I. Maibach, (eds), *Bioengineering of the Skin: Methods and Instrumentation,* CRC Press Boca Raton, Ann Arbor, London, Tokyo 41-51.

[38] Saur, R. (1993). Computergestützte Laser-Profilometrie Entwicklung eines neuen dermatologischen Verfahrens zur Objektivierung der Rauheitsstruktur der Haut", medical dissertation, Lübeck.

In: Paints: Types, Components and Applications
Editor: Stephanie M. Sarrica

ISBN: 978-1-61761-813-0
© 2011 Nova Science Publishers, Inc.

Chapter 3

PART 2: EXAMINATION OF UNTREATED AND TREATED ACRYLIC PAINT SURFACES BY 3D-MEASUREMENT TECHNOLOGY AT THE UNIVERSAL-MUSEUM JOANNEUM, GRAZ, AUSTRIA

Paul-Bernhard Eipper

Alte Galerie am Landesmuseum Joanneum,
Raubergasse, Graz, Austria

ABSTRACT

Due to the complex mixtures of acrylic paints it is obviously that grime and dirt are greater hazards to acrylic-paint then to others paint-surfaces. To remove these impurities, paintings are usually cleaned dry or wet with surfactants in aqueous medium. Modern acrylic paints produced by Schoenfeld Lukas and Hermann Schmincke, Düsseldorf were studied. To examine the effects of different cleaning methods, numerous acrylic paint surfaces were treated dry with a latex sponge and wet with (individually) water; demineralised water; magnetised water; saliva; methyl cellulose; carboxymethyl cellulose; Marlipal®1618/25; and SDS. The surfaces of the treated paints were examined by 3D-measuring technology based on micro mirrors. This technology provides measurements in seconds during the cleaning process and produces measurable images which can show changes on the surface.

Impurities on acrylic paintings cannot be easily removed. If they are removed dry, or wet with surfactants in aqueous media, it is very difficult to show the changes of the treated surface, even if abrasions of the surface have taken place. Particles of dirt and grime were absorbed in the paint film. Wet cleaning methods cause swelling of the paint surface. After drying the not visible changed surface contains absorbed particles and appeares quite similar to its previous state.

Due to the properties of the examined paints it can be concluded that 3D-exminations of oilpaints show more differences in the surfaces and are therefore more useful than 3-examinations of acrylic paints.

In some cases dry methods are not sufficient to clean surfaces completely. Therefore modifications of aqueous cleaning methods (such as: using mild non-ionic surfactants, thickening of the solutions used, reduction of contact humidity, and increasing temperature and pH) are possible but should be avoided.

DAMAGES ON ACRYLIC PAINT SURFACES

Meanwhile about 75 % of all new paintings are painted with acrylic paints. The necessity of developing adequate cleaning methods of these paint surfaces is given.

Taking in account that acrylic paint surfaces are very difficult to clean, these paints should better not be used for paintings: if these surfaces are covered with a varnish made of Cyklohexanon resin (e.g. Lukas Cyklohexanonharz (Art.nr. 2319) or acryl resin (e.g. Schmincke acrylic resin (Art.nr. 50585), or mixtures of these resins solved in white spirit it is easier to clean them. However these coatings change the apperance of the original surfaces.

If cleaning of varnished surfaces is needed this would not be without complications for such surfaces. If unvarnished acrylic surfaces have to be cleaned, the surface will undoubtedly be damaged [1-3]

The tendency of dirt accumulation at normal room temperatures (20° C) due to their porouse structure is acrylic paint surfaces immanent. The soft and sticky surface is caused by the used resins and their additives. The surface is electrical chargeable, even through streaming air as well as dry cleaning procedures. By temperatures over 20° C dirt can be "swallowed" by the surface. Temperatures beyond 15° C provokes craquelé.

Waterdrops cause water borders which can normally not be reduced. Fingerprints and saliva drops cause complications by restoration of such surfaces. Sometimes artists "restore" these damages by overpainting them (!).

Normally, restorers treat acryl paints the same way as oil paint surfaces. This is not useful, due to the high complexity of acrylic paints.

ACRYLIC EMULSIONS PAINTS

In 1843 acrylic acid was produced synthetic for the first time. The first solvent based acryl resin appears first in 1928 in Germany[1, 2]. Twenty years later some artists mixed colours by their own made of acrylic resin and dry pigments.

The first produced acrylic resin artist paints containing solvents appeared in 1947 named "Magna Plastic Artists Colours" by Leonhard Bocour Artist Colours, Inc., New York. These paints react not so senstitive to dry and wet cleaning than waterbased acrylic paints. However, the waterbased paints have displaced the solvent based acrylic paints.

The first waterbased arylic resin paints called "Liqiuitex" were available in 1955 being manufactured by Henry Levison, Permanent Pigments, Inc., Ohio. In Europe 1962 the "Cryla Artists' Acrylic Colour" manufactured by Rowney & Co., London, were the first waterbased acrylic paints. They were followed by "Lukascryl" manufactured by Lukas Künstlerfarbenfabrik Schoenfeld, "Primacryl" manufactured by Schmincke & Co, beide in Düsseldorf, and "Lascaux Acryl Künstlerfarben" manufactured by Alois K. Diethelm AG,

Brüttisellen, Switzerland and "Rembrandt Akrylfarbe" by Talens, Netherlands. 1970 appeares "Winsor & Newton Artists' Acrylic Colour" in London[1, 2].

In waterbased acrylic emulsion paints which are handled here, we found beneath alcalic binding media and pigments a huge range of materials, which do not disappear during drying of the paint. These materials have some influence to the dry paint film.

Waterbased acrylic paint can be easily detected from solventbased acrylic paints. The waterbased paints have little craters in the surface which were caused by the drying process. Solvent based paints appear more compact and quite similar to oil paints.

Ingredients of Acryl Emulsion Paints

The waterbased acrylic paints contained as binding media a copolymerisate of soft ethylacrylate and brittle methylmethacrylat. In addition we found initiatores (persulfate), polymerisation supporters (e.g. dodecylmercaptan), buffers (acids, alcalics (e.g. ammonium), hydrophilc surfactants (emulgators, nonionic e.g. alkylphenolethoxylate; anionic e.g. sodiumlaurylsulfate [SLS], dodecylbenzensulfonate); Schutzkolloide (polyvinylalkohol, polyvinylpyrrolidone, hydroxyethylcellulose, alginate, starc, caseic, alkalisalts of polymethacrylacid); conservationmedia (e.g. methylbezisothiazolin, chloromethylisothiazolin, bariummetaborat, formaldehyd); acrylmonomere[1, 2, 4]

It has to be mentioned, that not every acryl media called acryl media is one! The pure acrylic resin is quite expensive, so some producers dilute the copolymerisat of ethylacrylate/methylmethacrylate with some cheaper homopolymeric polyvinylacetate and/or polystyrole and/or polypropionate [1]. Terpolymeric polyvinylacetates can be found in exterior paints.

To formulate a paint, in addition to the binding media complex pigments have to be added. With the pigments further igredients appear, such as: surfactants (e.g. alkylphenolethoxylate, acetylendiol, alkyllarylsulfonate, sulfosuccionate); dispergers (e.g. oligophosphate, polycarboxylate, calium- or calziumsalts); softener (e.g. alcoholester, benzoatester, glykolether, glycoletherester, n-methyl-2-pyrrolidon); non foaming agents (e.g. acstor oil, silikonoil, alkoholes, glycole, wax); thickener (e.g. hydroxyethylcellulose, methylcellulose, carboxymethylcellulose, polysaccaride, ethoxylaturethan, attapulgite, silicate (e.g. aluminiumsilicate, magnesiumsilicate, rohagite; filling materials (e.g. calciumcarbonate, calcium-magnesiumcarbonate], aluminiumsilicate, siliciumdioxide; magnesiumsilicate, kalium-aluminiumsilicate, kieselgur, bariumsulfate, styrolacrylate; anti frozing agents (e.g. ethylen, propylenglycol); mikrobiozides, fungizides (e.g. chlorided phenole, formaldehyde, mercury derivates, tin derivates); antistaticas; solvents; antioxidantias; ammoniac. At last acrylic paints can contain 30 up to 50 % water. That's the reason because acrylic paints are much cheaper than oil paints: oil paints contains for three times more pigment [1-5].

Because conservators never know exactly what is in the paint they have to clean, it is quite difficult to find an adequate, save cleaning method for such paints.

Normally acryl paint medias are stable to alkalis but he polyvinylacetate-mixtures are not alkali resistant. Therefore the pH should not increase pH 6, or pH 8.

Dry Cleaning

If acrylic paint surfaces have to be cleaned the best way is to clean them dry, with compressed air and a vacuum cleaner to take the dust away.

If pressured air is without any positive effect, one can use dry latex sponges (e.g. "wishab", "akapad"), but on some soft paints (e.g. Schmincke) the remnants of these sponges are not easily to remove. One can only use very light pressure.

"Chemical sponges"[6, 7, 60] can cause higher frictional resistance. Normal erasers (e.g. "Staedtler Mars Plastic",) consists of polyvinylacetate, chalk and up to 30% of the softener dioctylphtalat[8] and should therefore not be used. The combination eraser plus water, performed by Hackney[9], can cause a mix of, dirt, paint and remnants of the eraser which ca - in monochrome parts of the painting – appear unattractive.

It is possible to dry up the surface with "groomstick" or the soft plasticine "Pelikan GE 20"[1].

It is also possible to cover soft pieces of cloth with Plextol D 360[1, 10] and remove impurities by dabbing the surface.

Up to now, the cleaning of acrylic paints with lasers is not developed well. One has to take the sensibility of acrylic paints into account. These surfaces react very sensitive to the pulse time and light emmission[11, 12]. If possible the short time pulsed Nd:YAG-laser with wave length under $\lambda = 1064$ nm can be used[13]. But therefore separate examinations have to be carried out.

If the surface have to be cleaned wet, one has to take damages into account.

Wet Cleaning

Most of the hundreds of surfactant solutions in use worldwide by conservators (e.g. Orvus WA, Surfynol 61, Triton-X 100, Synperonic N, etc.) as well as other substances (e.g. potassium oleate, ammonium citrate, desoxycholic acid, saponin, etc.) [14-26] have been shown to be harmful for treating art objects [27-34]. Paint surfaces should not be eroded, leached out, weakened, or discolored by the use of improper or unsafe materials. Because of the scarce financial resources available for such investigation in Europe, research at a sophisticated level is not possible for many of the numerous institutions, not even to mention private conservators. Conservators and curators are often not absolutely certain which materials/methods are safe and non damaging for surface cleaning these paintings. A cleaning procedure has to be developed and examined.

The wet cleaning of originally unvarnished acrylic paint surfaces is a frequently performed activity in conservation work. By the term "wet cleaning", only the removal of water soluble surface impurities is implied here (grime, particles, etc. which have accumulated over time).

It is well known that there are differences in the formulation of various acrylic paints.

A flexible solution for treating a given painted surface would enable conservators to make safe and confident decisions for handling works of art and thus prevent harm to treated paint surfaces.

The wet cleaning is due to the surfactants in the paint harmful to acrylic surfaces[35] and the "ultima ratio"[3]. If the surfactants which rest since the drying of the paint at the surface[36] are taken away, the surface became eroded and more brittle [1, 2, 37, 38]. Older acrylic paints are more forced by this procedure than younger ones.

If there is a craquelé, one have to consider the capillarity of such paints which cause a swelling of this paints[39]. To avoid this one have to use microporous sponges. These sponges guarantee short contact times. Cotton swabs have to be avoided strictly due to their strong abrasion[1, 28-34, 40]. In accordance to a wet cleaning, no demineralised water should be used because it is highly aggressive and therefore erosive to such paint surfaces[14, 17, 33, 41, 42]. The use of thickeners (e.g. Celluloseether) should be only carried out punctual (e.g. to remove fly spots).

Saliva

For wet cleaning saliva is not ideal. Saliva contains a number of ingredients, e.g. enzymes, digesting lipids, and carbohydrates, which can decompose binding media. Saliva solubilizes polar components from binding media [43 - 46]. Bacteria, fungi and microorganisms can also be detected in saliva. Saliva attracts flies which make the surface once again dirty [47].

Artificial saliva, which contains aromas and salts, should not be used [1, 2, 33, 48].

Water

There are different sorts of drinking water available. Water with ferrum, copper, and manganese ions cannot be recommended. It is better to use, depending on the pH-value of the dirt on the painted surface, water which includes carbon, calcium and magnesium ions. This means, it is possible to neutralize the dirt during cleaning the surface.

The practice of wet cleaning provoke a swelling of the upper parts of the paint film caused by the paint immanent surfactants which accumulate while the paint was drying [2, 4].

Unfortunately water dissolves this surfactants, too. That is the reason why sometimes no surfactant has to be added for cleaning acrylic surfaces: water takes the surfactant out of the paint.

Water with high ammounts of hydrogencarbonate can cause spots on acrylic surfaces and should therefore not be used [33]. Reducing the amount of water, by the use of microporous sponges is recommended.

Demineralised Water

Demineralised water has little or no cleaning effect. As already known [17, 42, 49], treatment of surfaces with demineralised water interacts unspefically with dirt and acrylic paints and can thereby harm the surface of paintings. In practical work, this means that it is not advisable to use demineralised water.

Magnetised Water

As it has been shown in the last years the cleaning effect of water can be optimized by magnetisation. This is caused by the lowered surface tension up to 8 %. While the size of the water clusters is minimized the surface of the water increased. For cleaning of oilpaints magnetized water is ideal but acrylic paints are affected by this, because the binding of the pigments in the acrylic binding media is not as high as the one in oil paints.

Sometimes the use of water did not guarantee clean surfaces. Then the use of additional substances has to be considered [33].

Water with Additions

Criteria for surface active substances include carbon chain length, critical micelle concentration (CMC), maximum foaming temperature (MFT), ethylene oxide number (EO/mol), hydrophobicity index (HI), hydrophilic lipophilic balance (HLB), and Zein Number. It must be taken into account that the selected surfactant should not tend to discolor the paint or catalyse the decomposition of the painting material.

A combination of 0,2 –0,3% methyl cellulose and 0,2 - 0,3% of linear, nonionic, pH-neutral, minimally hygroscopic, non-degradable, non-discoloring fatty alcohol ethoxylate Marlipal®1618/25 (C-chain length between C 16 and C 18), produced by Sasol (formerly Hüls, then CONDEA, Marl, Germany), in aqueous solution (100 ml water) was developed and found to be exceptionally good for cleaning oil paint surfaces. In the years following, we have attempted to demonstrate the safety and usefulness of this solution using SEM [18,19], ESEM [22], Laser profilometry [20, 21], and 3D-Stripe projection [22, 23] to detect any undesirable changes on the paint surface.

The use of ammoniac-, triammoniumcitrat-[50], such surfactants as Orvus WA Paste (= SDS) [21], Triton X-100[4], or cleaning agents can not be recommended [51, 52].

In a study it has been shown, that untreated an treated surfaces look quite similar. SEUFFERT points out that surfaces due to wet cleaning get swollen so much that ingredients and dirt are given away to the cleaning solution but are accumulated in the film too. After drying of the paint film the surface appears homogene again. Even when coloured cleaning materials show that there must have been abrasions on the surface macro- and SEM-measurements did not show any changes [1].

If solvents were used, surface changes appear immediately. Solvents dissolve the surface much faster then water. The surface gloss changes, pigments, extenders and additives were leached out [1,4].

In our examinations the combinations between thickeners and nonionic surfactants has been sufficient. Anionic surfactants should not be used because one never knows which surfactant was used during the fabrication of the paint. Anionic surfactants can produce salts on the surface.

Methyl Cellulose (MC)

The usual Na-methylcellulose is made of wood or cotton cellulose and caustic soda solution. MC soluble in cold water comes in several types, some of which are alkaline. MC is a bacteriocide and fungicide. In the cleaning of surfaces of paintings, MC is used to lower the surface tension of water, or to thicken surfactant solutions.

Carboxymethyl Cellulose (CMC)

CMC is, like MC, a derivate of cellulose. CMC can form complexes, that is, CMC can carry dirt in solution, even in absence of electrolytes; that is the reason that CMC is used in detergents. In the cleaning of paintings, CMC is used to lower the surface tension of water or to thicken surfactant solutions although it is not as effective for thickening surfactant solutions as MC.

Marlipal®16181/25 Powder

It is still problematic finding an ideal surfactant for conservators. According to information obtained from chemists working at Sasol, Germany's largest producer of surfactants, from the huge range of non-ionic surfactants only one can be used for cleaning paint surfaces. Marlipal®1618/25-powder, a sebum-fattyalcohol surfactant (Alkyl-polyethylene-glycolether) has a water-like character. Its specific chemical details are as follows: chain length of C 16 - C 18; EO/mol number 25; HLB number approximately 16; reaches critical micelle concentration at a concentration of 0.2% at pH 6.5-7; is 90% biologically degradable[53]; very low adipose effect; HI number approximates 1; therefore it is hygroscopic only to a small extent[33].

Sodium Dodecyl Sulphate (SDS)

SDS ($NaC_{12}H_{25}O_4S$, a monoester of sulphuric acid with dodecanol-1) is an anionic surfactant. It was used in this examination only for comparison. Its specific chemical details are as follows: the HLB number of the surfactant is approx. 35; SDS reaches its critical micelle concentration at 0,2%; its pH ranges between 6.5 - 7 (0.5 g/100ml); its Zein number is 640 (see appendix).

SDS can be considered an aggressive cleaning agent for paintings. Since SDS is a semi-ester of sulphuric acid, the ester binding may not be stable. It is known that catalytic agents, such as heavy metals found in paints, may hydrolyse SDS over time. The product is highly reactive and very hygroscopic sulphuric acid[33].

In general the results of water cleaned acrylic surfaces are satisfying. If not, small amounts of a mild nonionic surfactant can be added.

If this does not work it is possible to prolong the time of contact, increase the temperature of the solution, or change the pH-value of the solution (max. 6, max. 8) depending on the pH of the surface impurities to increase efficiency. This would be better than using a stronger, anionic surfactant [33].

3D-EXAMINATIONS OF CLEANED ACRYLIC PAINT SURFACES

Apparatus

This transportable technology, called MicroCad, was presented to the public for the first time in 2001 at Washington D.C., U.S.A. and Düsseldorf, Germany. The 3D-Stripe projection technique based on micro mirrors was developed at the GFMesstechnik, Berlin-Teltow, Germany.

The micro mirror projection can be differentiated from previous such stripe projections primarily through the use of digital projection. This measurement system produces and projects on the measured area different grey values, which may be regulated by the use of micromirrors. This permits local reflections from various areas on the surface of the object to be balanced, in contrast to the use of traditional laser profilometry which causes artefacts.

The micro mirror technology is based on a development by Texas Instruments, Dallas. 1024 x 768 micro mirrors of 16 x 16 micrometers each, are arranged in arrays (13 mm x 8 mm). Each of these micro mirror can be regulated in brightness and/or colour by a computer. In that way various colours and/or light intensities can be produced. The advantage of this technology is that the computer is calculates a pattern and this is then projected and finally an image is recorded with a CCD-camera. The difference between the recorded pattern which sent by the computer via the projector and that which was recorded by the camera and sent to the computer, gives the so called measurement effect. An additional important advantage of the micro mirrors is that one has a great deal of available light, for measuring the paint surface. These type of micro mirror projectors were developed for the use in Power Point-presentations projectors. In 1996/97 the idea for using the micro mirror projectors for measurement technology was developed by GFM.

The difference between the GFM-MicroCad-technology compared to usual stripe projection may be summarized as follows:

With micro mirror projection, one can project digitally, that means, that through specific control of the micro mirrors, measurement points of defined grey values between 0 and 255 Bits can be produced and projected. Thus differing local reflexions on surfaces can be equalized.

Through the use of micro mirrors for light projection, one observes (when compared to conventional projection systems such as LCD's) up to 90% of the light intensity from the lamp directed on the object to be measured. Thus one is the position to carry out optical measurement techniques in normally lighted rooms without the need to darken the room.

Through a special development pof GFM and Texas Instruments a so called high speed projection is feasible; that is, by using conventional lighting with structured white light, the video pulse time of the projectors/camera used are able to produce a 3D-profile in 68 milliseconds. In contrast to the 64 hour measuring time needed by laser- profilometry this is of great advantage.

The MicroCad projects stripe patterns on the surface using microscopic and macroscopic digitale stripe projections reflected from micro mirrors (referred to as digital micro mirror devices).

The light used, produced by a cold halogen source of 270 Watt turned the highest intensity level (about 20.000 Lux) is conveyed through the projector to the surface to be measured for 300 and 1000 milliseconds.

Following measurement of the surface with light projection only, stripe patterns are the projected on the surface of the object to be measured. These intersecting patterns are photographed by a CCD-camera of 1300 x 1024 Pixel and in seconds displayed on the computer screen. The measured surface is shown in the form of intersecting patterns of the projected lines on the object. The distance between measurement points is 1,5 micrometers pro pixel in the lateral (x-, y-) resolution and 0,3 micrometers in the vertical (z-) resolution with 1000 points/milimeter (ca 500 x enlargement). The photo technique delivers colour and black/white scaled (for orientation on the measuring field) camera photos, a photorealistic representation (which is very similar to SEM) and scaled colour micro topographic 3D-pictures. The last two, because based on 3D-point clouds, produce pictures with are both objectively measurable. Currently, the smallest measurement field is 2 x 1,5 mm^2.

The advantage of the MicroCad for conservators is that this device allows immediate in place examination and measurement in seconds during the cleaning process and producing measurable images which can show surface and craquelé changes before, during, and after treatment. This mobile measuring technique allows one to vary the surfactants solutions used during the cleaning process. The exact, scaled colour micro topographic images, give exact information concerning possible surface changes which can also be useful in legal court cases. The MicroCad can also be used for measuring whole picture surfaces, for determining the edges of overpainted areas, and comparing different brush strokes of questionable attributions. In the future no damaging paint sampling will need to be performed and no surface replicas will have to be produced.

EXPERIMENTAL

In accordance with our experiences using computer assisted laser profilometry [31, 32, 54-59] we evaluate the impact of cleaning methods (wet/dry) as well as various types of wet cleaning agents on the surfaces of paint samples, applying the MicroCad technology, because it requires no samples to be taken and it is easy to examine the same area before and after treatment. Effects provoked by various modifications (solution, viscosity, pH, time of contact, and temperature changes, etc.) will be evaluated for their effects to the surface. Measurements at ten to fifteen centimeter distance above the samples were carried out before and after treatments as described below.

MATERIALS

Paint Samples

The acrylic paints "Lukascryl dark ultramarine (Art. Nr. 4337)" from the firm of Schoenfeld, Düsseldorf, and "Primacryl dark krapp (Art.nr. 324)" were applied on glass. The dirt layer consisted of loose dirt and grime.

- untreated
- dry cleaning using latex-sponge ("wishab", "akapad") (duration 20 sec)
- wet cleaning using saliva on cotton swabs (duration 20 sec)
- wet cleaning using tap water on a microporous sponge (duration 20 sec)
- wet cleaning using magnetised tap water on a microporous sponge (duration 20 sec)
- wet cleaning using demineralised water on a microporous sponge (duration 20 sec)
- wet cleaning using methyl cellulose (0,2g/100ml demineralised water) on a microporous sponge (duration 20 sec)
- wet cleaning using carboxy methyl cellulose (0,2g/100ml demineralised water) on a microporous sponge (duration 20 sec)
- wet cleaning using Marlipal® 1618/25 (0,2g/100ml demineralised water) on a microporous sponge (duration 20 sec)
- wet cleaning using Marlipal® 1618/25 (0,2g/100ml demineralised water) and methyl cellulose (0,2 g/100ml demineralised water) on a microporous sponge (duration 20 sec)
- wet cleaning using SDS (0,2 g/100ml demineralised water) on a microporous sponge (duration 20 sec)
- wet cleaning using SDS (0,2%) and methyl cellulose (0,2 g/100ml demineralised water) on a microporous sponge (duration 20 sec)

For this examination solutions of 0,2% concentrations were used, because SDS and Marlipal® 1618/25 are reaching their critical micelle density at this point. If surfactant concentrations exceed the critical micelle concentration, the cleaning effect will not be more effective.

RESULTS

Due tue various compositions of acrylic paints and their different reactions it is quite difficult to formulate valid statements. The measurements illustrate the destruction of the surface, caused by swabs, in contrast to the visible abrasions and paint remnants on the swab, not enough. The effect of different waters on the surface can not seen clearly enough. But these findings verify previous studies[1, 3, 40]. But in general one could say:

Due to the sensitivity of acryl paints it is suitable to protect these paintings against soiling (e.g. protecting with glass sheets, backside protections, use of air cleaners, etc.)

The cleaning of acryl paints should take place at cool temperatures about 18° C +/- 1° C.

The Lukas acrylic paints are containing less and not so soft binding media than Schmincke acrylic paints. The Lukas paints are more hydrophilic and can be wettened more easy. The surface is sensitive to mechanical treatment. With Cyclohexanon resin varnished paints can be cleaned more easy.

The Schmincke acrylic paints are containing more and softer binding media than Lukas paints The Schmincke acrylic paints are more hydrophobe and can not be wettened easy. On mechanical treatment they react more sensitive than Lukas acrylic paints. With Cyclohexanon resin varnished paints can be cleaned more easy.

Dry cleaning with compressed air and a vacuum cleaner to take the dust away did not harm the surface.

Dry latex sponges (e.g. "wishab", "akapad"[60]) can cause remnants on soft paints. Therefore the use of "groomstick" or the soft plasticine "Pelikan GE 20" or coated soft pieces of cloth with Plextol D 360 for dabbing the surface can be recommended.

We found that all sorts of water erode the paint surfaces. The less harmful water seems to be tap water which does include carbonate, calcium and magnesium ions. Water with high ammounts of hydrogencarbonate can cause spots on acrylic surfaces and should therefore not be used.

Cotton swabs provokes scratches on the paint surface. Microporous sponges made of polyvinyl acetate (e.g. "blitzfix", "wondersponge") should be used. It has been observed throughout the cleaning process, the sponge should follow the direction of the brushstrokes of the paint. Tissue paper should not be used for drying the surfaces.

Combinations of anionic surfactants and methylcellulose should not be used.

We found that combinations of the nonionic surfactant Marlipal$^{®}$1618/25 (0,2g/100ml tap water) and methyl cellulose (0,2 g/100ml tap water) used on microporous sponge are the most sensitive cleaning agents for acrylic paints.

Working sheet for cleaning acrylic paint surfaces

1. Examination of paints: are there fixed well on the groundlayer? If not, the paint layers have to be consolidated first (e.g. with acrylate emlusion Plextol B 500).
2. Cleaning with compressed air and a vacuum cleaner to take the dust away. If the effect is a positive one: end of the cleaning procedure.
2a. Cleaning with compressed air a soft brush and a vacuum cleaner to take the dust away. If the effect is a positive one: end of the cleaning procedure.
2b. Dabbing the surface with "groomstick" or the soft plasticine "Pelikan GE 20", or with Plextol D 360 covered soft pieces of cloth to remove impurities on the surface. If the effect is a positive one: end of the cleaning procedure.
3. Cleaning the surface with dry latex powders (e.g. "wishab", "akapad"), but on some soft paints (e.g. Schmincke) the remnants of these sponges are not easily to remove. One can only use very light pressure. If the effect is a positive one: end of the cleaning procedure.
3a. Cleaning the surface with dry latex sponges (e.g. "wishab", "akapad"), but on some soft paints (e.g. Schmincke) the remnants of these sponges are not easily to remove. Sponges are not so soft than powder, one have to use only very light pressure. If the effect is a positive one: end of the cleaning procedure.
4. If there was no positive cleaning effect by dry cleaning one have to examine the paint: are the paints stable enough for a cleaning with aqueous solutions?

If the paint is stable one can examine the pH of the dirt on the surface with pH-indicator stripes (e.g. Merck). After that one can choose the suitable water in order to neutralize the dirt during cleaning. The water (max. pH 6, or max pH 8) has to be used with microporous sponges. If the effect is a positive one: end of the cleaning procedure.

5a. If there was no positive cleaning effect on e has to add surfactants to the used water (e.g. a combination of 0,2% Marlipal®1618/25 and 0,2% methyl cellulose) If the effect is a positive one: end of the cleaning procedure.

5b. If there was no positive cleaning effect one has to rise the temperature of the solution (max 25 °C) and /or to extend the contact time. It is possible to use a japanese paper fleece as compress. After this the possibilities of soft cleaning methods are exhausted.

Figure 1. Lukascryl Ultramarin Dunkel (4337), untreated

Blau_2 akapad-Schwamm Trockenreinigung

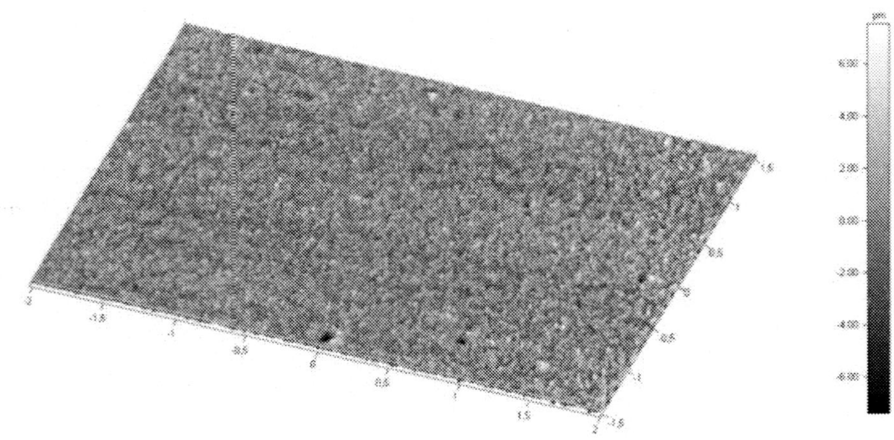

Figure 2. Lukascryl Ultramarin Dunkel (4337), dry cleaning with akapad (latex sponge)

Blau_3 Speichelreinigung mit Wattestäbchen

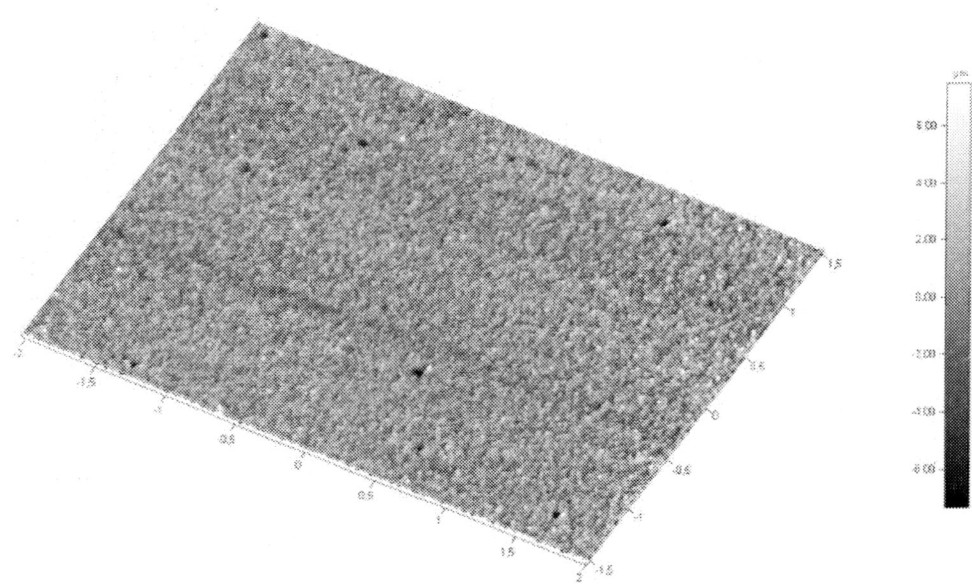

Figure 3. Lukascryl Ultramarin Dunkel (4337), saliva cleaning with cotton swabs.

Blau_4 Leitungswasser, entkalkt, Schwammreinigung

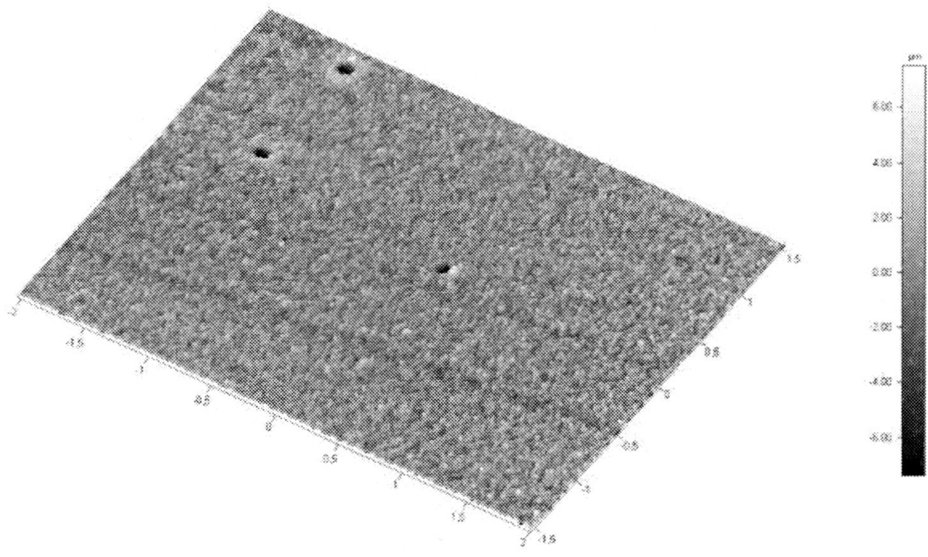

Figure 4. Lukascryl Ultramarin Dunkel (4337), tap water, decalcified, cleaning with microporous sponge.

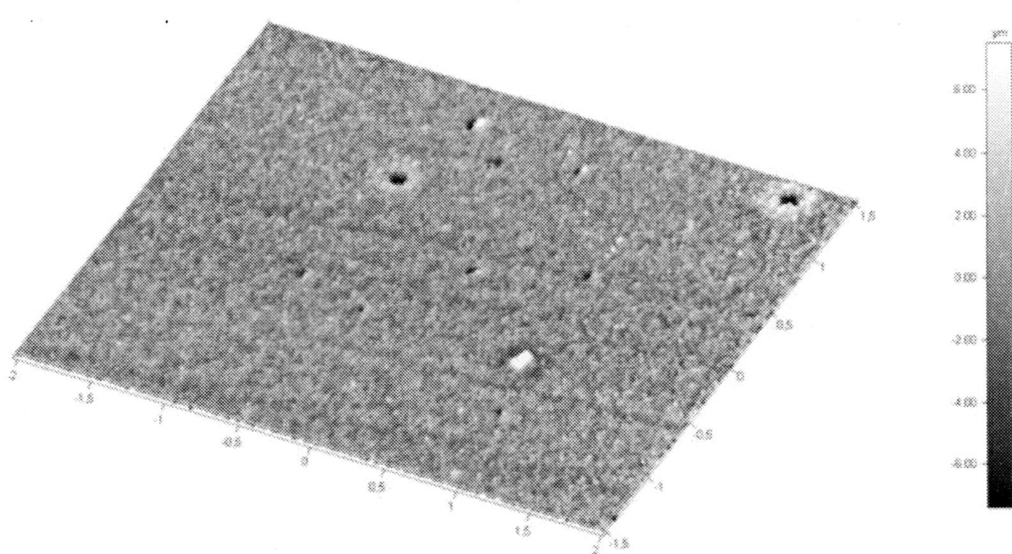

Figure 5. Lukascryl Ultramarin Dunkel (4337), tap water, decalcified, magnetized, cleaning with microporous sponge.

Blau_5 Demineralisiertes Wasser, Schwammreinigung

Figure 6. Lukascryl Ultramarin Dunkel (4337), demineralised water, cleaning with microporous sponge.

Blau_6 Methylcellulose in Leitungswasser, entkalkt

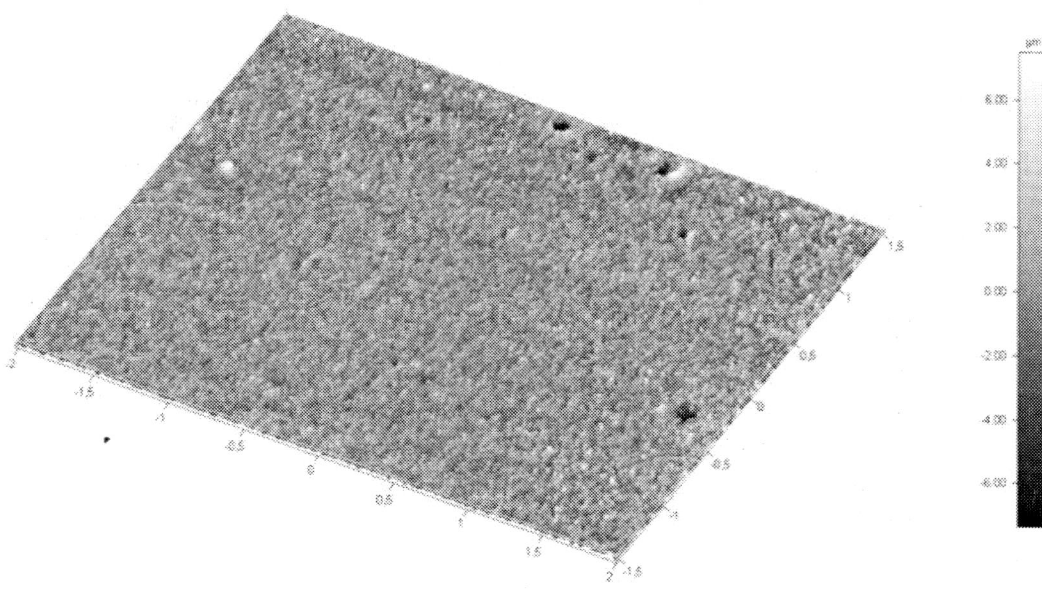

Figure 7. Lukascryl Ultramarin Dunkel (4337), Methylcellulose in tap water, decalcified, cleaning with microporous sponge.

Blau_7 Carboxymethylcellulose in Leitungswasser, entkalkt

Figure 8. Lukascryl Ultramarin Dunkel (4337), Carboxmethylcellulose in tap water, decalcified, cleaning with microporous sponge.

Blau_8 Marlipal 1618/25 in Leitungswasser, entkalkt

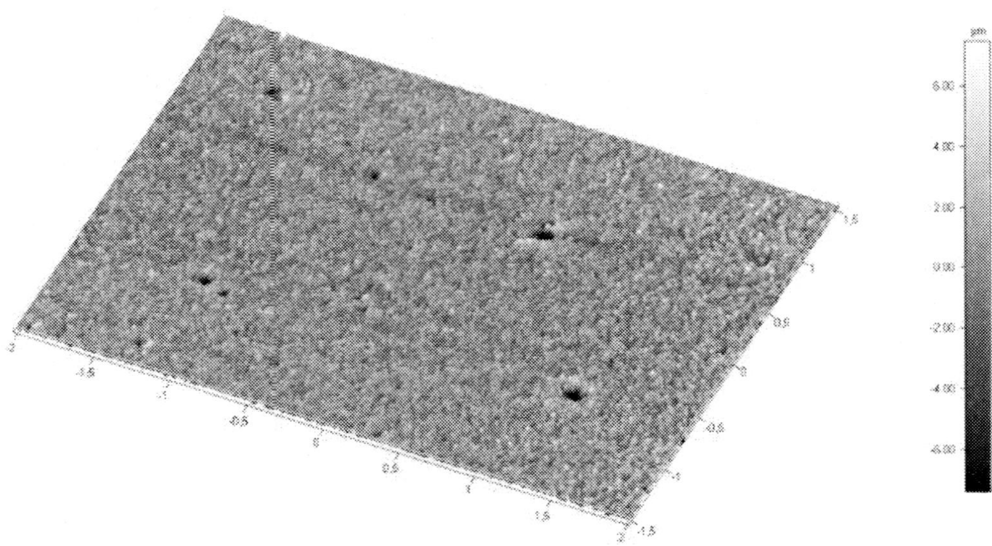

Figure 9. Lukascryl Ultramarin Dunkel (4337), Marlipal® 1618/25 in tap water, decalcified, cleaning with microporous sponge.

Blau_9 Marlipal 1618/25 in Methylcellulose in Leitungswasser, entkalkt

Figure 10. Lukascryl ultramarine Dunkel (4337), Marlipal® 1618/25 in Methylcellulose in tap water, decalcified, cleaning with microporous sponge.

Figure 11. Lukascryl ultramarine Dunkel (4337), SDS in tap water, decalcified, cleaning with microporous sponge.

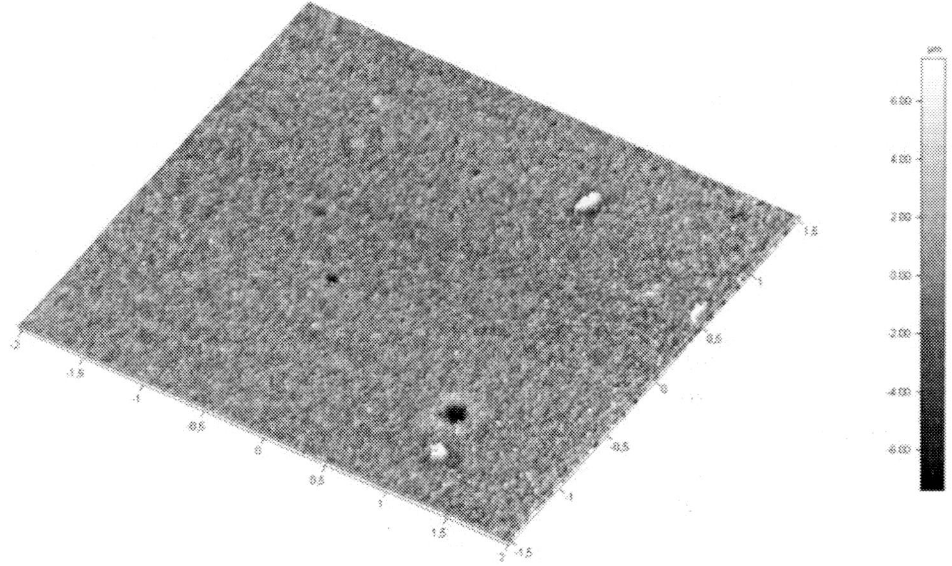

Figure 12. Lukascryl Ultramarin Dunkel (4337), SDS in Methylcellulose in tap water, decalcified, cleaning with microporous sponge.

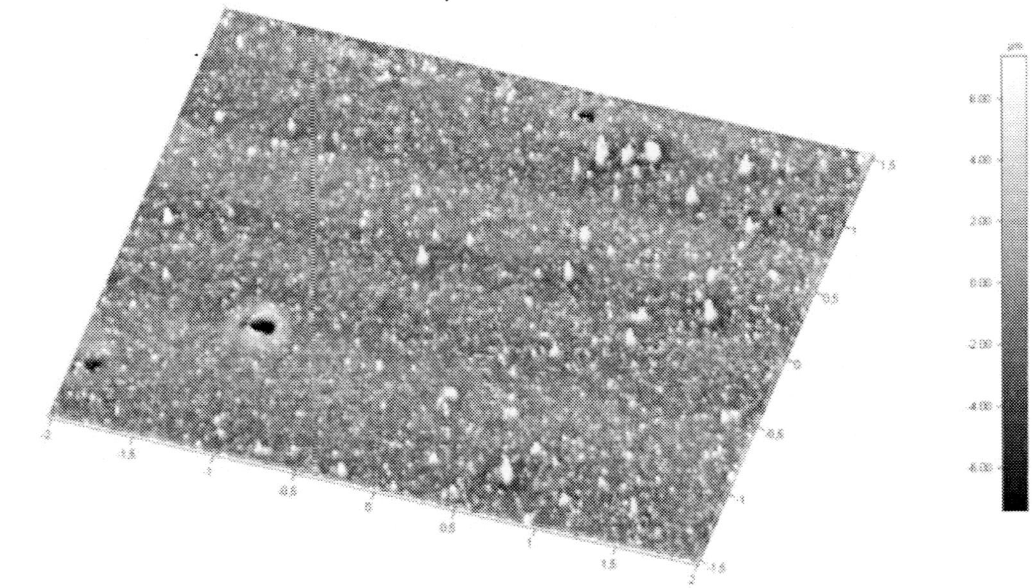

Figure 13. Schmincke Primacryl Krapp dunkel (324), untreated.

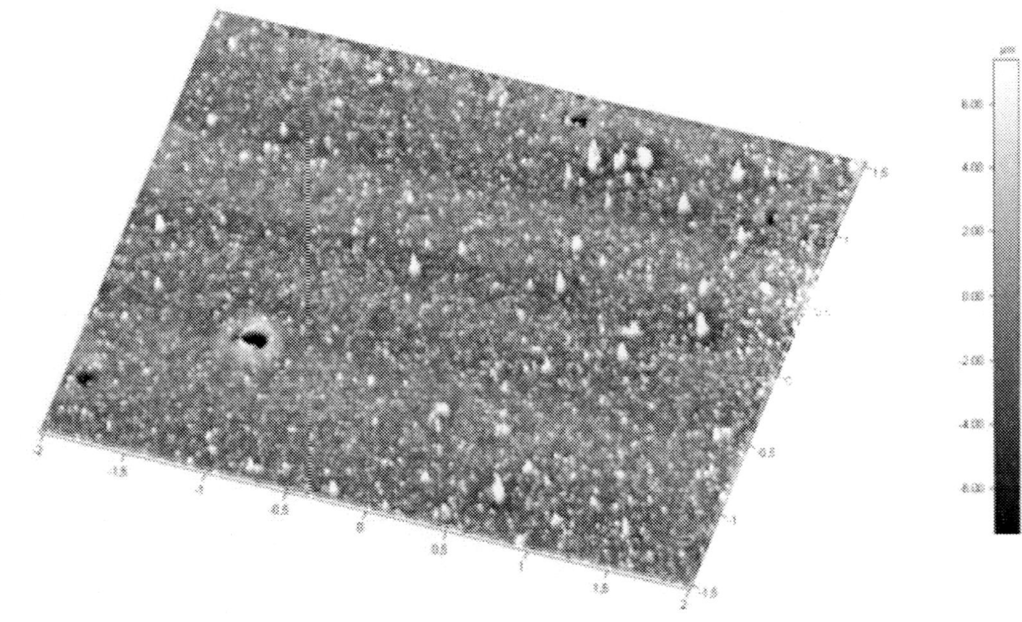

Figure 14. Schmincke Primacryl Krapp dunkel (324), dry cleaning with akapad (latex sponge).

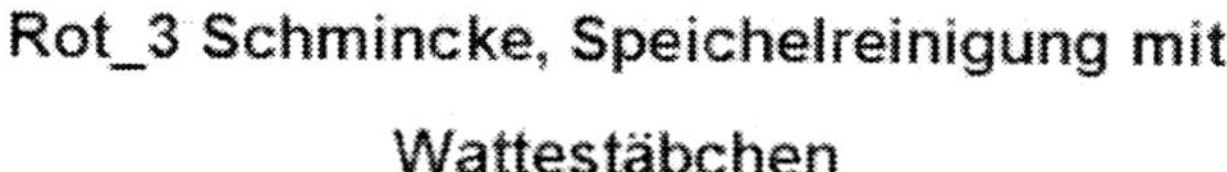

Figure 15. Schmincke Primarcryl Krapp dunkel (324), saliva cleaning with cotton swabs.

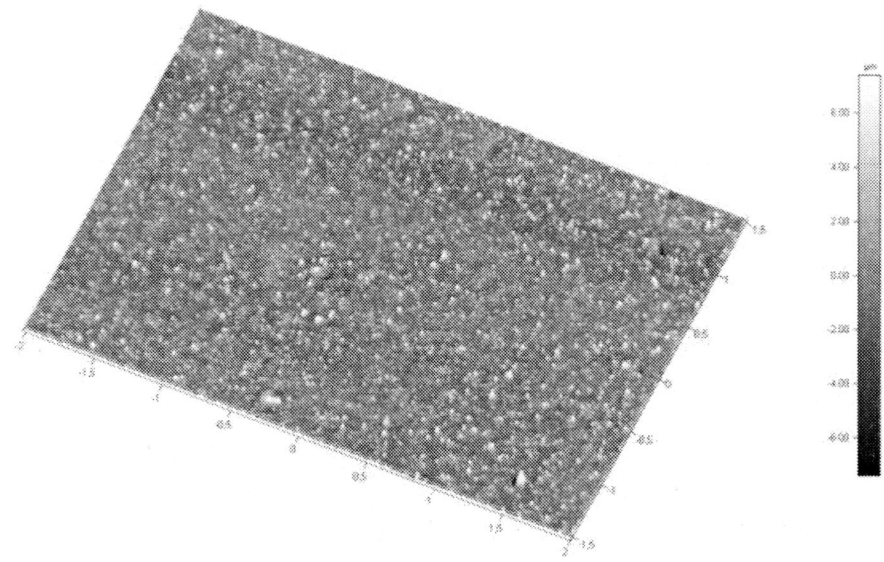

Figure 16. Schmincke Primacryl Krapp dunkel (324), tap water, decalcified, cleaning with microporous sponge.

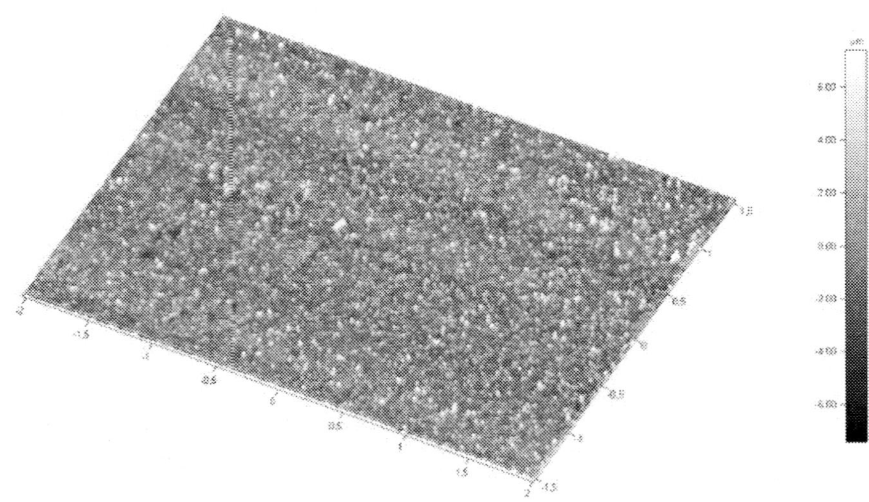

Figure 17. Schmincke Primacryl Krapp dunkel (324), tap water, decalcified, magnetized, cleaning with microporous sponge.

Figure 18. Schmincke Primacryl Krapp dunkel (324), demineralised water, cleaning with microporous sponge.

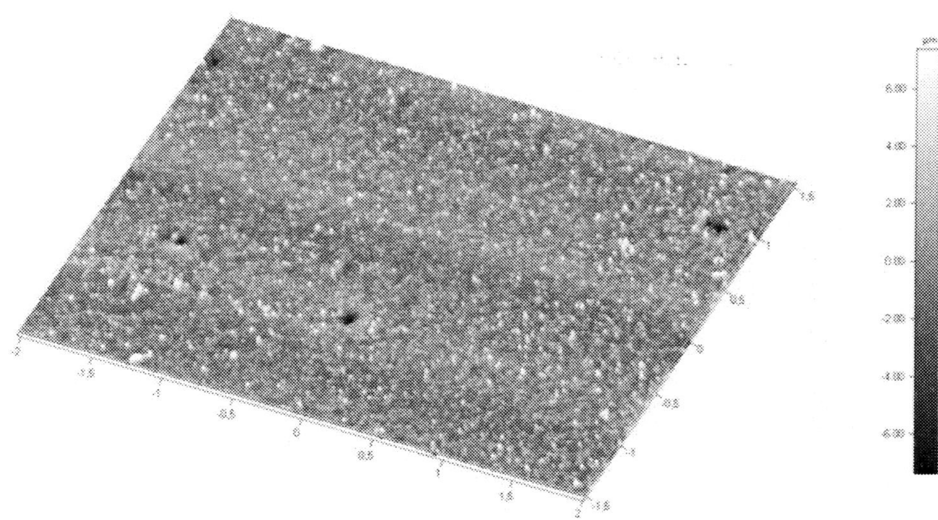

Figure 19. Schmincke Primacryl Krapp dunkel (324), Methylcellulose in tap water, decalcified, cleaning with microporous sponge.

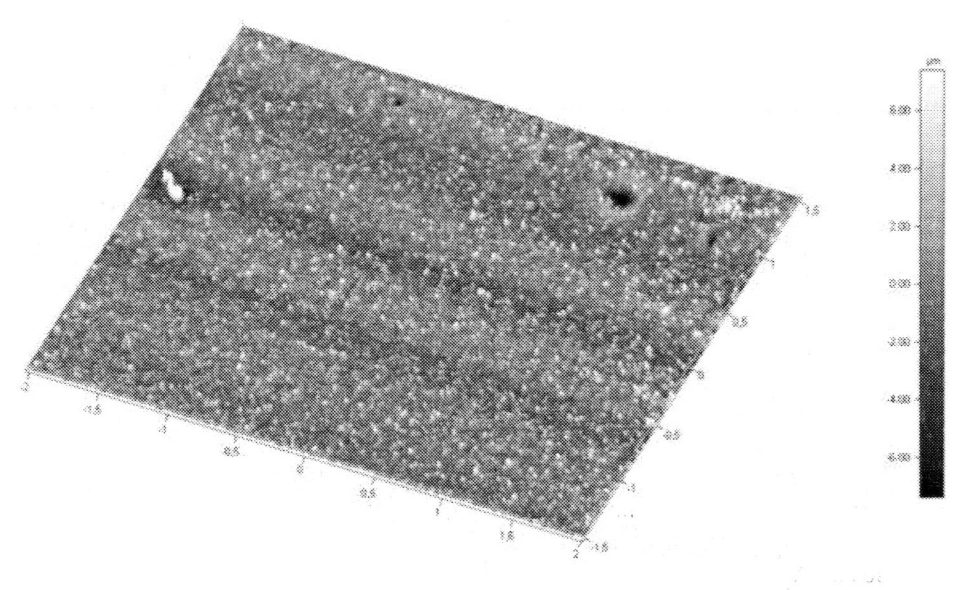

Figure 20. Schmincle Primacryl Krapp dunkel (324), Carboxymethylcellulose in tap water, decalcified, cleaning with microporous sponge.

Rot_8 Schmincke, Marlipal 1618/25 in Leitungswasser, entkalkt

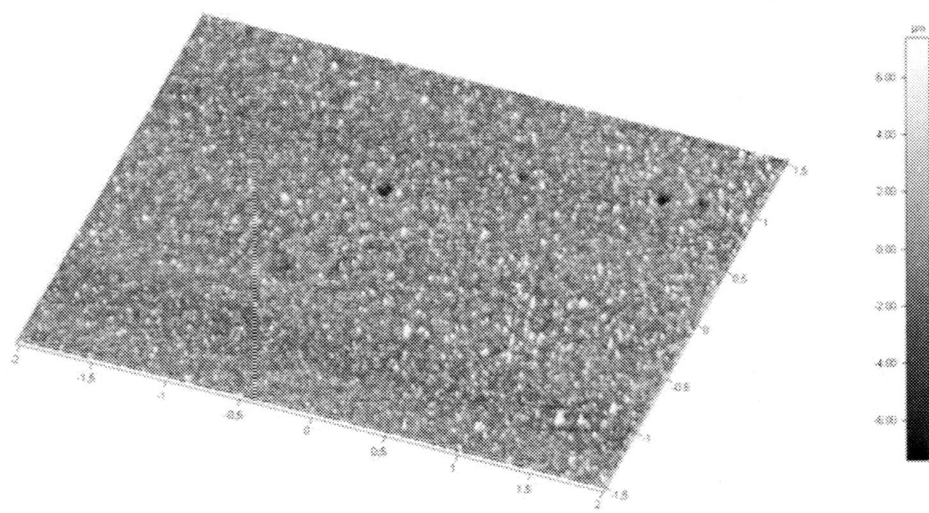

Figure 21. Schmincke Primacryl Krapp dunkel (324), Marlipal® 1618/25 in tap water, decalcified, cleaning with microporous sponge.

Rot_9 Schmincke, Marlipal 1618/25 in Methylcellulose in Leitungswasser, entkalkt

Figure 22. Schmincke Primacryl Krapp dunkel (324), Marlipal® 1618/25 in Methylcellulose in tap water, decalcified, cleaning with microporous sponge.

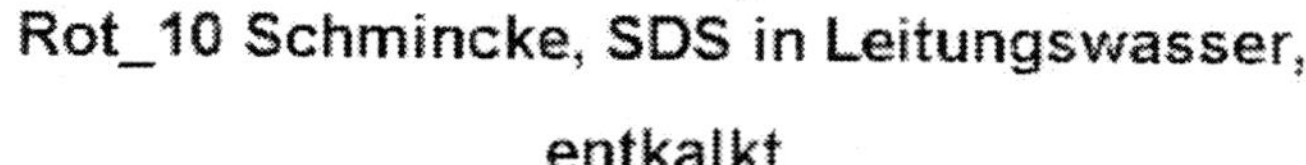

Figure 23. Schmincke Primacryl Krapp dunkel (324), SDS in tap water, decalcified, cleaning with microporous sponge.

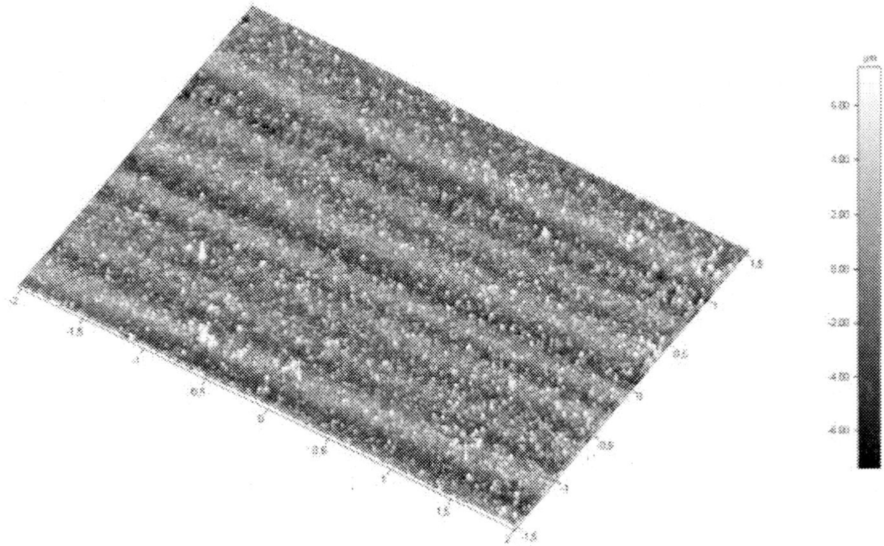

Figure 24. Schmincke Primacryl Krapp dunkel (324), SDS in Methylcellulose in tap water, decalcified, cleaning with microporous sponge.

ACKNOWLEDGMENTS

The author would like to thank Dr. G. Frankowski, Dipl.-Ing. K. Streibel, Dipl.-Ing. S. Zepke, Dipl.-Ing. T. Knoke (GFM-Messtechnik, Teltow/Berlin), Dipl.-Rest. C. Seuffert, Köln; Dipl.-Rest. D. Simmert, Riseby, Dr. Müller, Mrs. Solbisky-Thiel (Schmincke & Co. Künstlerfarbenfabrik, Düsseldorf), Dr. A. Kleine (Lukas Künstlerfarbenfabrik, Düsseldorf), Dr. C.-D. Hager (sasol, Marl), Dr. C. Krieb (Assindia Quelle, Essen) and Dr. E. Ringwald, Bonndorf;

ABOUT THE AUTHOR

Dr. Dipl.-Rest. PAUL-BERNHARD EIPPER received his Diploma as a Conservator in 1992 from the University of applied Sciences, Cologne, Germany. From 1990- 2005 he has been Head of Conservation at the August Deusser Museum, Zurzach, Switzerland. From 2000 -2009 he has been publicly appointed expert by the Chamber of Commerce, he serves as an expert in questions concerning art restoration under legal consideration in Germany. Since 2006 he has been Head of Conservation at the Universalmuseum Joanneum, Graz, Austria. In 2009 he received his PhD from the Faculty of Medicine at the University of Witten-Herdecke, D). He is the author of numerous books and articles. Author's address: Universalmuseum Joanneum, Weinzöttlstrasse 16, 8045 Graz, Austria, paulbernhardeipper@museum-joanneum.at

REFERENCES

[1]   Seuffert, C. (1994). 'Untersuchungen zum Verhalten von Acrylmalschichten bei der Oberflächenreinigung'. *Diploma Thesis*, University of applied sciences Cologne, 1-118.

[2]   Simmert, D. (1995). 'Acrylharzkünstlerfarben. Studien zu einem Malmaterial des 20. Jh.' *Zeitschrift für Kunsttechnologie und Konservierung, 1*, 78-105

[3]   Eipper, P. B. (2005). 'Acrylfarben-Oberflächenreinigung', *Museum aktuell, 117*, 10-22

[4]   Jablonski, E., Learner, T., Hayes, J. & Golden, M. (2003). 'Conservation Concerns for acrylic emulsion paints', *Reviews in Conservation 4*, 3-12

[5]   Chiantore, O., Scalarone, D. & Learner, T. (2003). 'Characterisation of artists' emulsion paints', *International journal of polymer analysis and characterisation, 8*, 67-82.

[6]   Sgoff, R. (2000). 'Das Kunstwerk im Kontext: Die Gesamterscheinung unterschiedlicher Oberflächen im Ensemble', in *Schmutz – Zeitdokument oder Schadensbild?*, 33-36.

[7]   Eipper, P. B. (2006). 'Mikroporöse Reinigungsschwämme zur feuchten Oberflächenreinigung: Kooperationsergebnisse des Landesmuseums Joanneum mit dem Bundesdenkmalamt', *Museum aktuell, 126*, 3898-3902.

[8]   Booth, P. (1990). 'Protecting paintings from dirt', in *Dirt and Pictures Separated*, 24-26.

[9]   Hackney, S., Townsend, J. & Eastaugh, N. (Ed), (1990). *'Dirt and Pictures Separated'*.

[10] Worch, M. T. (1989). 'Restaurierung eines frühchristlichen Grabes in St. Maximin zu Trier' *Restauro, 4*, 259-266.

[11] Truyen, A. (1995). 'Het verwijderen van overschilderingen door middel van laserstralen', *Mededelingenblad IIC Nederland, 2*, 11-15.

[12] Larson, J., Cooper, M. & Sportun, S. (2000). 'Current Developments in the Application of Laser Technology for Conservation', in *Tradition and Innovation IIC Melbourne Congress*, 107-110.

[13] Chappé, M., Hildenhagen, J., Dickmann, K. & Bredo, M. (2003). 'Pigmente unter Laserbestrahlung', *Restauro 1*, 27-31.

[14] Erhardt, D. & Bischoff, J. J. (1992). 'Resin soap components', in *Preprints to Restoration '92 Conservation. Training, Materials and Techniques: Latest Developments, held at the RAI International Exhibition and Congress Centre, Amsterdam* (1992), UK Institute of Conservation, London 77-79.

[15] Ford, B. & Byrne, A. (1991). 'The lipid stripping potential of resin soap gels used for cleaning oil paintings', Australian Institute for the *Conservation of Cultural Materials Bulletin, 17*, 51-60.

[16] Gilsa, B. (1991). 'Gemäldereinigung mit Enzymen, Harzseifen und Emulsionen', *Zeitschrift für Kunsttechnologie und Konservierung 1*, 48-58.

[17] Hedley, G., Odlyha, M., Burnstock, A., Tillinghast, J. & Husband, C. (1990). 'A study of the mechanical and surface properties of oil paint films treated with organic solvents and water', IIC Summaries of the Posters at the Brussels Congress (1990) *Cleaning, Retouching and Coatings*, 98-105.

[18] Khandekar, N. (2000). 'A survey of the conservation literature relating to the development of aqueous gel cleaning on painted and varnished surfaces', *Reviews in Conservation, 1*, 10-20.

[19] Mittal, K. L. & Bothorel, P. (eds.), (1986). '*Surfactants in Solution: Proceedings of two international symposia on surfactants in solution*', Plenum Publishing Corporation, 1-562.

[20] Urbaneja, M. A., Arrondo, J. R. L., Alonso, A. & Goñi, F. M. (1986). 'The Interaction of Triton X-100 with Multilammellar Phosphaticylcholine Liposomes', in *Surfactants in Solution: Proceedings of two international symposia on surfactants in solution*, 759-771.

[21] Watherston, M. M. (1976). '*The cleaning of colourfield paintings. The great decade of american abstract modernist art 1960-1970*', The Museum of Fine Arts, Houston, Texas

[22] Wolbers, R. C. (1990). '*Notes for workshop on new methods in the cleaning of paintings*', Getty Conservation Institute, 1-158.

[23] Wolbers, R. C. '*Slides for workshop on new methods in the cleaning of paintings. Laboratory exercises*', University of Delaware, Newark, DE, USA without year, without pages.

[24] Wolbers, R. C. (2000). '*Cleaning painted surfaces. Aqueous methods*', Archetype Publications, London, 1-198.

[25] Eipper, P. B. (2003). 'Oberflächenreinigung mit Tensiden', *Museum aktuell 93*, 4012-4019.

[26] Ashani, Y. & Catravas, G. N. (1980). 'Highly reactive impurities in Triton X-100 and Brij 35: partial characterization and removal', *Anal. Biochem. 109*, 55-62.

[27] Eipper, P. B. (1993). 'Die Reinigung von Gemäldeoberflächen mit Tensiden. Der Einsatz von modifizierten Polyvinylacetaten zur Konservierung von textilen Bildträgern, 1-134.

[28] Eipper, P. B. (1997). 'Vier Künstlerfarbenhersteller zwischen 1900 und 1970. Die Reinigung von Gemäldeoberflächen mit wässrigen Systemen, 1-115.

[29] Eipper, P. B. & Reiss, G. (1995). 'Vergleichende Versuche zur Reinigung von Gemäldeoberflächen mit Tensiden', Mitteilungsblatt, 49, 51-60.

[30] Eipper, P. B. & Reiss, G. (1996). 'Comparison of cleaning agents applied to surfaces of paintings', The Picture Restorer, 10, 5-10.

[31] Eipper, P. B. & Welzel, J. (1997). 'Die Überprüfung von mit Tensiden gereinigten Ölfarboberflächen durch die computergestützte Laserprofilometrie', Mitteilungsblatt, 54, 69-80.

[32] Eipper, P. B. & Welzel, J. (1999). 'Studies Towards an Examination of Surfactant-cleaned Oil Paint Surfaces by Computer-aided Laser Profilometry', The Picture Restorer, 15, 5-13.

[33] Eipper, P. B., Frankowski, G., Opielka, H. & Welzel, J. (2004). 'Ölfarben-Oberflächenreinigung. Die Überprüfung gereinigter Ölfarbenoberflächen durch das Raster-Elektronen-Mikroskop, das Niederdruck-Raster-Elektronen-Mikroskop, die Laser-Profilometrie und die 3D-Messung im Streifenprojektionsverfahren, 1-152.

[34] Eipper, P. B. & Frankowski, G. (2004). '3D-Measuring of surfactant cleaned oilpaint surfaces', The Picture Restorer, 26, 5-13.

[35] Hoppe, T. (2006). 'Acrylmalerei, 1-234.

[36] Digney-Peer, Burnstock, Learner, Khanjian, Hoogland & Boon, (2004). 'The migration of surfactants in acrylic emulsion paint films'. Lecture at the IIC Congress 'Modern Art, New Museums', Bilbao.

[37] Learner, T., Chiantore, & Scalarone, D. (2002). 'Ageing studies of acrylic emulsion paints', in ICOM Committee for Conservation 13th triennial meeting: Rio de Janeiro, 911-919.

[38] Hagan, Ploeger, Shurwell & Murray, (2004). 'Acrylic emulsion paint films and their properties before and after exposure to water: overview of recent developments'. Lecture at the IIC Congress 'Modern Art, New Museums', Bilbao.

[39] Pender, R. J. (2004). 'The behaviour of water in porous building materials and structures', Reviews in Conservation, 5, 49-62.

[40] Ormsby, B. (2006). 'The Effects of Surface Cleaning on Acrylic Emulsion Paintings - A Preliminary Investigation', in Oberflächenreinigung-Material und Methoden, 135-149.

[41] Burnstock, A. & White, R. (1990). 'The Effects of Selected Solvents and Soaps on a Simulated Canvas Painting' in IIC Summaries of the Posters at the Brussels Congress. Cleaning, Retouching and Coatings. Brüssel, 111-118.

[42] Michalski, S. (1990). 'A physical model of varnish removal from oil paint', in Cleaning, Retouching and Coatings. International Institute for Conservation of Historic and Artistic Works. London, 85-92.

[43] Romao, P. M. S., Alarcao, A. M. & Viana, C. A. N. (1990). 'Human saliva as a cleaning agent for dirty substances', in Studies in Conservation, 35, 153-155.

[44] Lehmann, R. R. (1992). "Ökologie der Mundhöhle', Dental Magazin 1, 102-106.

[45] Ramsay-O' Hoski, B. (1976). 'An investigation into the composition and properties of

saliva in relation to surface-cleaning of oil paintings', National Gallery of Canada, Ottawa.

[46] Girling, J. (1992). *'The use of human saliva in conservation'*, MA Thesis. Institute of Archaeology. University College, London.

[47] Nicolaus, K. (1998). *'Handbuch der Gemälderestaurierung'*, 1-427.

[48] Pietsch, A. (2002). *'Lösemittel'* 1-196.

[49] Phenix, A. & Burnstock, A. (1990). 'The deposition of dirt: a review of the literature, with scanning electron microscope studies of dirt on selected paintings', in *Dirt und Pictures Separated,* 11-18.

[50] Carlyle, L., Townsend, J. H. & Hackney, S. (1990). 'Triammonium Citrate: An investigation into its application for surface cleaning', in: Hackney, S., Townsend, J. H. & Eastaugh, N. (Hrsg.): *Dirt and Pictures Separated,* 44-48; Mansmann, K., 'Oberflächenreinigung mit Ammoniumcitraten' in *Zeitschrift für Kunsttechnologie und Konservierung* 2 220-237.

[51] Bjarnhof, M. (1984). 'Removal of damp blotches by the aid of low pressure table'. *ICOM Preprints*, Kopenhagen, 84.2.10 - 84.2.11.

[52] Southall, A. (1990). *'Detergents, soaps, surfactants,* 29-34.

[53] Bockmühl, D. & Weyer, C. (2006). 'Bekämpfung und Prävention mikrobieller Schäden an Lederoberflächen durch herkömmliche Methoden der Lederreinigung', in *Oberflächenreinigung-Material und Methoden,* 126-130.

[54] Saur, R., Schramm, U., Steinhoff, R. &Wolff, H. H. (1991). 'Strukturanalyse der Hautoberfläche durch computergestützte Laser-Profilometrie', *Der Hautarzt, 42,* 499-506.

[55] Rohr, M. & Schrader, K. (1994)., 'Surfactant-induced Skin Roughness. Quantitative Analysis of the Structure of the Skin Via Automated Non-touch Laser Scanning', *Euro Cosmetics, 24-28.*

[56] Rohr, M. & Schrader, K. (1995). 'Hautphysiologische In-vivo-Untersuchungen verschiedener Tenside' *Skin Care Forum, 12,* 12-16.

[57] Efsen, J., Handsen, H. N., Christiansen, S. & Keiding, J. (1995). 'Laser profilometry', in Serup, J. and Jemec, G. B. E. (eds.), *Handbook of Non-Invasive Methods und the Skin,* 97-105.

[58] Müller, U. (1995). 'Roughness Measured by Profilometry: Mechanical, Optical and Laser', in Berardesca, E., Elsner, P., Wilhelm, K.P. and Maibach, H. I. (eds.), *Bioengineering of the Skin: Methods and Instrumentation,* 41-51.

[59] Saur, R. (1993). *'Computergestützte Laser-Profilometrie Entwicklung eines neuen dermatologischen Verfahrens zur Objektivierung der Rauheitsstruktur der Haut'*, MA Thesis, Lübeck.

[60] Eipper, P. B. (2003). 'Trocken-Schwämme zur Oberflächenreinigung', *Museum aktuell, 91,* 3898-3902.

In: Paints: Types, Components and Applications
Editor: Stephanie M. Sarrica

ISBN: 978-1-61761-813-0
© 2011 Nova Science Publishers, Inc.

Chapter 4

COOL PAINT AS URBAN HEAT ISLAND MEASURE TECHNOLOGY[*]

Hideki Takebayashi
Kobe University, Kobe, Japan

ABSTRACT

Urban heat island measure technologies are classified roughly improvement of the building and pavement coating, promotion of ventilation in the urban area, reduction of the exhaust heat from building and vehicle. Cool paint is one of the technologies of improvement of the building and pavement coating, with green roof, water-holding material, etc. Because the performance of cool paint depends on quantity of reflective solar radiation and those of green roof and water-holding material depend on quantity of evaporation, it is hard to compare their performance directly. They are compared based on the surface heat budget model, and surface temperature on each material is the index for the performance. The effect by the improvement of building and pavement coating depends on surface heat budget of each material, the effect by the promotion of ventilation in the urban area depends on heat flux from nearby surface to upper air, and the effect by the reduction of the exhaust heat from the heat exchange device depends on the heat budget of each machine. So, it is hard to compare their performance directly. They are compared based on surface boundary layer model, and air temperature nearby ground surface is the index for the performance by each heat island measure technology. Japanese engineers have developed several improvement technologies related as above. Through various technical evaluation, the effect by cool paint as heat island measure technology is compared with the other technologies.

---

[*] A version of this chapter was also published in Cool Paint as Urban Heat Island Measure Technology, by Hideki Takebayashi, published by Nova Science Publishers, Inc. It was submitted for appropriate modifications in an effort to encourage wider dissemination of research.

INTRODUCTION

In Japan, the urban heat island phenomenon, i.e., increased air temperature that occurs in the urban area has been recognized to be a serious problem. Various heat island measure strategies are developed, and the effectiveness is confirmed in various applications, for example, cool roofs, green roofs, cool pavements, effective HVAC systems, etc. The characteristic of each technology has been clear (AIJ 2007; Takebayashi & Moriyama 2007), but an evaluation method for the effectiveness of these measure strategies has not been established; thus, there are only a few examples of the implementation of measure strategies for the urban heat island phenomenon. Therefore, it is necessary to develop an evaluation method for measure strategies so that the residents and industries can make further technological advancements by implementing these strategies. In general, the evaluation of the effectiveness of heat island measure strategies is carried out by using meso– or small–scale numerical weather simulation models (Taha 1997). Example of the calculation results of meso-scale numerical weather simulation model is shown in figure 1 (Kitao, Moriyama, Nakajima, Tanaka & Takebayashi 2009). Air temperature and wind vector distributions in Osaka region, Japan at 14:00, August 3, 2006 are shown in figure 1. The calculation results in the case of present land cover condition and potential natural land cover condition are shown in left and right figures. Horizontal air temperature profile at the height of 2 m on A - A' line in figure 1 is shown in figure 2. In the area of around 15 to 20 km from Osaka bay, air temperature rise is restrained by the cooler sea breeze. The air temperature difference between coastal area and inland area is about 5 degrees. Air temperature rise by the land cover change from natural to artificial land cover is about 2 degrees. So, air temperature reduction effect by the sea breeze is very large. The target of air temperature reduction by the heat island measure technologies is about 2 degrees at the maximum.

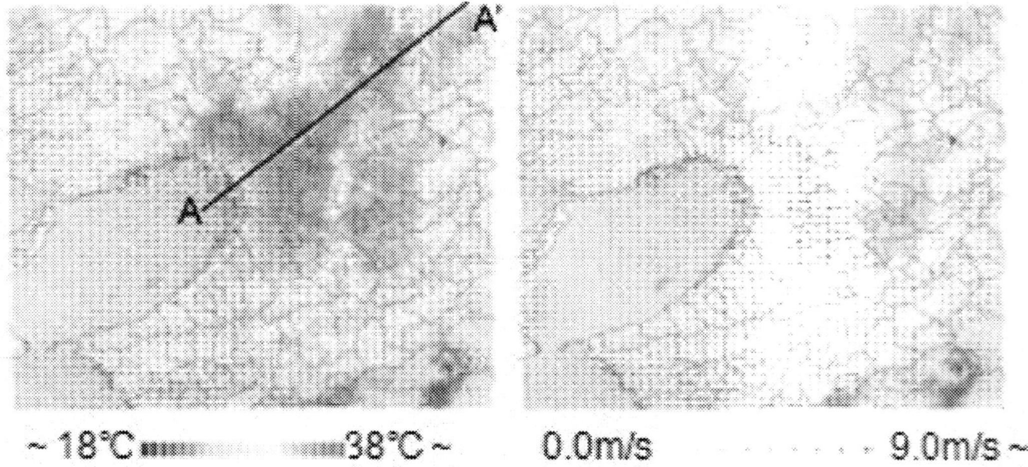

Figure 1. Example of the calculation results of meso-scale numerical weather simulation model Air temperature and wind vector distribution in Osaka region, Japan at 14:00, August 3, 2006. Left: Present land cover condition, Right: Potential natural land cover condition.

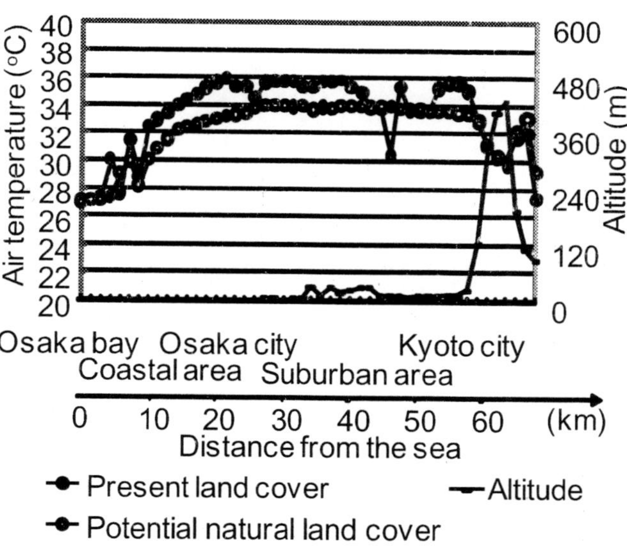

Figure 2. Horizontal air temperature profile at the height of 2m on A - A' line in figure 1.

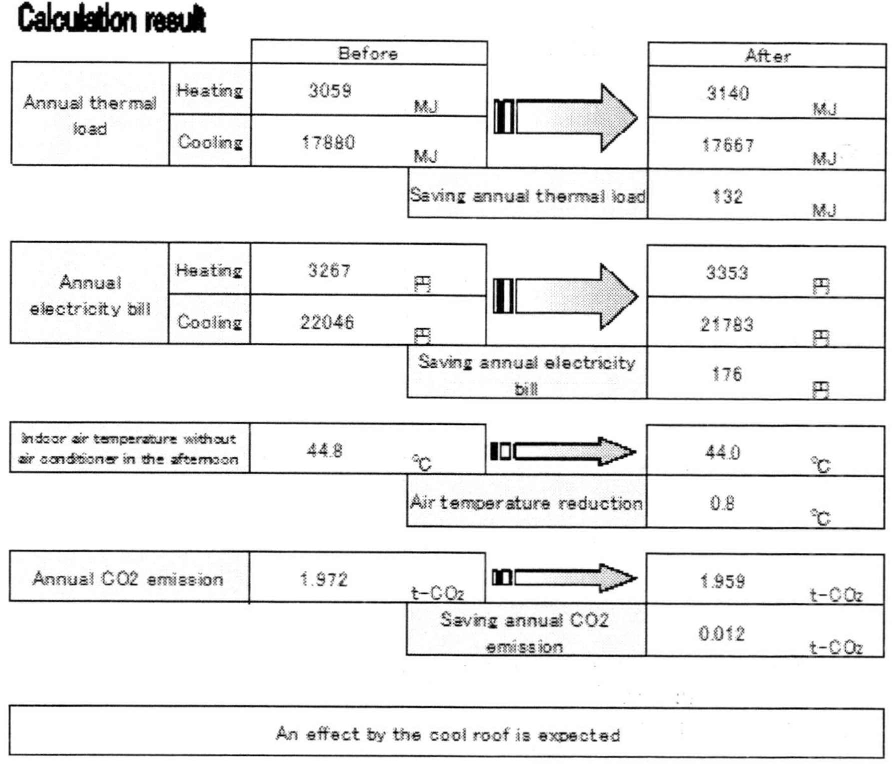

		Before			After	
Annual thermal load	Heating	3059	MJ		3140	MJ
	Cooling	17880	MJ		17667	MJ
			Saving annual thermal load		132	MJ
Annual electricity bill	Heating	3267	円		3353	円
	Cooling	22046	円		21783	円
			Saving annual electricity bill		176	円
Indoor air temperature without air conditioner in the afternoon		44.8	℃		44.0	℃
			Air temperature reduction		0.8	℃
Annual CO2 emission		1.972	t-CO₂		1.959	t-CO₂
			Saving annual CO2 emission		0.012	t-CO₂

Figure 3. An example of the calculation results by the evaluation tool for the private benefit, which is made by Prof. Kondo.

However, it is difficult to carry out numerical simulations for the individual heat island measure strategy. In this paper, a simple evaluation method of surface air temperature reduction by the heat island measure technology is explained and the evaluation results by various measure technologies are presented.

The working group for the appropriate use of cool roof in Architectural Institute of Japan has been carried out several activities such as collection and rearranging the technical information, the evaluation and measurement method, making of the evaluation tool, for the appropriate use of cool roof which is related to the coating of the building surface, while various measure technologies are suggested to the urban heat island phenomenon. The group has developed the evaluation tool for calculating the reduction of the air conditioner load, the increase and decrease of the annual energy consumption by changing the solar reflectance of the building roof, which was examined by the multiple regression analysis for the typical detached house and gymnasium. The calculation model is based on the heat load calculation program used widely in the field of the building environmental engineering conventionally (Udoh, Kondo & Takeda 2009). The working group names this tool the evaluation tool for "the private benefit." An example of the calculation results by the evaluation tool for the private benefit is shown in figure 3. The author translated Japanese Excel sheet into English. When a climatic condition, a building condition, a housing condition, an air-conditioning condition are set up, calculation results of annual thermal load, annual electricity cost, indoor air temperature without air conditioner in the afternoon, annual CO_2 emission are output by the heat load calculation. The system is built by a multiple regression analysis based on many calculation results performed beforehand. The parameters are the area and the completion year of the building, setting room temperature, driving time and heat source of air conditioner, color of the roof before and after complete change, and the electricity company. A large energy saving effect is expected in the building which has not enough insulation such as factory or warehouse, but it is not expected so much in the general building such as office or apartment house.

The group has also developed the evaluation tool for "the public benefit." On this evaluation tool, the improvement effect of the urban thermal environment such as the heat island measure effect and outdoor human thermal environment mitigation effect are calculated.

OUTLINE OF URBAN HEAT ISLAND MEASURE TECHNOLOGY

Several research papers concerning the technical performance evaluation of various kinds of heat island measure technologies are reported in the past studies, and their results are reviewed in some documents and books (AIJ 2008; SHASEJ 2009; Osaka HITEC 2007). Based on those results, the table on the effect of the heat island measure technology is arranged by the working group for the appropriate use of cool roof in Architectural Institute of Japan. The members of the working group are specialists on this field in both university and private company. The quantitative evaluation results are uneven in the same technology category, so the representative characteristics are shown in the table. All technologies except the exterior Insulation and finishing system (outside insulation) are developed for the purpose of heat island measure mainly, so the heat island measure effect by each technology is

Table 1. Rearranging table on the effect of urban heat island measure technology by Cool Roof Committee of Architectural Institute of Japan (Takebayashi, et. al., 2010)

Target	Menu	Heat island measures		Outdoor thermal environment		Energy saving		Other effects	Consideration matters
		Day time	Night time	Day time	Night time	Cooling	Heatng		
Roof	Green	○○	○○	○○	○○	○	○	Improvement of landscape, ecosystem, real estate value, customers service, flood relaxation, environmental education	Maintenance
	High reflectance	○○	○○	-	-	○	△	Improvement of indoor thermal environment	Aging deterioration of performance, reflection for circumference, heating load increases
	Sprinkling, Retaining water	○○	○	-	-	○	-		
	Outside insulation	△	○	-	-	○○	○○		
Wall	Green	○○	○	○○	○○	○	○	Improvement of landscape, ecosystem	Maintenance
	High reflectance	○	○	△	○	○	△	Improvement of indoor thermal environment	Aging deterioration of performance, reflection for circumference, heating load increases
Wall	Sprinkling	○○	○	○	-	○	-	Improvement of landscape	
	Outside insulation	○○	○	△	○	○○	○○		
Road, Pavement, Parking, Open space	Green	○○	○○	○○	○○	-	-	Improvement of landscape, ecosystem	Maintenance
	High reflectance	○○	○○	△	○○	-	-		Aging deterioration of performance, reflection for circumference
	Retaining water	○○	○○	○○	○○	-	-	Rainwater penetration	Water reserving condition
	Sprinkling	○○	○○	○○	○○	-	-		

○○: Superior effect is expected. ○: Good effect is expected. △: Opposited effect may happen. -: Not intended.

judged to be good. On the other hand, the effects of the energy saving by those technologies are not so larger than that by the outside insulation, so enough insulation is given priority to over cool roof and green roof in the case that energy saving is recognized as a top priority purpose. As a supplementary matter, the other effects and consideration matters are also shown in the table. The maintenance of green roof and the aging performance (solar reflectance) deterioration of cool roof are important consideration matters. For green roof, many effects such as an improvement of landscape, ecosystem, real estate value, etc., are pointed out, so green roof is different from cool roof at a point that a general effect is expected. On the basis of these characteristics, it is expected that the appropriate technical choice accepted the local climate or the building use is carried out.

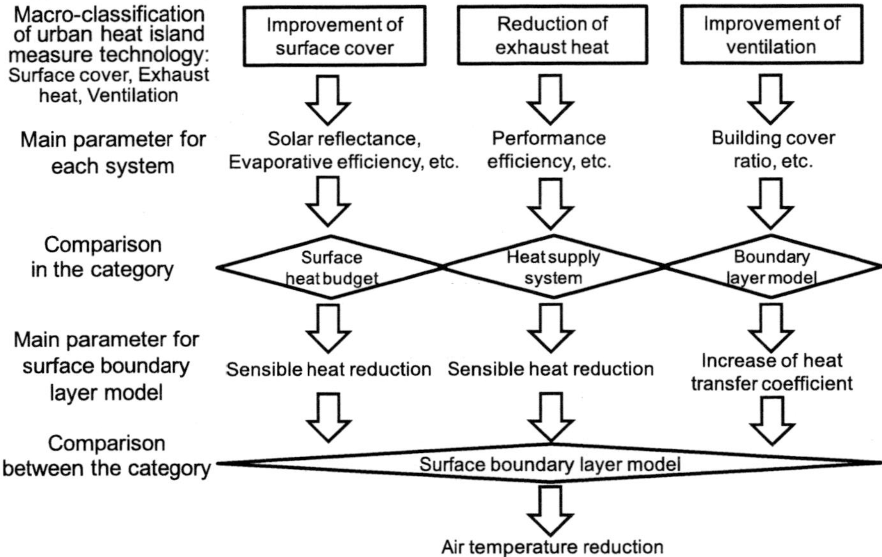

Figure 4. General evaluation process of the urban heat island measure effect.

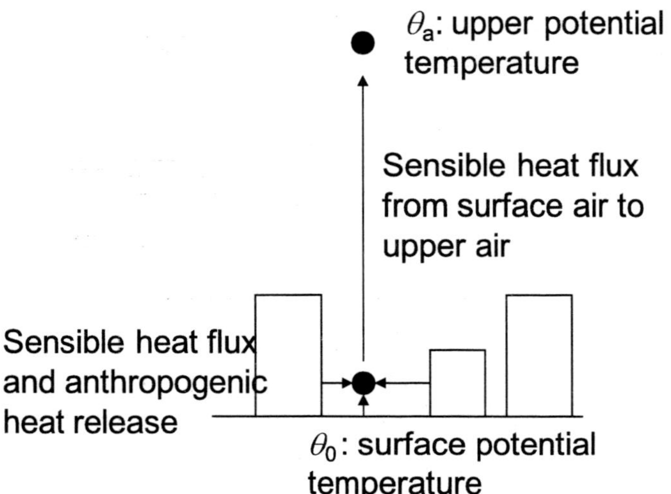

Figure 5. Surface boundary layer model.

EVALUATION METHOD OF THE URBAN HEAT ISLAND MEASURE EFFECT

The general evaluation process of the urban heat island measure effect is shown in figure 4. The heat island measure technology is classified in three categories; the improvement of surface cover, the reduction of exhaust heat and the improvement of ventilation. Cool roof and green roof belong to the category of the improvement of surface cover. The heat island measure effects are obtained by reducing the sensible heat flux in the case of the improvement of surface cover and the reduction of exhaust heat, and by increasing the heat transfer from surface to upper air in the case of the improvement of ventilation. Heat budget nearby ground surface indicating the status of surface air temperature is represented by the surface boundary layer model which is shown in figure 5. In this model, if the sensible heat flux or exhaust heat released to nearby ground surface air is smaller or the sensible heat flux transported from the nearby ground surface to upper air is larger, air temperature nearby ground surface becomes lower. In fact, air temperature nearby ground surface is evaluated by the three dimensional meso-scale numerical weather simulation model such as MM5, WRF (University Corporation for Atmospheric Research; WRF Community), because air temperature is also affected by the geographical features i.e. close to the sea or the mountain, located in the urban or the suburbs, etc. If the company which develops cool roof, green roof, water-holding material pavement, and artificial exhaust heat reduction technology, etc. evaluates the heat island measure effect of their own product in the case that they are applied to the real building or the city, it is required for the complicated calculation such as the above. Therefore, the simple evaluation method is suggested.

SURFACE HEAT BUDGET ON VARIOUS TECHNOLOGIES

Surface heat budget model on the roof, wall, road, pavement, etc. is shown figure 6 and surface heat budget equation is shown in table 2. Main parameters are solar reflectance (ρ) and evaporative efficiency (β). And the other parameters are emissivity (ε), thermal conductivity (λ) and heat capacity ($C_p\gamma$). Emissivity (ε) is approximately constant except metal roofs, and thermal conductivity (λ) and heat capacity ($C_p\gamma$) have a small sensitivity for surface temperature in comparison with solar reflectance (ρ) and evaporative efficiency (β). Therefore, default values are set in emissivity, thermal conductivity and heat capacity, and surface heat budget is calculated in the case that main two parameters (solar reflectance and evaporative efficiency) are changed. Calculation results of sensible heat flux and surface temperature at 13:00 and 21:00 averaged in summer season from 18 July to 15 September, 2005 are shown in figures 7 and 8. Observation data of solar radiation, air temperature, relative humidity, wind velocity and cloudiness that were observed at Osaka observatory are used for this calculation. Calculation results of s sensible heat flux and surface temperature are shown by the counter lines. We can read the calculation value when solar reflectance (ρ) of the horizontal axle and evaporative efficiency (β) of the vertical axle are selected. Outside insulation means the effect by lower heat storage of the surface material e.g. tiled roof. The sets of solar reflectance (ρ) and evaporative efficiency (β) that are extracted from several observation results by authors and the other Japanese researchers are plotted. Examples of several kinds of surface heat budget experimental sites are shown in figure 9. Surface heat budget of roof, road, pavement, parking lot, etc. have been studied and main parameters are estimated based on these results.

From figures 7 and 8, when solar reflectance (ρ) or evaporative efficiency (β) is large, sensible heat flux and surface temperature is small and heat island measure effect is large. For example, we can recognize that solar reflectance (ρ) 0.5 and evaporative efficiency (β) 0.15 are almost equal for the daytime heat island measure effect. In the nighttime, the contribution of solar reflectance is relatively smaller than that of evaporative efficiency because solar radiation is 0. In this way, we can compare several materials with different characteristics and get an indicator for the development of new materials.

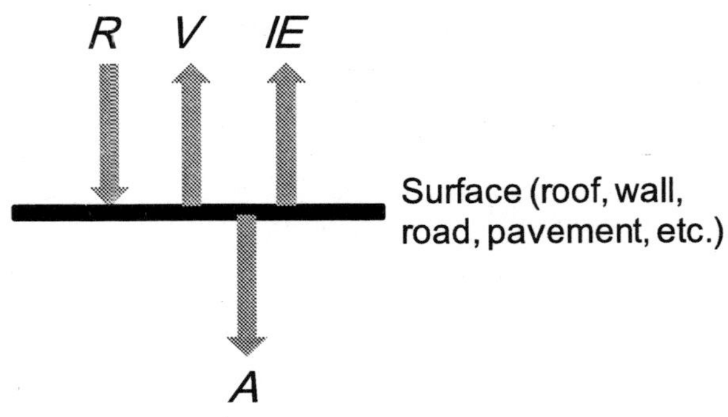

Figure 6. Surface heat budget model.

Table 2. Surface heat budget equation

$$R = V + lE + A$$

$$R = (1 - \rho)S{\downarrow} + \varepsilon L{\downarrow} - \varepsilon\sigma T_s^4$$

$$V = \alpha(T_s - T_a)$$

$$lE = l\beta(\alpha/C_p)(X_s - X_a)$$

$$A = \lambda/\Delta z (T_s - T_{z1})$$

R: Net radiation (W/m²)
V: Sensible heat flux (W/m²)
lE: Latent heat flux (W/m²)
A: Conduction heat flux (W/m²)
ρ: Solar reflectance (Albedo)
$S{\downarrow}$: Solar radiation (W/m²)
$L{\downarrow}$: Long wave radiation (W/m²)
ε: Emissivity (=1.0)
σ: Stefan-Boltzmann constant (=5.67*10⁸W/m²K⁴)
α: Convection heat transfer coefficient (=4.2v+6.2W/m²K) v: wind velocity (m/s)
T_s: Surface temperature (C)
T_a: Air temperature (C)
l: latent heat (=2,512kJ/kg)
β: Evaporative efficiency (-)
C_p: Specific heat of air (=1.0kJ/kgK)
X_s: Saturated absolute humidity(kg/kg)
X_a: Absolute humidity of air (kg/kg)
λ: thermal conductivity (W/mK)
Δz: Distance from surface to first layer (m)
T_{z1}: Temperature in the first layer (C)

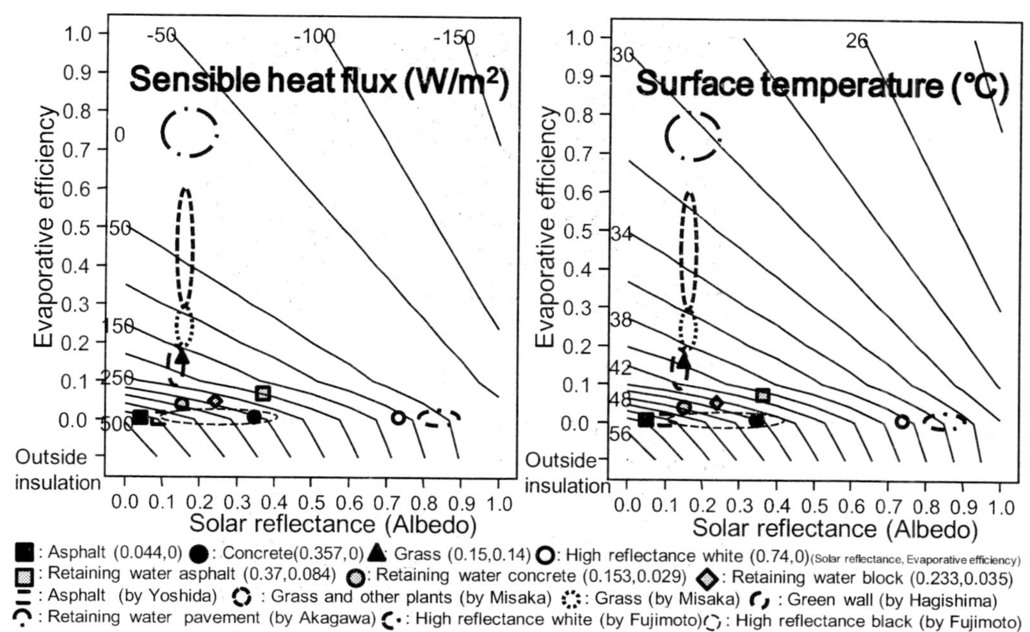

Figure 7. Calculation result of sensible heat flux and surface temperature at 13:00 averaged in summer (from 18 July to 15 September, 2005, at Osaka, Japan).

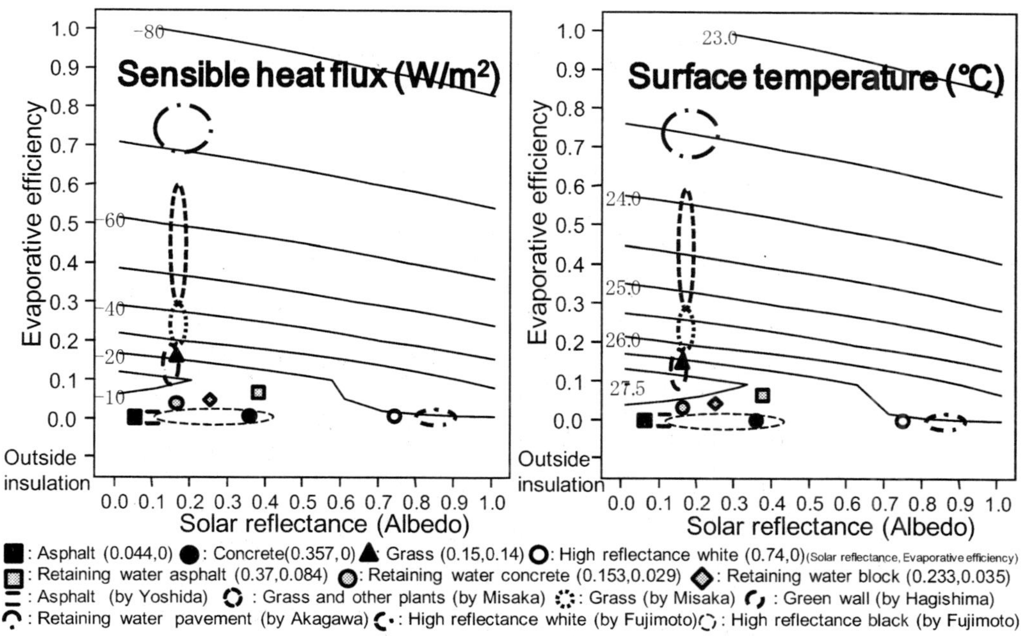

Figure 8. Calculation result of sensible heat flux and surface temperature at 21:00 averaged in summer (from 18 July to 15 September, 2005, at Osaka, Japan).

Figure 9a. Experimental roof of Kobe University (Takebayashi & Moriyama 2007).

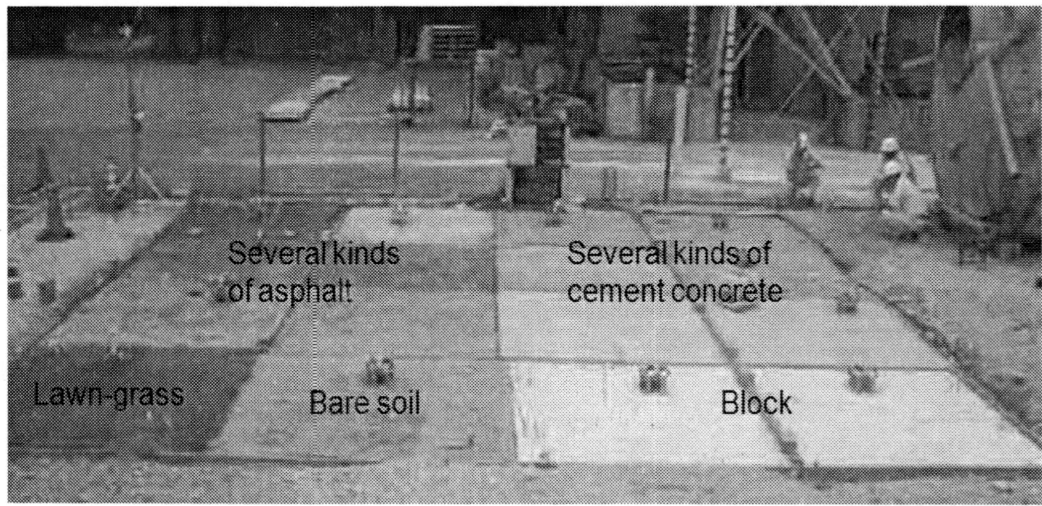

Figure 9b. Experimental road and pavement surface of Sumitomo Osaka Cement Co., Ltd. (Takebayashi et.al. 2008).

Figure 9c. Experimental parking surface of Hyogo Prefecture (Takebayashi & Moriyama 2009).

Figure 9d. Experimental incline roof of the industrial guild of Sekisyu Kawara (Sugihara, Moriyama & Takebayashi 2006).

Figure 9. Several kinds of surface heat budget experimental sites.

SIMPLE EVALUATION METHOD OF SURFACE AIR TEMPERATURE REDUCTION

In the meso–scale weather simulation model, which is used for evaluating the effectiveness of heat island measure strategies, the lower layer of the one-dimensional surface–boundary layer model is coupled with the upper layer of the three-dimensional hydrodynamic turbulence model. The amount of fluxes exchanged between the ground surface and the atmosphere, such as momentum, heat, and moisture fluxes, is estimated by calculation results of the ground surface heat budget. Recently, an urban canopy model is incorporated in the surface–boundary layer model for precise estimation of the amount of fluxes (Masson 2000; Kusaka, Kondo, Kikegawa & Kimura 2001; Kondo, Genchi, Kikegawa, Ohashi, Yoshikado & Komiyama 2005).

The structure of the meso–scale weather simulation model, which consists of the three-dimensional hydrodynamics model and the surface–boundary layer model, is shown in figure 10. The surface air temperature and wind velocity are calculated using the one-dimensional surface–boundary layer model; then, these results are set as the lower boundary condition of the three-dimensional hydrodynamics model. The amount of sensible heat flux released, which is estimated from the surface heat budget of the ground surface, is set as the lower boundary condition. Therefore, if we can obtain the upper air data across the entire calculation domain, we can calculate the surface air temperature and wind velocity by using only the one-dimensional surface–boundary model with precision similar to that of general meso–scale weather simulation models. This method is simple, i.e., it does not require three-dimensional calculations. In fact, it is difficult to acquire the upper air data with a higher space resolution; we acquire the upper air data at four observation points.

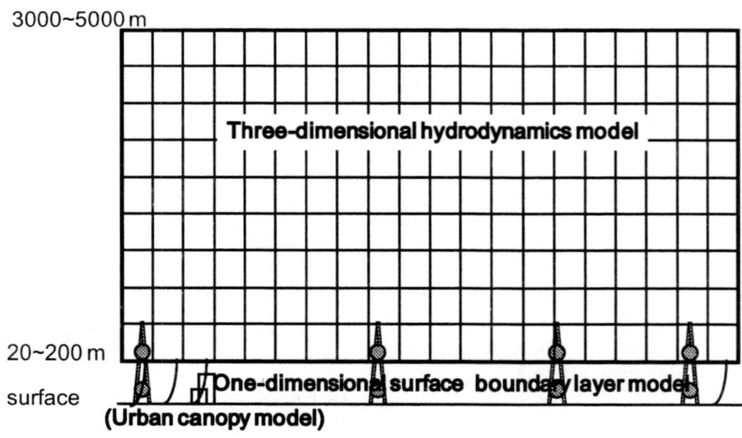

Figure 10. Structure of general meso-scale weather simulation model.

As shown in figure 10, the upper boundary condition of the surface–boundary layer model, in other words, the lower boundary condition of the three-dimensional hydrodynamics model is obtained directly. We install observation instruments on four steel towers with the cooperation of an electric power company and a phone company. The instruments are set from the west end to the east end of Osaka plains. Observation data recorded by the instruments is used for calculating the upper boundary condition of the surface–boundary layer model, and the change in the surface air temperature with the change in the amount of sensible heat flux released is particularly examined.

Outline of upper weather observation is shown in figure 11, and the installation situation of measuring instruments is shown in figure 12. The wind direction and velocity are recorded using a two–dimensional ultrasonic anemometer. The air temperature is measured using a thermistor–type thermometer. The observation height is determined to be located in the constant flux layer. Nanko is a coastal area, Namba is an urban area, and Aramoto and Ishikiri are suburbs. Each observation instrument is installed on the south side of the steel tower for reducing the effect on the observation data by the tower, because land and sea breeze are flowing in the east–west direction in the summer. Further, ten years' data collected at Osaka tower, which is located in central Osaka city, are also used for secular variations. It is located in urban area, and its distance from the seashore is the same as that of Namba tower.

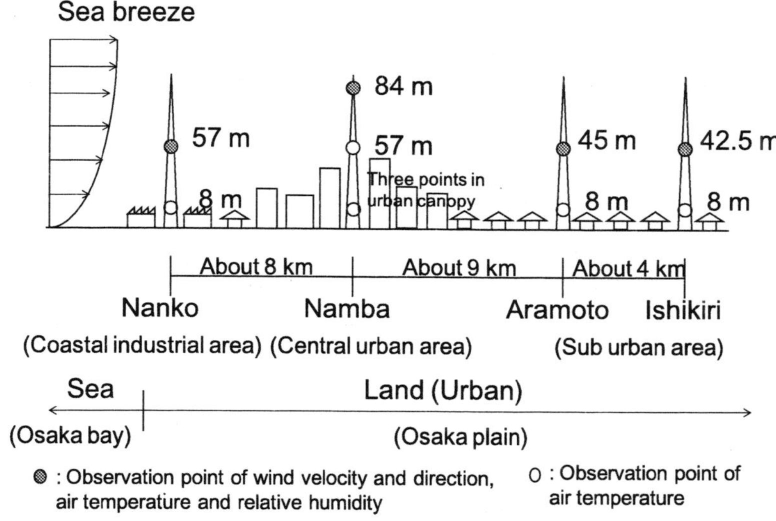

Figure 11. Outline of upper weather observation.

The sensible heat flux in the one-dimensional surface–boundary layer model is defined as shown in (1).

$$V = \alpha_c \left(\theta_0 - \theta_a \right)$$

(1)

where V is the sensible heat flux (W/m^2), α_c is convection heat transfer coefficient (W/m^{2}K), θ_0 is surface air potential temperature (K) and θ_a is upper air potential temperature (K) (see figure 5).

The convection heat transfer coefficient is expressed as shown in (2) on the basis of the Monin–Obukhov similarity theory.

$$\alpha_c = \frac{C_p \gamma k^2 u_a}{F_m F_h}$$

(2)

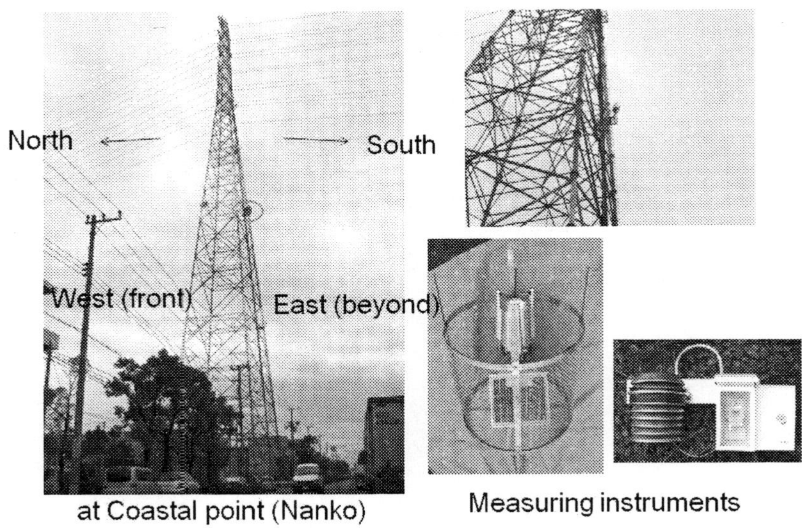

Figure 12. Installation situation of measuring instruments.

where C_p is the specific heat of air (=1.0 kJ/kgK) γ is density of air (=1.2 kg/m^3), k is von Karman constant (=0.35) and u_a is upper wind velocity (m/s). F_m and F_h are integral values of the universal function (-). Businger's experimental expressions are used for the universal function.

It is noted that this similarity theory can be effectively used for explaining heat transportation phenomenon in urban areas (Moriwaki & Kanda 2006). However, in the case of unstable weather conditions, the constant flux layer is not formed. Since the similarity theory is generally used in meso-scale weather simulation models and the main purpose of this study is to develop a simple evaluation method for the effectiveness of the heat island measure strategies, we use this similarity theory, although it has certain limitations.

Parameters used in this model are the upper wind velocity u_a, roughness parameter z_0, and difference in the upper and surface air potential temperatures $\theta_0-\theta_a$. The upper wind velocity u_a and the upper and surface air potential temperatures θ_a, θ_0 are measured on the steel towers, and z_0 is assumed to be 1.5 m, because Osaka city is almost urbanized. α_c is estimated using the above mentioned values. Some estimation methods for z_0 are suggested, for example, using geometric characteristics such as building height and density and turbulence statistics such as friction speed, but, in this study, it is assumed according to previous studies (Moriyama & Takebayashi 1999).

The hourly mean values of the convection heat transfer coefficient, which are estimated by the above mentioned method using the observation data recorded at the four towers, are shown in figure 13. The convection heat transfer coefficient is approximately 100 W/m^{2}K in the daytime and approximately less than 50 W/m^{2}K in the nighttime. In the daytime, sea breeze blows into the urban area, which makes the climatic conditions unstable, and therefore, a constant flux layer may not be formed. It is believed that heat is transported due to relatively large-scale turbulence in the sea breeze.

When additional amount of sensible heat flux ΔV (W/m^2) is released into the surface air, the change in the surface air temperature $\Delta\theta$ (K) is estimated. In this case, the sensible heat flux transported from the surface air to the upper air is expressed as shown in (3). Upper air temperature θ_a changes also as a result of changing sensible heat flux, however its quantity is considerably small in comparison with $\Delta\theta$, because the upper air exists in the constant flux layer.

$$V + \Delta V = \alpha_c' \left(\theta_0 + \Delta\theta - \theta_a \right) \tag{3}$$

where $\Delta\theta$ causes α_c to change to α_c'. Thus, it is necessary to recalculate α_c' by convergence calculation. However, if $\Delta V < V$, then $\alpha_c' = \alpha_c$. So, (4) can be derived from (1) and (3).

$$\Delta\theta = \Delta V / \alpha_c \tag{4}$$

Here, the evaluation of the effectiveness of heat island measure strategies is carried out by assuming the negative ΔV value and $\alpha_c' < \alpha_c$. Therefore, (4) gives an underestimate of $\Delta\theta$, implying that the evaluation is incorrect.

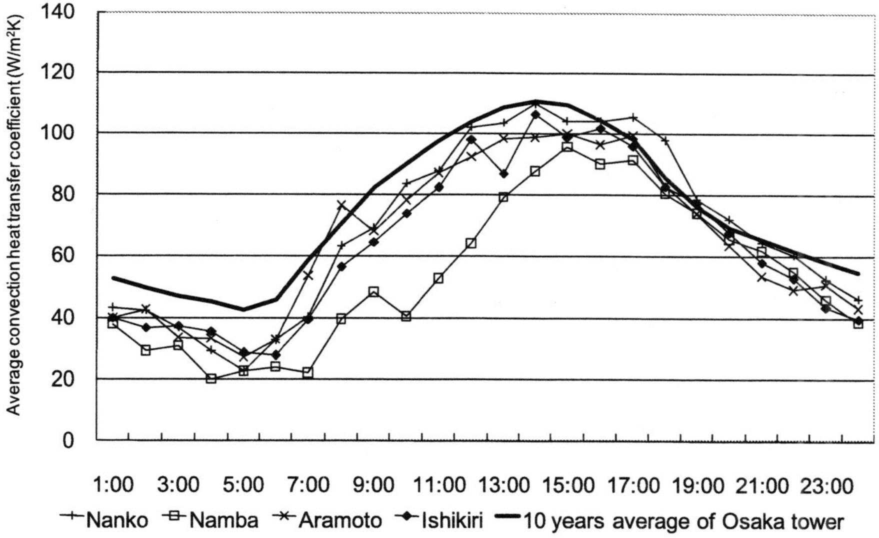

Figure 13. Hourly mean values of convection heat transfer coefficient recorded at four towers and Osaka tower from August 1 to 31.

When additional sensible heat flux of 10 W/m^2 is released into the surface air, the change in the surface air temperature which is estimated from (4) using the convection heat transfer coefficient of figure 13 is shown in figure 14. When the convection heat transfer coefficient is small, the change in the surface air temperature is large. For the same amount of additional sensible heat flux released, the difference of the surface air temperature rise between daytime and nighttime changes by a factor of two or three. However, because the standard deviation in the convection heat transfer coefficient is large in the nighttime, i.e., the average convection heat transfer coefficient of 50 W/m^{2}K changes to 20 W/m^{2}K, so $\Delta\theta$ changes from 0.2 °C/(10W/m^2) to 0.5 °C/(10W/m^2). Because the atmosphere is stable in the nighttime, it may be said that the surface air temperature rises greatly with a small change in the additional sensible heat flux released.

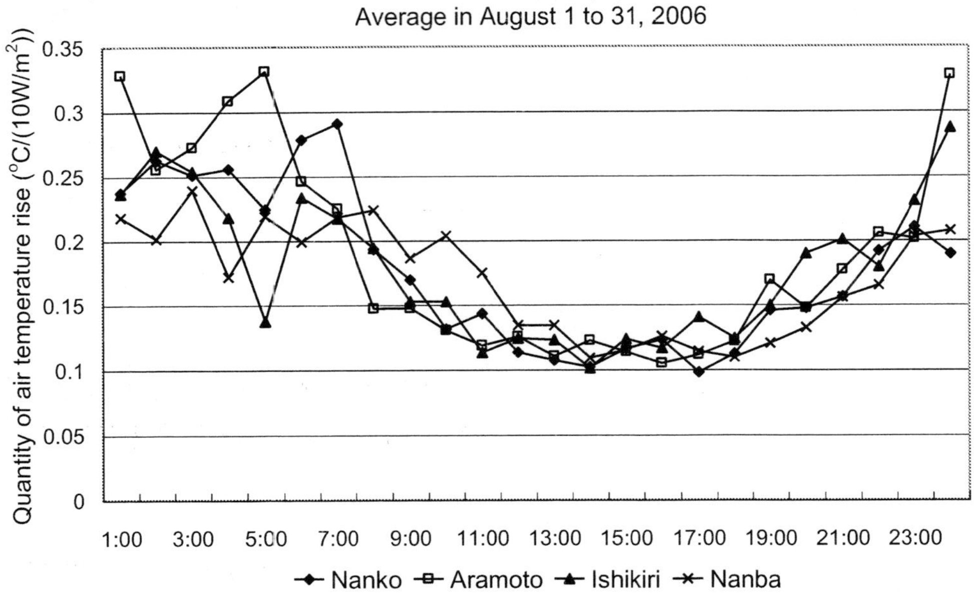

Figure 14. Quantity of surface air temperature rise in the case of additional thermal road of 10 W/m² estimated by observation results at each tower.

EVALUATION RESULTS OF VARIOUS TECHNOLOGIES

Urban heat island measure strategies implemented on buildings and road surface coatings are selected for the evaluation of the heat island measure effect. Parameters are set according to previous studies, as shown in table 3 (Takebayashi & Moriyama 2007). Calculation result of surface temperature and sensible heat flux averaged in each hour from 1 to 31 August 2006, at Osaka are shown in figures 15, 16. Sensible heat flux reduction averaged in daytime and nighttime based on surface heat budget model (averaged from 1 to 31 August 2006, at Osaka) is shown in table 4. The decrease in the amount of sensible heat flux released is calculated by assuming that the heat island measure strategy is implemented on all concrete surface area (i.e., green roofs, white cool roofs, gray cool roofs) or all asphalt surface area (i.e., water keeping asphalt, water keeping concrete, water keeping block). The base condition is assumed to be consisted of concrete surface 60% and asphalt surface 40%. But the relationship between each other building that means the sunshade by the nearby buildings and mutual radiation exchange is not considered here.

Surface air temperature reduction is calculated by the sensible heat flux released reduction and the changing values in the surface air temperature with the release of additional sensible heat flux. The reduced surface air temperatures recorded at Nanko (coastal) and Namba (urban) towers are shown in tables 5 and 6. Because the effect of the measure strategies in reducing the amount of sensible heat flux released is small in the nighttime, the decrease in the surface air temperature is small, too. If 10 W/m² of anthropogenic heat is reduced uniformly in the nighttime, the surface air temperature is reduced by 0.21 °C in Nanko and by 0.19 °C in Namba.

Table 3. Parameter of surface heat budget

	Asphalt	Concrete	Green roof	Cool roof - white	Cool roof - gray	Water keeping asphalt	Water keeping concrete	Water keeping block
Albedo (-)	0.044	0.357	0.15	0.74	0.36	0.37	0.153	0.233
Evaporative efficiency (-)	0	0	0.14	0	0	0.084	0.029	0.035
Thermal conductivity (W/mK)	0.84	1.59	0.6	1.59	1.59	0.87	0.99	0.65
Thermal capacity (J/m^3K)	700000	100000	2900000	100000	100000	1500000	500000	300000

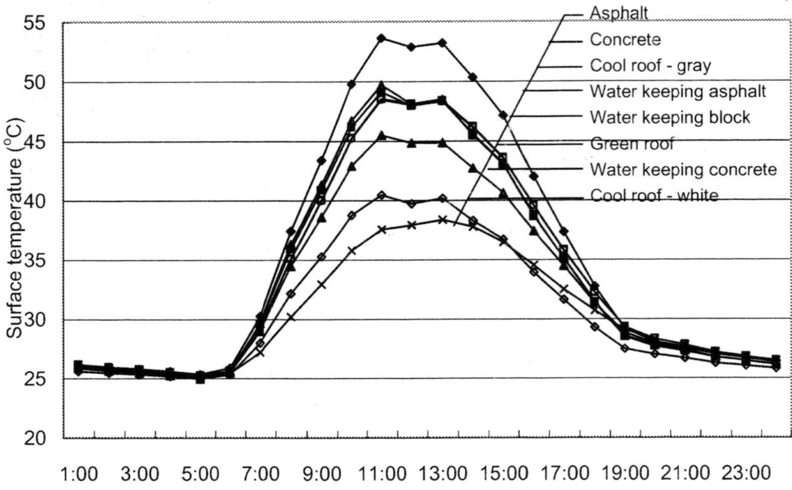

Figure 15. Calculation result of surface temperature averaged in each hour from 1 to 31 August 2006, at Osaka.

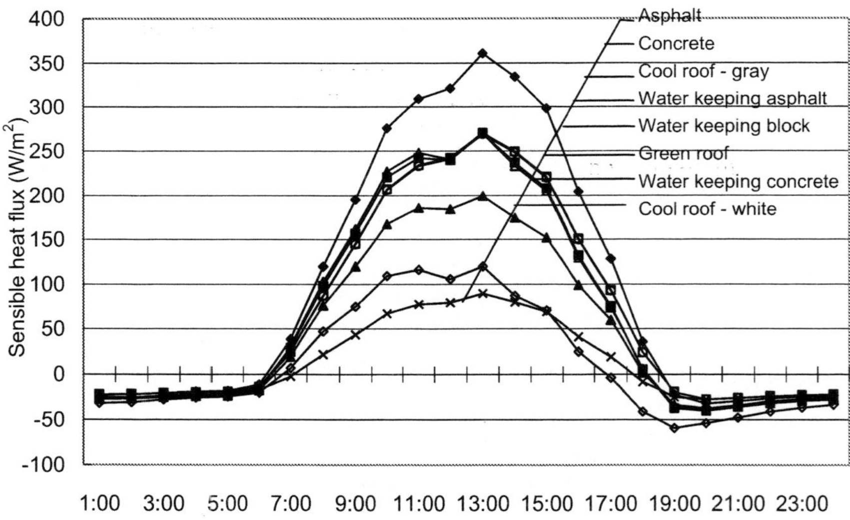

Figure 16. Calculation result of sensible heat flux averaged in each hour from 1 to 31 August 2006, at Osaka.

Table 4. Sensible heat flux reduction averaged in daytime and nighttime based on surface heat budget model (averaged from 1 to 31 August 2006, at Osaka). The difference with concrete in the case of roof, the difference with asphalt in the case of road

W/m^2	Green roof	Cool roof - white	Cool roof - gray	Water keeping asphalt	Water keeping concrete	Water keeping block
daytime	34.5	92.6	0.73	30.1	11.2	11.0
nighttime	4.21	0.58	0.00	2.57	1.02	1.10

Table 5. Surface air temperature reduction averaged in daytime and nighttime at Nanko (Coastal) (averaged from 1 to 31 August 2006, at Osaka)

°C	Green roof	Cool roof - white	Cool roof - gray	Water keeping asphalt	Water keeping concrete	Water keeping block
daytime	0.42	1.19	0.01	0.38	0.14	0.13
nighttime	0.08	0.01	0.00	0.05	0.02	0.02

Table 6. Surface air temperature reduction averaged in daytime and nighttime at Namba (Central urban) (averaged from 1 to 31 August 2006, at Osaka)

°C	Green roof	Cool roof - white	Cool roof - gray	Water keeping asphalt	Water keeping concrete	Water keeping block
daytime	0.48	1.37	0.01	0.43	0.16	0.15
nighttime	0.07	0.01	0.00	0.04	0.02	0.02

ESTIMATION METHOD OF SOLAR REFLECTANCE

Main parameters of surface heat budget are solar reflectance (Albedo) and evaporative efficiency. Measurement methods of solar reflectance are classified two categories; using the pyranometer, e.g. ASTM E-1918 and the spectrum meter, e.g. ASTM E903, JIS K5602. An example of the measurement of solar reflectance using the pyranometer is shown in figure 17. In this measurement, 4 white sheets that solar reflectance is given are set on the objective surface under the pyranometer. In the next measurement, the white seats are replaced on the black seats that solar reflectance is given. Then, more accurate solar reflectance on the objective surface except neighboring influence is found with the following equations, using the measurements results of solar radiation on the objective surface, white and black sheets.

$$\rho_{white\ measure} = \phi * \rho_{white\ right\ value} + \rho_{other} \tag{5}$$

$$\rho_{black\ measure} = \phi * \rho_{black\ right\ value} + \rho_{other} \tag{6}$$

$$\rho_{objective\ measure} = \phi * \rho_{objective\ right\ value} + \rho_{other} \tag{7}$$

where, ϕ is the view factor between the objective area and the pyranometer. ϕ and ρ_{other} (solar reflectance of neighboring area) are obtained from the equations (5) and (6). Then, $\rho_{objective\ right\ value}$ (solar reflectance of objective area) is obtained from the equation (7).

Figure 17. Example of the measurement of solar reflectance using the pyranometer.

If the surface of objective area is not flat e.g. typical Japanese tile, solar reflectance is reduced by the multiple reflection of each surface material. We can get the reduction value from the flat surface using the chart in figure 18. The chart is for the equal diffusion reflection surface, and in the case of the mirror reflection surface we have another chart sheets for several incident angles. The triangle roof which is used for the factory is assumed as the roof shape. The shape is expressed by the view factor between two surface materials. If the shape has a deep valley and the view factor between two surface materials is large, the reduction of solar radiation from the flat surface is large because of the multiple reflection of each surface material. And, the reduction of solar reflectance in the case of large solar reflectance of the material is larger than that in the case of small solar reflectance of the material. Reduction of solar reflectance from the flat surface is shown in figure 19. In the case that solar reflectance of the material is large, the reduction potential by the multiple reflection is also large, so the maximum point in the reduction of solar reflectance from the flat surface moves to larger side in the solar reflectance of the material.

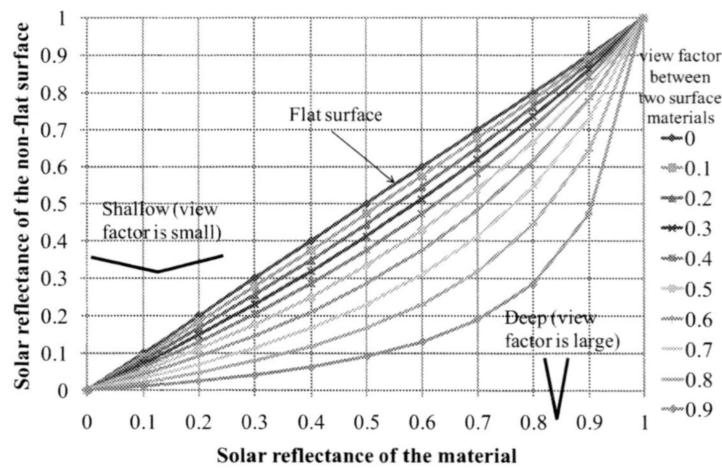

Figure 18. Solar reflectance of the non-flat surface in the case of the equal diffusion reflection surface.

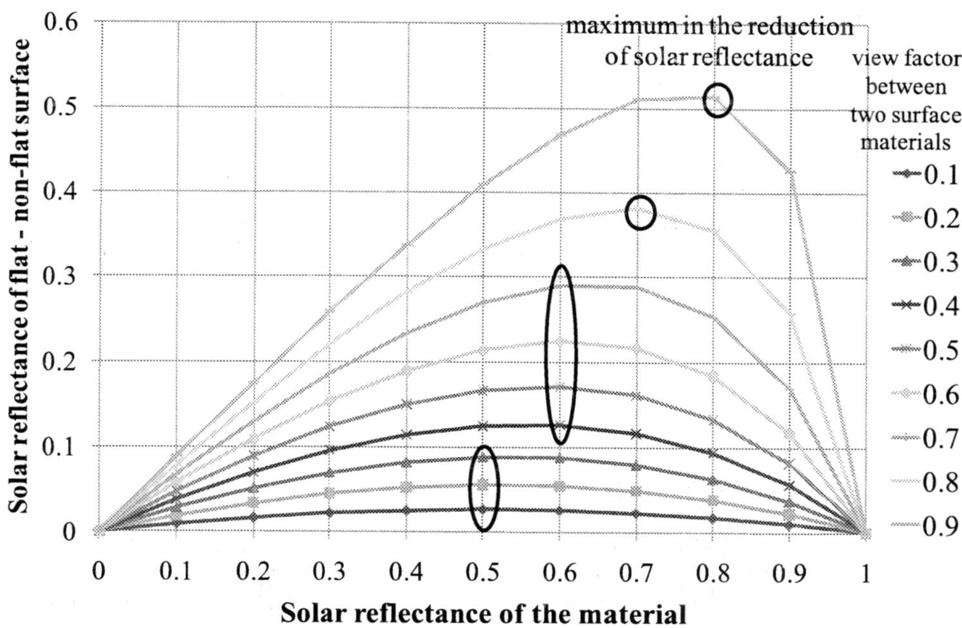

Figure 19. Reduction of solar reflectance from the flat surface.

ESTIMATION METHOD OF EVAPORATIVE EFFICIENCY

Evaporative efficiency β (see table 2) is also main parameter of surface heat budget model. β is estimated from the observation result of the evaporative flux using the fourth equation in table 2. Several methods are suggested for the measurement of evaporative flux. However, some of those need a large objective area and a special experiment for the measurement. Here, for the comparatively easy on-site application at the limited area such as the green roof, an example of the measurement is explained as follows. Measurement image of evaporative flux from the green roof is shown in figure 20. The weight of some samples for gravimetry is measured by an electronic balance and the implication water rate is measured by TDR (Time Domain Reflectometry) sensors which are buried in the soil. Quantity of the water evaporated during the period of the two measurement time is estimated from the changing weight and the implication water rate. If evaporation flux is provided, evaporative efficiency β is calculated according to the fourth equation in table 2. Example of measurement results of evaporative flux from grass cover is shown in figure 21. Latent heat flux is also shown in figure 21. In the daytime relatively large latent heat flux is transferred from grass surface to atmosphere. However, the variation between measurements methods is large, so the improvement of the precision is necessary. Estimation results of evaporative efficiency β on grass cover surface with different kinds of soils are shown in figure 22. It is about 0.2 to 0.3 and its change by time is small.

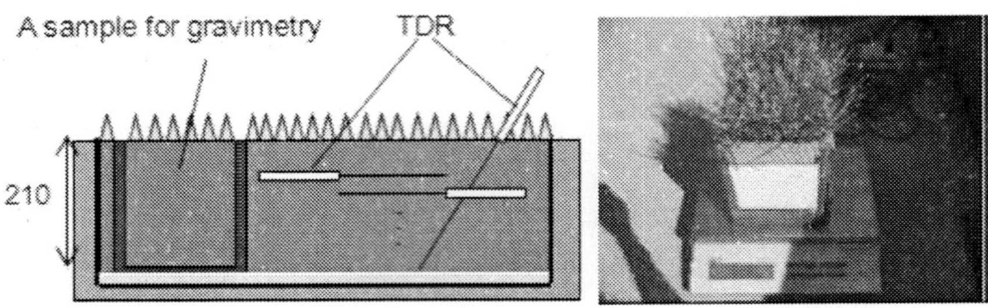

Figure 20. Measurement image of evaporative flux from the green roof.

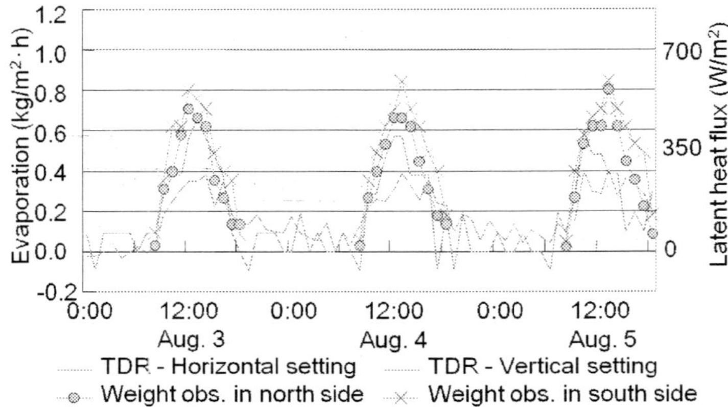

Figure 21. Example of measurement results of evaporative flux from grass cover.

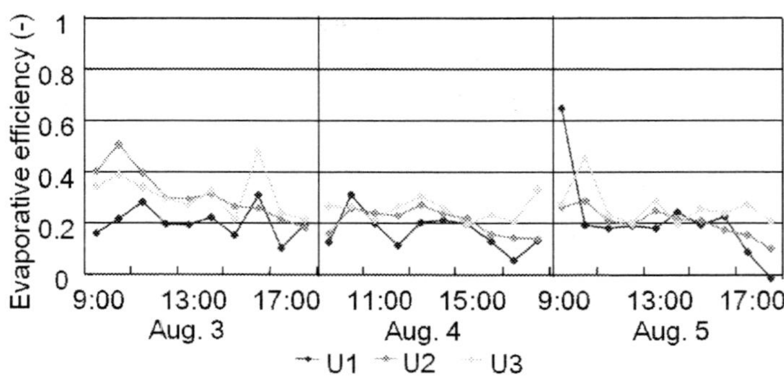

Figure 22. Estimation results of evaporative efficiency β on grass cover surface with different kinds of soils.

EFFECT ON RADIATION ENVIRONMENT BY URBAN FORM

Reflection of solar radiation and evaporation are main roles for the heat island measure technology. When quantity of reflective solar radiation or evaporation is large, the heat island measure effect is large. It is also important point of view where the priority to the heat island measure technology should be applied. Relationship between urban form and sensible heat flux is shown in figure 23. The measure technology for the roof surface is given priority in large building coverage ratio area; e.g. downtown area. The measure technology for the road surface is given priority in small building coverage ratio; e.g. residential area. Priority of the measure technology for the wall surface is not so high.

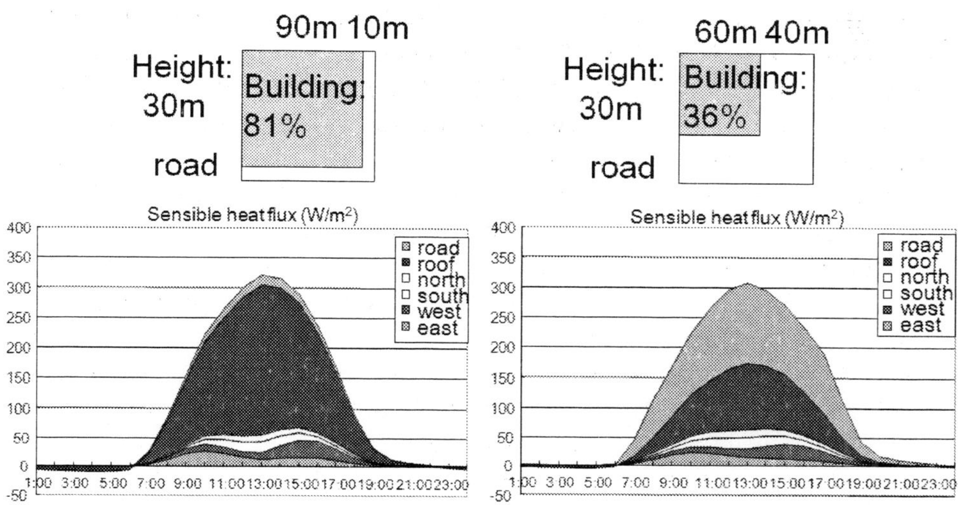

Figure 23. Relationship between urban form and sensible heat flux which is averaged in each hour from 18 July to 15 September 2005, at Osaka.

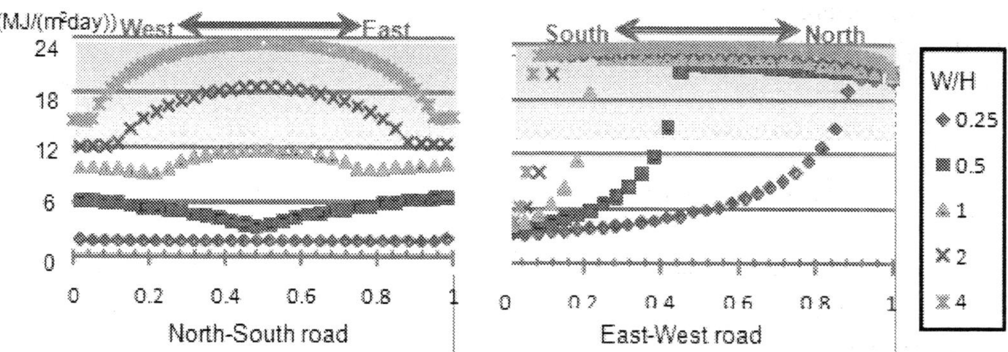

Figure 24. Calculation results of the distribution of day multiplication insolation in north-south and east-west road at 4 August 2009, at Osaka.

Priority is different by a place in the road space. Calculation results of the distribution of day multiplication insolation in north-south and east-west road are shown in figure 24. In the north-south direction road, when the building height of both sides is low for the road width, in other words W / H is big, the day multiplication insolation is large around the center of the road. In the east-west direction road, the day multiplication insolation is small in the area near to the south side building influenced by the shadow of the south side buildings, and when the building height of south side is high for the road width, in other words W / H is small, the shadow area is large. In the crossing, both characteristics of east-west and north-south road are shown, and the day multiplication insolation is large because the shadow area is few.

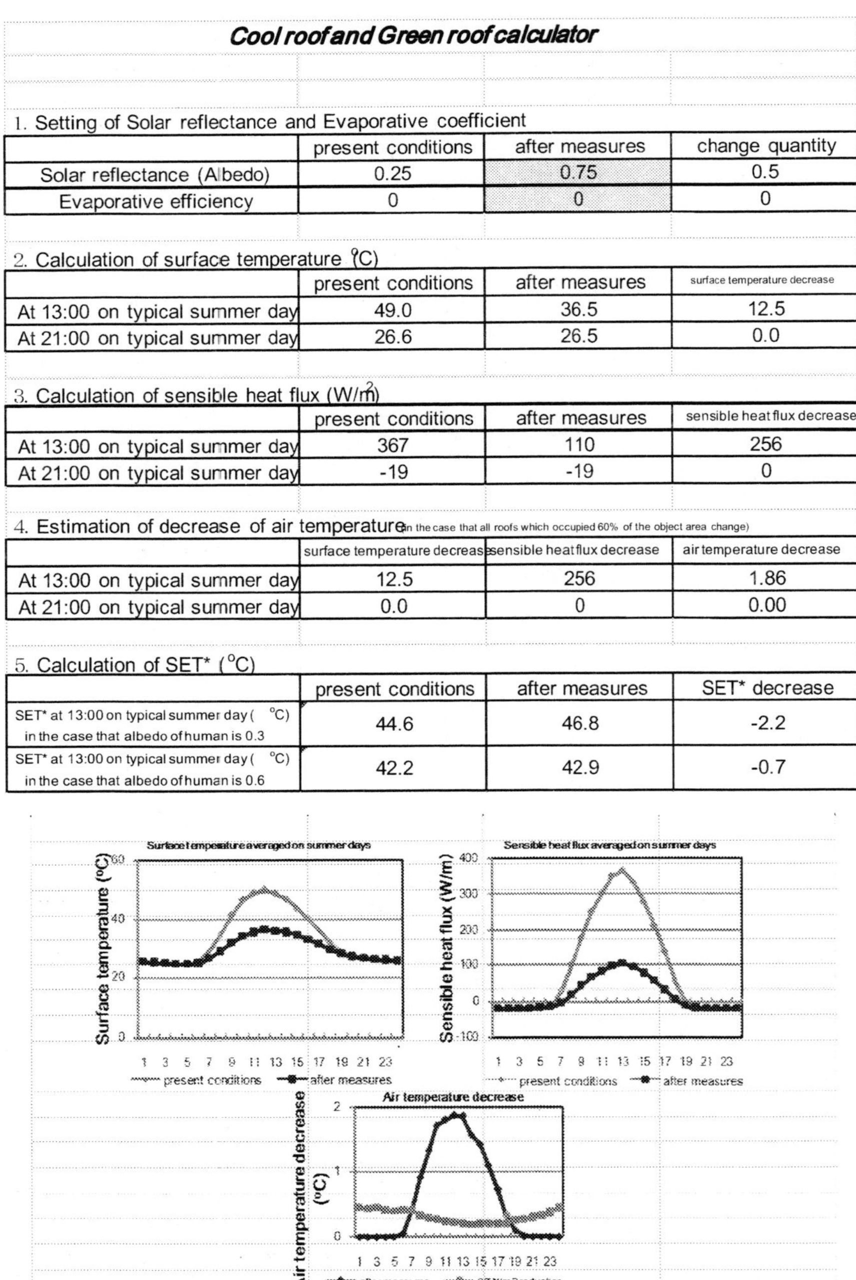

1. Setting of Solar reflectance and Evaporative coefficient

	present conditions	after measures	change quantity
Solar reflectance (Albedo)	0.25	0.75	0.5
Evaporative efficiency	0	0	0

2. Calculation of surface temperature (°C)

	present conditions	after measures	surface temperature decrease
At 13:00 on typical summer day	49.0	36.5	12.5
At 21:00 on typical summer day	26.6	26.5	0.0

3. Calculation of sensible heat flux (W/m²)

	present conditions	after measures	sensible heat flux decrease
At 13:00 on typical summer day	367	110	256
At 21:00 on typical summer day	-19	-19	0

4. Estimation of decrease of air temperature (in the case that all roofs which occupied 60% of the object area change)

	surface temperature decrease	sensible heat flux decrease	air temperature decrease
At 13:00 on typical summer day	12.5	256	1.86
At 21:00 on typical summer day	0.0	0	0.00

5. Calculation of SET* (°C)

	present conditions	after measures	SET* decrease
SET* at 13:00 on typical summer day (°C) in the case that albedo of human is 0.3	44.6	46.8	-2.2
SET* at 13:00 on typical summer day (°C) in the case that albedo of human is 0.6	42.2	42.9	-0.7

Figure 25. Simple evaluation sheet of the effect by the heat island measure technology.

SIMPLE EVALUATION TOOL

The simple evaluation excel sheet of the effect by the heat island measure technology is shown in figure 25. The sheet is one page and a user can print it easily. The database of the calculation results is stored in another seat and a user can also confirm the database. When a user changes solar reflectance ρ or evaporative efficiency β, quantity of change of surface temperature, sensible heat flux, air temperature and SET* are calculated based on the above-mentioned method. A concrete surface roof that solar reflectance is 0.25 and evaporative efficiency is 0 is assumed for the based condition before the heat island measure technology is introduced. Calculation results at 13:00 and 21:00 are shown in the table and the hourly change of each element is confirmed in the graph under the table. SET* is calculated in the case of 0.3 and 0.6 solar reflectance of the cloths of the human body at the daytime. Examples of the calculation results of the effect by the heat island measure technology are shown under the table in figure 25. In this case, an application of the high solar reflectance paint which is 0.75 is assumed. Calculation results of air temperature decrease by measures to reduce artificial heat release of 30 W/m^2 all day long is shown in the figure, too. It is confirmed that the air temperature decrease by the reduction of the artificial heat release is larger in the nighttime in comparison with the daytime.

Calculation results of reduction of sensible heat flux and air temperature at 13:00 averaged from 18 July to 15 September 2005 at Osaka, Japan are shown in figure 26. By the outside insulation (exterior insulation and finishing system) that thermal conductivity and heat capacity are set up based on the conventional study results, surface temperature rises and sensible heat flux increases in the daytime, but the changes of them are not so large in the nighttime. It is difficult to distinguish the difference of calculation results in the nighttime because they are averaged values of summer season. The heat island measure effect by the outside insulation is not so large in comparison with the effect by changing solar reflectance or evaporative efficiency, but the energy-saving effect by the outside insulation is generally large.

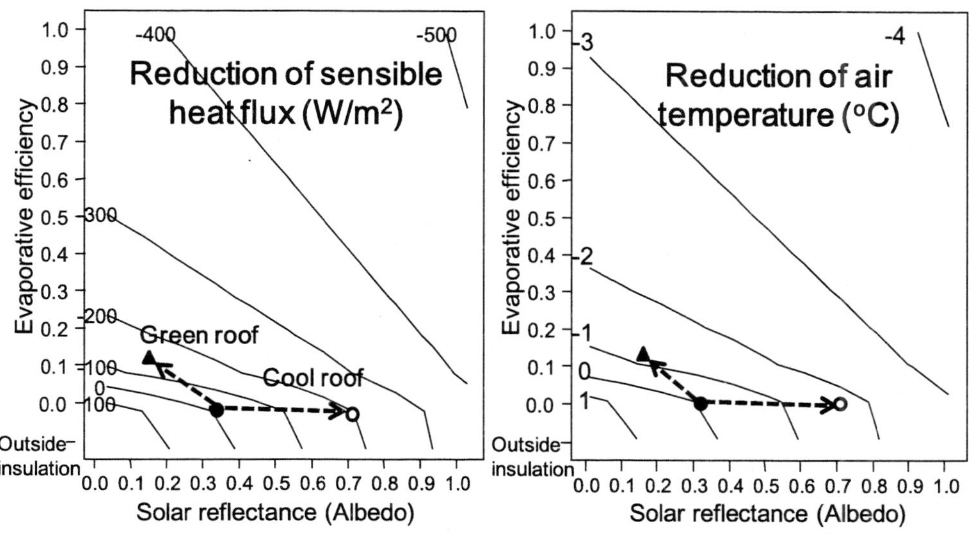

Figure 26. Calculation results of reduction of sensible heat flux and air temperature at 13:00 averaged in summer (from 18 July to 15 September, 2005, at Osaka, Japan).

Calculation results of surface temperature and reduction of SET* at 13:00 averaged from 18 July to 15 September 2005 at Osaka, Japan are shown in figure 27. The human being on the cool roof feels hot by the reflective solar radiation. However, its influence is relaxed if the human wears the clothes with higher solar reflectance.

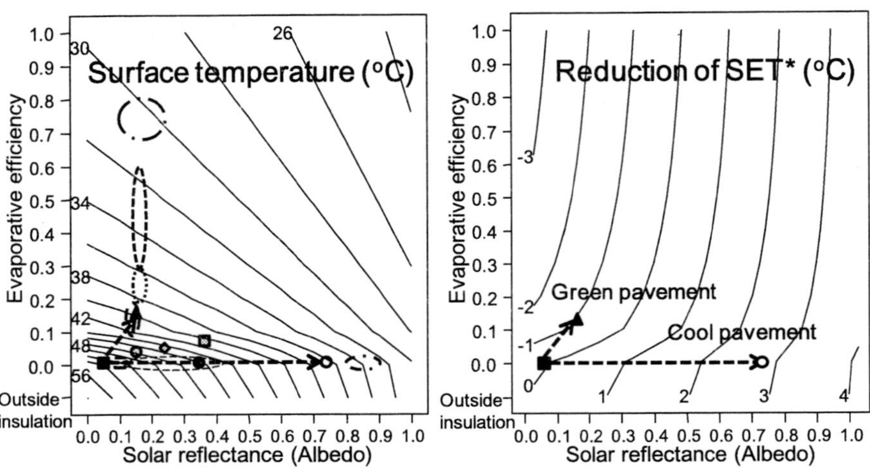

Figure 27. Calculation results of surface temperature and reduction of SET* at 13:00 averaged in summer (from 18 July to 15 September, 2005, at Osaka, Japan).

CONCLUSION

As a background of the introduction of heat island measure technology, it is explained that the target of air temperature reduction nearby ground surface by the heat island measure technologies is about 2 degrees based on the calculation results of meso-scale numerical weather simulation model. The working group for the appropriate use of cool roof in Architectural Institute of Japan has been carried out several activities concerning the cool roof technology. The group has developed the evaluation tools for "the private benefit" and "the public benefit" and rearranged the table on the effect of the heat island measure technology.

The heat island measure technology is classified in three categories; the improvement of surface cover, the reduction of exhaust heat and the improvement of the ventilation in urban area. General evaluation process of the heat island measure effect consists of the local heat budget of each technology and surface boundary layer model. Surface heat budget model and its application examples on several kinds of the heat island measure technologies are described. Simple evaluation method of surface air temperature reduction using surface boundary layer model with upper weather observation data is suggested and its evaluation results on various technologies are explained.

Main parameters of surface heat budget are solar reflectance (Albedo) and evaporative efficiency. Estimation methods of solar reflectance and evaporative efficiency are described in detail on-site measurement as mind. The effect by urban form to the radiation environment should be considered too. The priority of measures is different by the position on the road. Finally, a simple evaluation tool for "the public benefit" is explained. When a user changes solar reflectance ρ or evaporative efficiency β, quantity of change of surface temperature, sensible heat flux, air temperature and SET* are calculated.

REFERENCES

Architectural Institute of Japan. 2007. Urban heat island and buildings / city, A vision of measure and problem, Tokyo: Architectural Institute of Japan (in Japanese).

Architectural Institute of Japan. 2008. Proceedings of symposium for the appropriate use of cool roof, Tokyo: Architectural Institute of Japan (in Japanese)

Kitao, Nanako, Moriyama, Masakazu, Nakajima, Shunsuke, Tanaka, Takahiro and Takebayashi, Hideki. 2009. "The characteristics of urban heat island based on the comparison of temperature and wind field between present land cover and potential land cover." Proceedings of the seventh international conference on urban climate.

Kondo, Hiroaki, Genchi, Yutaka, Kikegawa, Yukihiro, Ohashi, Yukitaka, Yoshikado, Hiroshi and Komiyama, Hiroshi. 2005. "Development of a multi-layer urban canopy model for the analysis of energy consumption in a big city: Structure of the urban canopy model and its basic performance." Boundary Layer Meteorology, 116, 395-421.

Kusaka, Hiroyuki, Kondo, Hiroaki, Kikegawa, Yukihiro and Kimura, Fujio. 2001. "A simple single-layer urban canopy model for atmospheric models: Comparison with multi-layer and slab models." Boundary Layer Meteorology, 101, 329-358.

Masson, Valéry. 2000. "A physically-based scheme for the urban energy budget in atmospheric models." Boundary Layer Meteorology, 94, 357-397.

Moriyama, Masakazu and Takebayashi, Hideki. 1999. "Making method of 'Klimatope' map based on normalized vegetation index and one-dimensional heat budget model." Journal of Wind Engineering and Industrial Aerodynamics, 81, 211-220.

Osaka Heat Island Measure Technology Consortium. 2007. Information of heat island measure technology, Osaka: Osaka HITEC (in Japanese)

Sugihara, Teruaki, Moriyama, Masakazu and Takebayashi, Hideki. 2006. "Measurement results of solar reflectivity and surface temperature on several kinds of roof tiles." AIJ, 671-672.

Taha, Haider. 1997. "Urban climates and heat islands: albedo, evapotranspiration, and anthropogenic heat." Energy and Buildings, 25, 99-103.

Takebayashi, Hideki and Moriyama, Masakazu. 2007. "Surface heat budget on green roof and high reflection roof for mitigation of urban heat island." Building and Environment, 42, 2971-2979.

Takebayashi, Hideki and Moriyama, Masakazu. 2009. "Study on urban heat island mitigation effect achieved by converting to grass-covered parking." Solar Energy, 83, 1211-1223.

Takebayashi, Hideki, Kondo, Yasushi and Working group for the appropriate use of cool roof. 2010. "Simple evaluation system of cool roof for proper promotion (part 2) Development of evaluation tool for public benefit." AIJ J. Technol. Des., 33. (in Japanese)

Takebayashi, Hideki, Moriyama, Masakazu, Nishioka, Masatoshi, Kimishima, Takeyuki, Nabeshima, Minako, Kagata, Mamoru and Honda, Takashi. 2008. "Study on urban heat island mitigation effect based on surface heat budget on environmental consideration type pavements" J. Environ., Eng., AIJ, 623, 73-83. (in Japanese)

The Society of Heating, Air-Conditioning and Sanitary Engineers of Japan. 2009. Heat island measure, How to think and lead for planning the city normal air temperature, Tokyo: Ohmsha (in Japanese)

The University Corporation for Atmospheric Research. The PSU/NCAR mesoscale model (MM5), http://www.mmm.ucar.edu/mm5/.

The WRF Community. The weather research and forecasting model (WRF), http://wrf-model.org/index.php.

Udoh, Kunihiro, Kondo, Yasushi and Takeda, Hitoshi. 2009. "Simple evaluation system of cool roof for proper promotion." AIJ J. Technol. Des., 31, 849-854. (in Japanese)

In: Paints: Types, Components and Applications
Editor: Stephanie M. Sarrica

ISBN: 978-1-61761-813-0
© 2011 Nova Science Publishers, Inc.

Chapter 5

TECHNIQUE ASSESSMENT OF COATING PROCESSES USING MULTI-CRITERIA DECISION SUPPORT

J. Geldermann and S. Wiedenmann
Chair of Production and Logistics, University of Goettingen,
Goettingen, Germany

ABSTRACT

Since the integration of new techniques affects the total process of the industrial production, an isolated view of an individual emission reduction measure is not reasonable. Rather it requires an integrated technique assessment, which considers conflicting objectives between economic, technical, social and ecological aspects adequately with respect to a sustainable development. Therefore, it is not only important that the material and energy flows of different techniques are classified in an environmental impact assessment, but also that they are evaluated in a techno-economic assessment.

This paper describes two case studies of Multi-Criteria Decision Support for the selection of a coating technique by companies, aiming for a long-term orientation towards sustainability. Coating techniques are of importance because the conventionally used materials have a high solvent content and are thus of environmental relevance. Therefore, a changeover to low-solvent painting processes to avoid VOC emissions is currently under discussion, while at the same time low-solvent aqueous or powder coating systems are being developed. A new challenge is the use of renewable resources for which an outlook is given.

Keywords: Technique Assessment, Multi-Criteria Analysis, Coating, Automotive Industry, VOC. Multi-Criteria Decision Support, Life Cycle Analysis (LCA), Monte Carlo Simulation, Principal Component Analysis.

1. INTRODUCTION

Substantial reductions of VOC emissions and more general environmental improvements are possible in the metal coating industry. In some sub-sectors, notably job coating, VOC emissions can be reduced by two-thirds using coating technologies already available [1,2]. However, in environmentally advanced sub-sectors such as passenger cars and agricultural machines, VOC emissions have already been reduced so far that other environmental aspects have become more important, requiring explicit choices on trade-offs between them.

Conventional, solvent-based paints contain approx. 30–80% (by weight) organic solvents for regulation of viscosity and film formation. From an environmental perspective, emissions of organic solvents, mostly VOCs (Volatile Organic Compounds), are problematic because they are subject to photochemical reactions in the presence of sunlight and nitrogen oxides (NO_x) leading to formation of toxic photo-oxidants, especially ozone. These substances have negative effects on the health of humans and animals due to their neurotoxicological properties, and also affect the health of plants. Various legislative initiatives of the European Commission are aimed at reducing VOC emissions in order to decrease the number of days in the year with excessive ozone levels and the resultant acute and chronic effects of ozone exposure [3,4]. As a consequence, paint producers have developed innovative products: **Waterborn coatings** are self-emulsifying resins and self-cross linking acrylic and polyurethane emulsions. Nevertheless, they still have a solvent content of up to 15%, which is necessary for satisfactory adhesion and flow of the material after its application. Furthermore, the drying of waterborne coatings requires more energy due to the higher vaporization enthalpy of water in comparison to solvents. **High-Solids** have a solids content of more than 65%. The film forming agents used are mainly based on epoxy resins or 2-component polyurethanes. Products containing 85% solids by volume have been introduced into the market recently; however, the solids content during actual application does not generally exceed 72%, so that the solvent content is generally still around 30%. Solids content in coatings can be increased to a limited extent by replacing part of the solvents with super-critical carbon dioxide. Therefore different application devices, e.g., hot sprayers, must be developed and used. **Powder coatings** have penetrated widely into some sectors such as the can coating or white goods sectors. The employment of combined environmentally friendly technologies such as powder/water-borne or powder/radiation curing is increasing [5]. Innovative approaches seen in new powder spray guns or high speed powder coating processes are being developed to open and expand new market segments [6]. The powder coatings are mainly duromers and thermoplastics and are dry materials which do not contain any solvents at all. Hence, neither VOC-emissions nor waste water are released during application, which is mostly done with electrostatic spray guns. Due to the recycling of the overspray, transfer efficiencies of up to 98% are possible. But, very often a lower surface quality can be observed as well as higher energy consumption mainly to cure the coating. Radiation-curing paints (containing less than 5% organic solvents), Infrared radiation curing, UV radiation curing and electron beam hardening are other innovative products with environmental advantages, but not yet in wide use in industrial coating. The comparison of these different types of paints in terms of sustainability is difficult.

Since the integration of new techniques for the cross-media environmental protection affects the total process of the industrial production, an isolated view of an individual emission reduction measure is not reasonable [7,8]. Rather it requires an integrated technique assessment, which considers conflicting objectives between economic, technical, social and ecological aspects adequately with respect to a sustainable development. Therefore, it is not only important that the material and energy flows of different techniques are measured in a material balance and are classified in an impact assessment as suggested by a Life Cycle Assessment (LCA; cf. ISO 14040), but also that they are evaluated in a techno-economic assessment. Consequently, a comprehensive catalogue of criteria must be included in the assessment considering dynamic- and progress-oriented concepts such as *Best Available Techniques* (BAT) as introduced in national and international guidelines (for example [2] for the sector of metal coating using organic solvents). This process can be supported, structured and transparently modelled using Multi-Criteria Decision Support Methods. An integrated technique assessment therefore provides a holistic evaluation as a decision basis with consideration of the aspects of all involved groups ("stakeholder participation"). Stakeholders share, or perceive that they share, the impacts arising from a decision. They have a claim, therefore, that their perceptions and values should be taken into account [9]. Consequently, any sustainability strategy of a company should be constantly re-appraising the needs of its shareholders and stakeholders and working continuously with all parties to arrive at a new balance of interests.

This chapter illustrates a decision-making process for the development of a concept for the implementation of integrated coating techniques at a car manufacturing company. Starting point of the decision process was the elaboration of process models and their Life Cycle Assessment. For the simultaneous evaluation of ecological, technical, economic and social criteria, various moderated workshops can be conducted. The quantitative analysis of the collected information and data can be supported by approaches for Multi-Criteria Decision Support. Thus, the various aspects can be consolidated with consideration of all involved groups and thereby it is ensured that the results of the Eco-balance are considered systematically in a techno-economic analysis and additional criteria (future safety, aspects of quality etc.) are incorporated in the decision.

This chapter has the following structure: Section 2 gives the background on technology management and innovation and explains the need for Multi-Criteria Decision Support. The first case study for technique assessment is described in Section 3, where seven alternatives for automobile coating are briefly introduced and analysed within different scenarios. Monte Carlo Techniques and the Principal Component Analysis are applied to investigate the robustness of the decisions. A further example is given in a case study of a paint supplier in Section 4. Challenges for practitioners and researchers arising from the use of renewable materials are investigated in Section 5. Finally, Section 6 summarises the chapter and draws conclusions.

2. DECISION SUPPORT IN THE CONTEXT OF TECHNOLOGY MANAGEMENT AND INNOVATION

Environmentally compatible products and production processes would be unthinkable without innovation processes. Such processes not only revolutionise the materials and technologies employed, they also lead to new management methods. Technology management has been coined in the last two decades and originally stems from the planning and controlling of research and development. It aims to understand, shape and relate the interactions between technology, organisations, and society and their impact on the sustainable use of natural and human resources. Especially the keeping pace with technological progress is of outmost importance for industrial firms [10-13]. Mainly two characteristics of technological progress can be observed:

- The creation of improved or novel products is called product innovation. Examples could be waterborne coating systems or solvent free adhesives for the automobile production.
- The changeover to improved or novel production processes is known as process innovation, like the use of coating robots for the reduction of overspray losses.

Moreover, qualitative improvements of well established products and processes count as technological process. Due to the advanced state of the art in automobile production (as in many other industrial sectors), a clear distinction between product and process innovations is not possible. In fact, any change in the production process comprises various types of innovation.

From a methodological viewpoint, the application of different scientific methods is required for technology management, taking into account both experiences in management and economics and combining various perspectives. In the literature on technology and innovation management (for example [14-17], numerous phase models for decision making are described, starting from the generation of ideas until their final implementation. Such systematic approaches aim to structure and to speed-up the innovation process.

Multi-Criteria Decision Support Models are considered to be helpful thanks to their graphical representation of the investigated alternatives with respect to the evaluation criteria taken into account. In the context of technology management and innovation, process models help to disseminate any economic and emission related consequences of innovative products and processes. They can also illustrate a complete view of shop internal material flows or even decision relevant parts of the supply chain. Thus, various technological solutions and scenarios can be compared with each other, identifying possible environmental and economic benefits [18].

CASE STUDY 1: COMPARISON OF COATING TECHNIQUES IN AUTOMOBILE PRODUCTION

The interpretation of the obtained results of process models is difficult, since they often come in incommensurable units of measurements and often show goal conflicts requiring an approach considering multiple criteria simultaneously. In the following, a hypothetical case study on the selection of a coating process is presented, in order to demonstrate the course of Multi-Criteria Decision Support and especially the graphical representation of the obtained results. Since the actual data of the several underlying case studies are subject to trade secrets, only indicative results can be shown. In the scope of the assessment of several coating techniques seven alternatives were compared and the strengths and weaknesses of the different alternatives were discussed and weighted against each other. As alternatives the following options were available:

(a) Standard process material (supplier A)
(b) Standard process material (supplier B)
(c) Integrated technology (existing facilities) (supplier C)
(d) Integrated technology (existing facilities) (supplier D)
(e) Integrated technology (new facilities) (supplier E)
(f) Integrated technology (new facilities) (supplier F)
(g) Integrated technology (new facilities) with powder slurry clearcoat (supplier G)

The Eco-balance of the *standard process* and the simulated Life Cycle Analysis of the different alternatives are used for the ecological characterisation consolidated in the different relevant impact categories, for example the *Global Warming Potential* (GWP 100) or the *Photochemical Ozone Creation Potential* (POCP). In general, the Eco-balance is a "preferably comprehensive comparison of the impact on the environment of two or more different products, product groups, systems, procedures or behaviours" over its entire life cycle as the term "Life Cycle Assessment" (LCA) illustrates.

Therefore four working steps are defined in the DIN /ISO standard 14040 containing (1) the definition of the objective and the investigation framework, (2) the material balance, (3) the estimation of all possible effects and (4) the data evaluation. A Multi-Criteria Decision Analysis (MCDA) is particularly for the interpretation of the results and the weighting of the relevance of useful assistance. It enables the consideration of additional criteria (for example economic key data or technical parameters), which are not included in the Eco-balance.

For strategic decision making in a company, moderated workshops help to gather expertise from the various departments. Figure 1 shows a selection of criteria, which were at first collected in a brainstorming session and subsequently structured in a criteria hierarchy (also sometimes called 'decision tree'). It indicates that the respective round of experts spent much effort on considering and defining criteria which characterise the product quality (i.e. the quality of the coated car body). It should be noted that other companies may have other criteria hierarchies.

Decision trees systematically structure the decision problem and transparently model the subjective weighting of the various criteria. The avoidance of weighting is not a real option, since implicitly an equal weighting of all criteria is assumed, which can however be a first starting point.

To support the weighting of the mentioned criteria the software WebHIPRE was used [19;20], but also other MCDA software tools are helpful [21;22;43]. In order to include different views on the decision problem, the workshop participants for example can make a weighting according to the SWING method for the sub-objectives of the first level. These weights could be used as **baseline scenario** in the Multi-Criteria Decision Analysis. Furthermore a **scenario ecology** (100% of the weighting on the sub-objective ecology) and a **scenario economics** (100% of the weighting on the sub-objective economics) can be evaluated. This systematic and transparent development of a criteria hierarchy and the weighting of the individual partial aspects of the decision are made possible by methods of the Multi-Criteria Decision Support and moderated workshops.

Based on the collected data different scenarios can be analyzed and ranked with the help of Multi-Attribute Value Theory (MAVT) and PROMETHEE, whereby the theory and proceeding of the two Multi-Criteria Decision Methods are not described in detail but the main characteristics of the modelling and the results are discussed.

Using PROMETHEE a partial ranking of the different techniques can be constructed favouring the integrated coating techniques on new facilities over integrated techniques on existing facilities finally over the existing standard technique (see Figure 3). The additionally regarded powder slurry procedure shows incomparableness to the integrated procedures in the summary of all aspects, it however shows advantages compared with the remaining procedures.

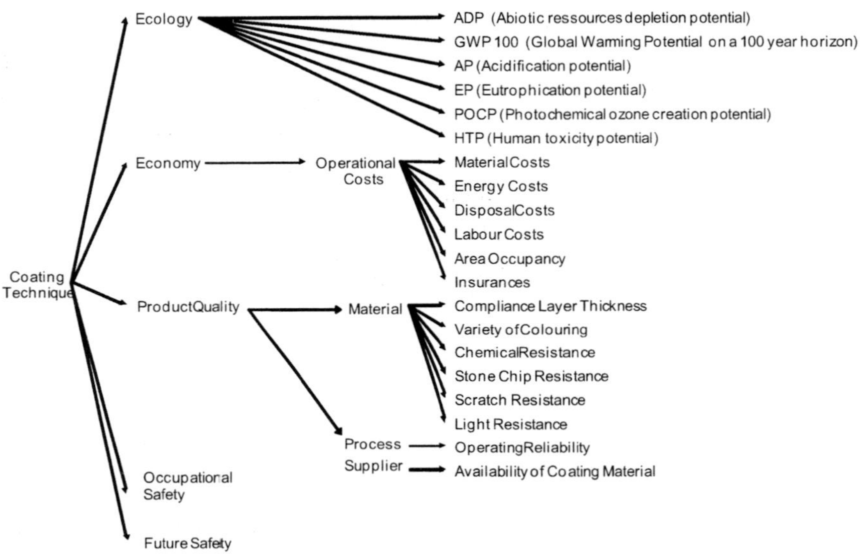

Figure 1. Criteria hierarchy including the weighting of all criteria

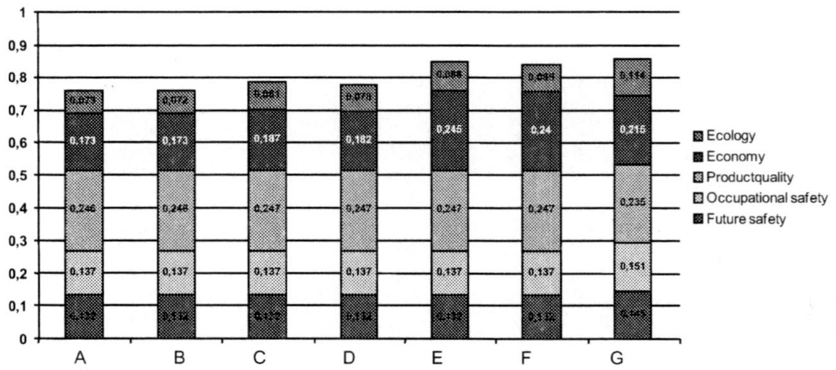

Figure 2. Single criterion values for the baseline scenario using MAVT

The majority of environmental decision problems involve uncertainty and risk. By their very nature, estimates and long-term forecasts, as required in LCA and for the evaluation of emerging technologies, are obviously uncertain; and a technology considered optimal on the basis of particular assumptions made today is highly unlikely to turn out optimal in the actual situation in a decade. For reviews discussing different types of uncertainty, variability, and risk see, for example [21;23;24].

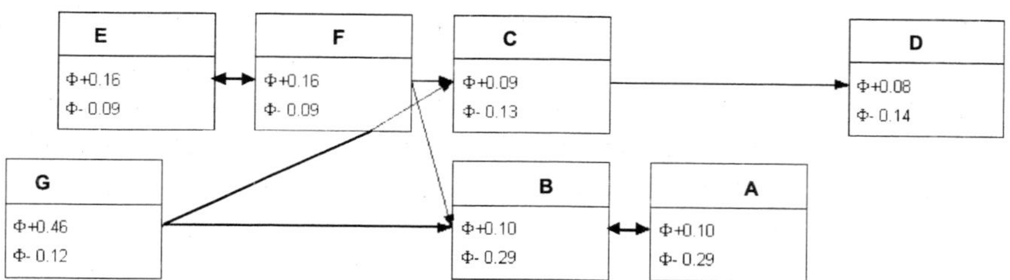

Figure 3. PROMETHEE I Ranking of the baseline scenario

Considering the above mentioned difficulties in modelling value judgements and normalisation, it is important to carry out sensitivity analyses as proposed in the ISO 14044. Sensitivity analyses investigate how the results of a model change with variations in the input variables and are therefore essential in any Multi-Criteria Decision Analysis Problem. Using PROMETHEE several local and global [25] sensitivity analyses can be carried out [26].

Especially the use of *Monte Carlo Simulation* (MCS) and the *Principal Component Analysis* (PCA) [27] allow a simultaneous consideration of the uncertainty of the process data and the value judgements of the decision maker. Besides investigating the robustness of the decision, the strengths and weaknesses of each alternative can be addressed and so, the *distinguishability* of all alternatives can be evaluated [28]. Consequently, sensitivity analyses are important to iteratively re-model the decision problem and to facilitate learning about the given problem.

The Principal Component Analysis is based on the matrix M of the single criterion net flows in which the strengths and weaknesses of an alternative can be analysed for each criterion individually. By calculating the eigenvectors of the covariance matrix M'M and building a matrix of all eigenvectors sorted by the magnitude of the eigenvalues, the axes can be transformed [27]. The transformation represents a rigid rotation of the old axes into the new principal axes based on the matrix of the sorted eigenvectors. The component scores in the new coordinate system are uncorrelated and have maximum variance. By projecting the cloud of alternatives from the R^n onto the plane of the two eigenvectors with the largest eigenvalues (henceforth called 1st and 2nd principal component), the so-called GAIA plane [29], as much information as possible is preserved. Apart from alternatives and criteria axes, the weighting vector (the PROMETHEE *'decision stick'* or weighting vector π) can be projected on the GAIA plane. By defining upper and lower bounds for each weight, the convex hull of all valid weighting combinations can be projected on the GAIA plane. [30] call the projection of the set of all extreme points of the unit vectors associated to all allowable weights the "Human Brain" or the PROMETHEE VI sensitivity tool. While this idea has been roughly sketched by [30], the implementation in Matlab is its first software realization [26].

An example shows the analysis of the baseline scenario by the PCA, whereby the decision problem is very well illustrated due to the structuring and consideration of all criteria (see Figure 4). All criteria axes are star-shaped distributed on the plain of the first and second principal component, i.e. all alternatives can be well described by the criteria. Since the alternatives distribute themselves based on the characteristics of the attributes on the plain, alternatives with similar attribute characteristics lie close together on the plain. In this case, this is to be recognized very clearly for the two different materials in the respective techniques. For an assumed data uncertainty of 10% for all attribute values it is no longer possible to differentiate between the individual coating materials A and B, however between the individual techniques. By application of the MCS variations in the attribute values are considered and aggregated over the different attributes. Thus, a scatter plot instead of individual data points is the result of the analysis. The distinction between alternative A (red) and alternative B (yellow) is then no longer possible. This applies also to alternatives C and D (turquoise and green) and alternatives E and F (light blue and dark blue). The individual techniques can still be differentiated from each other even with considered data uncertainty.

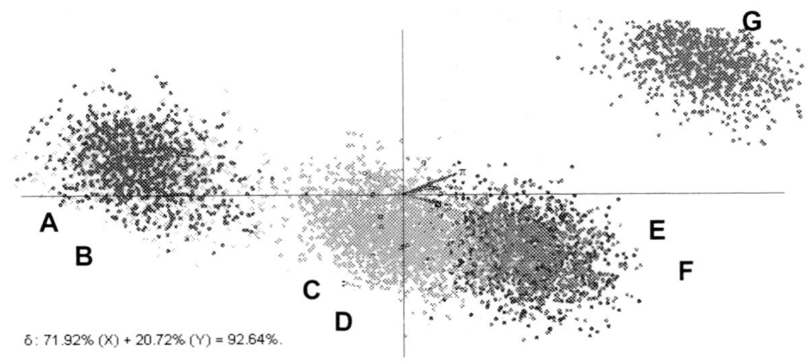

Figure 4. Distinguishability Analysis of the baseline scenario

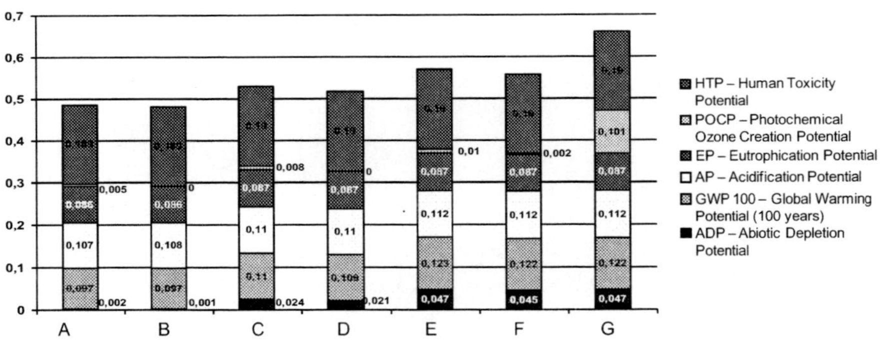

Figure 5. Single criterion values for the scenario ecology using MAVT

Summarising the results it can be concluded for the baseline scenario that integrated procedures on new facilities are better than the integrated procedures on existing facilities and these again are better than the existing standard technique. Main factor is the economic sub-objective. Although for economic reasons (above all material costs play an important role here) the integrated techniques are better, these are however surpassed due to ecological reasons by the integrated techniques in combination with a powder slurry clearcoat in the overall view of the baseline scenario. Furthermore, regarding the sub-objective economics it can be stated that the integrated techniques on new facilities are as good as the integrated techniques on existing facilities or the existing standard technique in all attributes, even though the integrated procedures on new facilities are identified as incomparable with the integration technique using a powder slurry clearcoat. However a conclusion can be drawn that these concepts are better than the techniques (standard or integrated) on existing facilities.

Basically the differences between the different coating concepts are not very significantly pronounced. With a 100% weighting of the sub-objective ecology (scenario ecology), the relative differences between the different concepts however become clearer, although the ranking of the different alternatives remains constant for all weighting factors (see Figure 5).

Altogether the weighting within the ecological criteria for the ranking of the alternatives is unimportant since the existing facilities are dominated regarding the respective coating material by the integrated technique on existing facilities and these again by the integrated coating technique on new facilities. Thus, in this case the weighting of the individual attributes is insignificant for the ranking of the alternatives.

3. CASE STUDY 2: AUTOMOTIVE REFINISH PRIMERS

While the previous case study presented the strategic decision problem of an automobile producer and his choice of the most suitable coating product, this case study outlines the decision problem of paint suppliers who are developing various environmentally friendly innovations for surface treatment in the car repair sector. Six types of primers and their application techniques in vehicle refinishing are being compared on the basis of data delivered by [31;32] with an Eco-efficiency Analysis. The starting point for the recalculation of the case study was an initiative by an informal working group of various European coating producers on environmental assessment. The six alternatives [31;32]:

(i)   One-Component UV Curing Primer (1K UV),
(ii)   One-Component UV Curing Primer, using an aerosol can (1K UV-A).
(iii) Conventional Two-Component Urethane Primer with thermal drying (2K U-T),
(iv)   Two-Component Urethane with infrared drying (2K U-IR),
(v)   Epoxy with thermal drying (E-T),
(vi)   Epoxy with infrared drying (E-IR)

The details of the decision problem are given in [33] and will not be repeated in this paper. Just the graphical representation of the decision analysis is reported in the GAIA plane in Figure 6, projecting all alternatives, criteria and the weighting vector. It can easily be seen that the criteria are not mutually independent (as claimed by [34] and [31]) since the projections of the criteria do not span the decision space equally, but rather point into one direction.

Figure 6 also shows the variation in value judgments with the grey area around the weighting vector π (as the PROMETHEE VI sensitivity tool) and the uncertainty in the attribute data with the scatter plot of 1,000 points created by changing all attribute values in a Monte Carlo Simulation using a standard deviation of 5 % and a normal distribution. This plot makes clear that a differentiation between the 1K UV-A alternative and the two IR alternatives (E-IR and 2K U-IR) might be difficult if the uncertainty in the underlying data is taken into account. This also becomes obvious when the small differences of their projections on the weighting vector π are compared. But even under these uncertain conditions, the 1K UV curing coat is always the best ranked alternative.

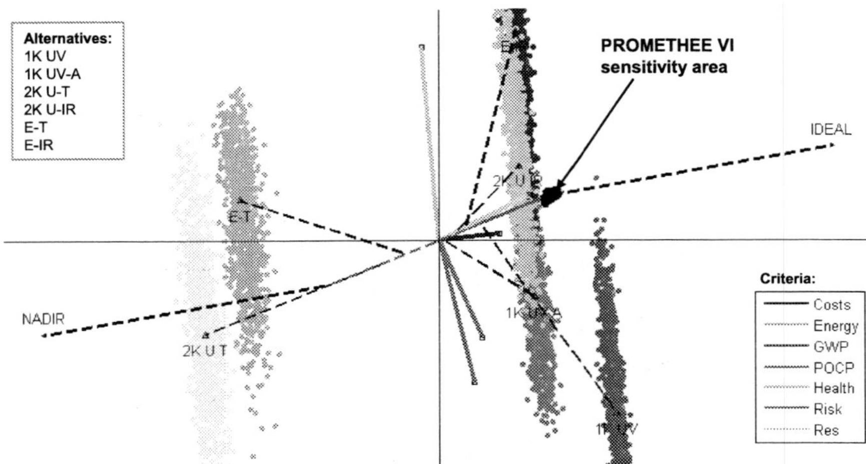

Figure 6. Illustration of alternatives using PCA and Monte Carlo Simulation

Thus, the PROMETHEE Multi-Criteria Analysis helped to gain further insights into the given selection or ranking problem concerning different techniques, and enabled the decision maker to understand his own preferences and transparently model and communicate them. The sensitivity and uncertainty analyses in particular elucidate the variability in the underlying data and the subjective value judgments.

4. Outlook on the Use of Renewable Resources

Due to the ongoing strive to develop and use more environmentally friendly products and to increase resource efficiency, science and industry searches for innovative technological solutions. The paint industry currently heavily depends on crude oil, but is seeking for alternatives.

Since crude oil, the raw material for many coating ingredients, is limited, exploring the options for material use of renewable resources is important. The requirements for the industrial use of renewable products are a competitive price, constant quality and uninterrupted supply throughout the year. Binding agents and additives have been identified as the most promising application area for materials derived from renewable resources [35]. The substances can be derived from carbohydrates, starch, celluloses and their derivates as well as plant oils. In Germany 93.000 t of vegetable oil have been used for these applications in the year 2007 [36].

Plant oils consist of triglycerides, which are esters of glycerol with longchain fatty acids. The composition of the fatty acids varies with the plant, the crop, the season and the growing conditions. Drying oils are characterized by an iodine value > 170, which indicates a high amount of double bonds. Besides the carboxyl group the double bonds are the second important functional group for the industrial application of fats and oils [37]. Modification towards e.g. epoxy resins, alkyd resins and polyurethane resins are possible [38]. Epoxidized vegetable oil with high contents of linolenic acid, e.g. linseed oil, can be used as crosslinker for powder coatings [39]. Epoxidized plant oils are used as monomers for photoinitiated cationic polymerization. This reaction can be carried out without the use of VOCs [40]. Life Cycle Assessment proved such a binder based on linseed epoxide to be more environmental friendly then a comparable petrochemical product [41].

It should be paid attention, however, that the consideration of further criteria does not overshadow conventional important criteria. Case study 2 gives an example that the strict focus on environmental criteria lead – in mathematical terms – to a deteriorated decision space, which the graphical representation of the GAIA plane within the Multi-Criteria Decision Support Software of PROMETHEE revealed.

For a comprehensive base for decisions on the use of coatings produced from renewable materials, the consideration of further criteria is required, e.g., economic criteria and options for cascading use [42].

5. CONCLUSION

The paper described the use of moderated workshops and Multi-Criteria Decision Support for technique assessment when a company chooses an innovative production process, taking both environmental and techno-economic criteria into account.

The results show the advantages of the integrated technology on new facilities compared to the standard process that is currently implemented. In addition, the available sensitivity analysis allows a valuation of the robustness of the results. In this case the result depends on the preferences of the decision maker. Thus, considering the complex task of assessing coating concepts for serial coating of passenger cars, the described approach provides a beneficial contribution for organising the decision process, mutual understanding and transparency as well as traceability of the decision.

Especially the Multi-Criteria Decision Support Methods allow us to represent all information on the preference behaviour transparently. Thus, the preferences can be better communicated. From the numerous sensitivity and uncertainty analyses, it is possible to examine the variability of the underlying data.

ABOUT THE AUTHORS

Jutta Geldermann is full professor for production and logistics at the Georg-August-University of Goettingen (Germany). She holds a diploma in Industrial Engineering and a PhD in Business Administration, both from the University of Karlsruhe (TH). Her major research areas are Multi-Criteria Decision Support Systems and assessment and optimisation of the economic performance of emission reduction strategies on regional, national and supranational levels.

Susanne Wiedenmann is research assistant at the chair of Production and Logistics at the Georg August University of Goettingen. She holds a diploma in Industrial Engineering from the University of Karlsruhe (TH). Her major research interests include Multi-Criteria Decision Making, production planning with renewable resources and risk management.

REFERENCES

[1]   Geldermann, J; Rentz, O. Multi-Criteria Analysis for Technique Assessment: Case Study from Industrial Coating. *Journal of Industrial Ecology*, 2005, 9, 127-42.

[2]   Rentz, O; Peters, NH; Nunge, S; Geldermann, J. Bericht über Beste Verfügbare Techniken (BVT) in den Bereichen der Lack- und Klebstoffanwendung in Deutschland / Reports on the Best Available Techniques (BAT) in the Sectors of Paint and Adhesive Application in Germany (German / English). *Düsseldorf: VDI-Verlag*, 2003.

[3]   Geldermann, J; Schollenberger, H; Rentz, O; et al. An integrated scenario analysis for the metal coating sector. *Technological Forecasting and Social Change*, 2007, 74, 1482-1507.

[4]   May, T. Chemical leasing concepts in industries applying coatings. International Workshop on *"Sustainable Chemistry - Integrated Management of Chemicals, Products and Processes"*. 2004. Dessau, Umweltbundesamt (UBA Berlin), Organisation for Economic Co-operation and Development (OECD) and the Federal Institute for Occupational Safety and Health (FIOSH, German acronym BAuA). 27-1-2004.

[5]   Busato, F. Powder and Waterborne Coatings 2000-2010 - Is Past Growth Sustainable? In: Adler H-J, Potje-Kamploth K, eds. *Quo Vadis- Coatings*. Weinheim: Wiley-VCH, 2002, 17-21.

[6]   Strohbeck, Ulrich, Cudazzo, Markus, Domnick, Joachim, Pulli, Karlheinz, and Ye, Qiaoyan. Advanced Powder Coating Technologies Open New Markets for VOC-free Metal Coating. Geldermann, Jutta, Schollenberger, Hannes, Hubert, Isabelle, and Rentz, Otto. 209-219. 2004. Karlsruhe, French-German Institute for Environmental Research. *Inegrated Scenario Analysis and Decision Support for the Modern Factory.*, 23-9-2004.

[7]   Geldermann, J; Jahn, C; Spengler, T; Rentz, O. Proposal for an Integrated Approach for the Assessment of Cross-Media Aspects Relevant for the Determination of "Best Available Techniques" BAT in the European Union. *International Journal of Life Cycle Assessment*, 1999, 4, 94-106.

[8]   Basson, L; Petrie, JG. An Integrated Approach for the Consideration of Uncertainty in Decision Making Supported by Life Cycle Assessment. *Environmental Modelling & Software*, 2007, 22, 167-76.

[9]   French, S; Geldermann, J. The varied contexts of environmental decision problems and their implications for decision support. *Environmental Policy and Science*, 2005, 8, 378-91.

[10]   Brockhoff, K. *The dynamics of innovation: strategic and managerial implications.* Heidelberg: Springer, 1999.

[11]   Milling, P. Der technische Fortschritt beim Produktionsprozeß: ein dynamisches Modell für innovative Industrieunternehmen. (*The Technological Progress of the Productionprocess: A Dynamic Model for Innovative Industrial Firms*) Wiesbaden: Gabler, 1974.

[12]   Zahn, E. Handbuch Technologiemanagement. (*Handbook of Technology Management*)

Stuttgart: Schäffer-Poeschel, 1995.

[13] Tschirky, H; Koruna, S. Technologie-Management: Idee und Praxis (*Technology Management: Idea and Praxis.*) / Hugo Tschirky, 1998. Zürich: Verlag Industrielle Organisation, 1998.

[14] Strebel, H. Innovations- und Technologiemanagement. (*Innovations Management and Technology Management.*) Wien: WUV, 2003.

[15] Schwarz, EJH. Nachhaltiges Innovationsmanagement. (*Sustainable Innovations Management.*) Gabler: Wiesbaden, 2004.

[16] Bullinger, HJ. Technologiemanagement: forschen und arbeiten in einer vernetzten Welt. (*Technology Management: Researching and Working in a Cross-linked World.*) Heidelberg: Springer, 2002.

[17] Gelbmann, U; Vorbach, S. Strategisches Innovations- und Technologiemanagement. (Strategic Innovations Management and Technology Management.) In: Strebel H, ed. *Innovations- und Technologiemanagement.* Wien: WUV-Univ.-Verl, 2003, 93-210.

[18] Geldermann, J; Rentz, O. Decision support through mass and energy flow management in the sector of surface treatment. *Journal of Industrial Ecology*, 2004, 8, 173-87.

[19] Hämäläinen, RP; Mustajoki, J. *Web-HIPRE- Java Applet for Value Tree and AHP Analysis.* Computer software, Systems Analysis Laboratory, Helsinki University of Technology, 1998.

[20] Geldermann, J; Bertsch, V; Treitz, M; French, S; Papamichail, KN; Hämäläinen, RP. Multi-criteria Decision Support and Evaluation of Strategies for Nuclear Remediation Management. OMEGA - The *International Journal of Management Science*, 2009, 1, 238-251.

[21] Bertsch, V. *Uncertainty handling in multi-attribute decision support for industrial risk management.* Karlsruhe University Press, 2008.

[22] Geldermann, J; Schöbel, A. On the similarities of various MCDA methods (submitted). *Journal of Multi-Criteria Decision Analysis,* 2010.

[23] French, S. Modelling, making inferences and making decisions: the roles of sensitivity analysis. *TOP*, 2003, 11, 229-52.

[24] Huijbregts, MAJ; Breedveld, L; Huppes, G; de Koning, A; Suh, S. Normalisation figures for environmental life-cycle assessment - The Netherlands (1997/1998), Western Europe (1995) and the world (1990 and 1995). *Journal of Cleaner Production*, 2003, 11, 737-48.

[25] Saltelli, A; Chan, K; Scott, EM. *Sensitivity Analysis.* Chichester: John Wiley and Sons, 2000.

[26] Treitz, M. *Production Process Design Using Multi-Criteria Analysis.* Karlsruhe: Karlsruhe University Press, 2006.

[27] Timm, NH. *Applied multivariate analysis.* New York: Springer, 2002.

[28] Basson, Lauren. Context, Compensation and Uncertainty in Environmental Decision Making. 2004. *PhD Thesis,* Department of Chemical Engineering, University of Sydney, Australia.

[29] Brans, JP; Vincke, P; Mareschal, B. How to select and how to rank projects: The PROMETHEE method. *European Journal of Operational Research*, 1986, 24, 228-38.

[30] Brans, JP; Mareschal, B. PROMETHEE Methods. In: Figueira J, Greco S, Ehrgott M, eds. *Multiple Criteria Decision Analysis - State of the Art Surveys.* New York: Springer, 2005, 163-95.

[31] Wall, Charlene, Richards, Bradley, and Bradlee, Christopher. *The Ecological and Economic Benefits of UV Curing Technology.*, 2004 RadTech Report 2004.

[32] Richards, Bradley and Wall, Charlene. *Automotive Refinish Primers for Small Surface Damage Repair.* http://corporate.basf.com/file/15376.file5 . 2005. 1-7-2005.

[33] Geldermann, J; Treitz, M. *Quantifying Eco-Efficiency with Multi-Criteria Analysis.* Research Paper der Georg-August-Universität Göttingen, Wirtschaftswissenschaftliche Fakultät, Schwerpunkt Unternehmensführung, Professur für Produktion und Logistik, 2008.

[34] Saling, P; Kircherer, A; Dittrich-Krämer, B; et al. Eco-Efficiency Analysis by BASF: The Method. *International Journal of Life Cycle Assessment*, 2002, 7, 203-18.

[35] Institut für Farben und Lacke e.V. Machbarkeitsstudie zum Einsatz von nachwachsenden Rohstoffen in der Lackindustrie. (Feasibility Study on Application of Renewable Resources in the Coating Industry.) Fachagentur Nachwachsende Rohstoffe (FNR).16. 2000. Münster, Landwirtschaftsverlag GmbH. Schriftenreihe *"Nachwachsende Rohstoffe"*.

[36] Carus, M; Raschka, A; Piotrowski, S. The development of instruments to support the material use of renewable raw materials in Germany (Summary). Market volumes, structure and trends – Policy instruments to support the industrial material use of renewable raw materials. , 2010. nova-Institut für politische und ökologische Innovation GmbH.

[37] Brüse, F. Tenside, Polymerbausteine und Polymere aus Nachwachsenden Rohstoffen durch Epoxidringöffnungen mit Aminen und enzymatische Polykondensationen. (Polymer Building Blocks and Polymers from Renewable Resources through Epoxy Ring Opening with Amines and Encymatic Polycondensation.) 2003. Fakultät für Mathematik, Informatik und Naturwissenschaften der Rheinisch-Westfälischen Technischen Hochschule Aachen, Aachen.

[38] Meier, MAR; Metzger, JO; Schubert, US. Plant oil renewable resources as green alternatives in polymer science. *Chemical Society Reviews*, 2007, 36, 1788-802.

[39] Overeem, A; Buisman GJH; Derksen, JTP; et al. Seed oil rich in linolenic acid as renewable feedstock for environment-friendly crosslinkers in powder coatings. *Industrial Crops and Products*, 1999, 10, 1999, 157-65.

[40] Crivello, JV; Narajan, R. Epoxidized Triglycerides as Renewable Monomers in Photoinitiated Cationic Polymerization. *Chemistry of Materials*, 1992, 4, 692-9.

[41] Bartmann, D; Peters, H; Lübker, U; Lott, A; Sack, W; Metzger, JO; Diehlmann, A; Kreisel, G. Strahlenpolymerisierbare lösemittelfreie Schutz- und Dekorationsbeschichtungen für Holz und Holzwerkstoffe auf Basis nachwachsender heimischer Rohstoffe. (Radiation-polymerizable Solvent-free Protective and Decorative Coatings for Timber and Timber Derived Products Based on Renewable Resources), 2000, Osnabrück, Deutsche Bundesstiftung Umwelt. Abschlussbericht Az 08150.

[42] Oertel, D. Industrielle stoffliche Nutzung Nachwachsender Rohstoffe. Sachstandsbericht zum Monitoring "Nachwachsende Rohstoffe". (*Industrial Material Use of Renewable Materials. Report on Monitoring "Renewable Resources".*) [Arbeitsbericht Nr.114]. 2007, Berlin, Büro für Technikfolgen-Abschätzung beim Bundestag.

[43] Geldermann, J; Schöbel, A. On the similarities of various MCDA methods. *Journal of Multi-Criteria Decision Analysis*, Submitted in April 2010.

In: Paints: Types, Components and Applications
Editor: Stephanie M. Sarrica

ISBN: 978-1-61761-813-0
© 2011 Nova Science Publishers, Inc.

Chapter 6

ENVIRONMENTALLY FRIENDLY PAINTS

Heba Abdelrazek Mohamed

Polymers and Pigments Department, National Research Center,
Dokki, Giza, Egypt

1. PAINTS

The terms 'paint' and 'surface coating' are often used interchangeably. Surface coating is the more general description of any material that may be applied as a thin continuous layer to a surface [1]. Paints do not only used to color surfaces and make them attractive, they can protect and extend lifetime of the coated substrate. Paints are complex mixtures of organic and inorganic ingredients. Among these ingredients that might be present in paint formulations are polymer particles (binder), pigments and extenders, diluents (solvents or water) and additives [2]. The composition of paints is simplified with quantitative percent of ingredients in Figure (1).

The adhesion of paints to substrate like metal (steel) is purely physically due to hydrogen bonds that develop when two surfaces are brought closely together. Binders with polar groups like epoxies and alkyds have good wetting properties and show excellent physical adhesion characteristics. Adhesion takes place when the coating and substrate separation is not more than approximately 0.5 nm. Any contaminant on the substrate will increase the separation and mask reactive sites on the steel and consequently decrease the paint film adhesion. So, surface preparation is very important step before paint application [3-5].

1.1. Pigment Volume Concentration

Beside the importance of the chosen binder, there is pigment volume concentration (PVC), which represents the most important parameter in characterizing paints specially physical charecterization. According to Equation (1), PVC is the mathematical ratio of the volume of pigments and fillers to the total volume of dried paints. Pigment distribution in latex paints improved by reducing the latex particle size. From purely geometric considerations, the best results should be obtained with infinitely small latex particles. A series of model paints pigmented with titanium dioxide showed that the results were valid for latex/pigment size ratios from 0.75 to 1.5 Um [6].

$$PVC\% = \frac{\text{Volume of Pigments and fillers}}{\text{Volume of binders + Volume of pigments and fillers}} \times 100 \qquad \text{Equation (1)}$$

Critical PVC (CPVC) is the point at which binder wets the pigments and binds them all together without voids [7,8].In other word CPVC is prerequisite for the formulation of pigmented paints. The CPVC values of a pigment were calculated from the measured values of pigment density ρp and from the identified value of pigment absorption by linseed oil (g/100 g of a pigment) according to Equation (2).The CPVC value represents conditions under which the pigment–binder system contains the right amount of the binder allowing the tightly organized pigment particles to be still soaked with the binder [9]. Figure (2) represents the relation between PVC and the required output characteristics from paint formulations.

$$CPVC = \frac{10\ 000/\ \rho p}{(100/\ \rho p) + (slo\ /0.93)} \qquad \text{Equation (2)}$$

Where: ρp, pigment density (gcm−3);
    slo, oil absorption (g/100 g);
    0.93, linseed oil density (g/cm^{-3}).

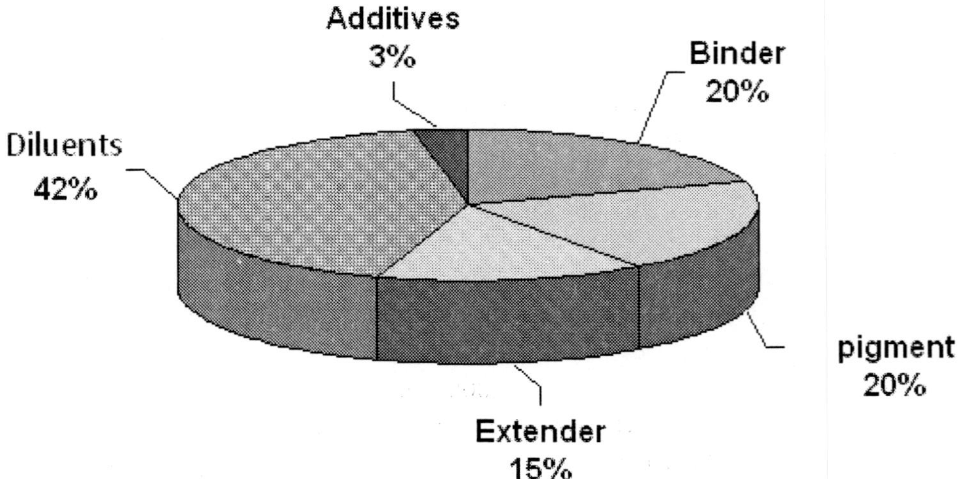

Figure 1. Composition of paints

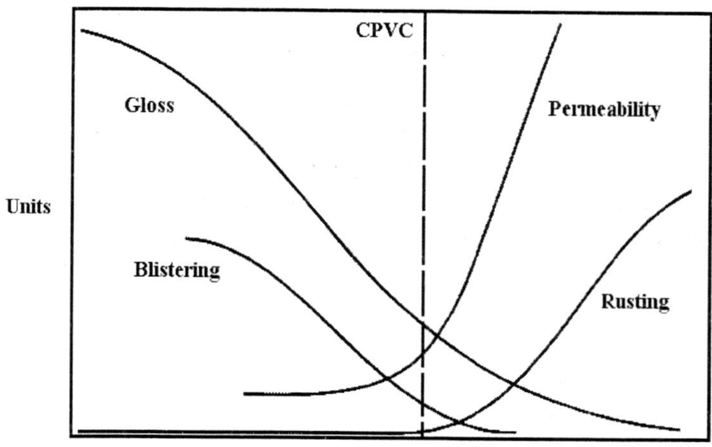

Figure 2. The relation between pigment volume concentration and paint characteristics

1.2. Environmentally Friendly Paints

Paints involve high content of volatile organic compounds (VOC's), which evaporate during their application, drying and curing. VOC's may also react with nitrogen oxides under natural condition of sunlight and heat to generate lower atmospheric ozone which is harmful for both environments and living organisms.

In the current wave of environmental prudence and cost-cutting consciousness, national and international environmental management systems are now in place to allow companies to have formal structures to demonstrate such awareness and compliance [10], consequently, paints manufacturers have found themselves caught between the anvil of environmental regulations and the hammer of the consumers whom will not settle with a lesser level of performance. Great efforts have been exerted in order to switch from using solvent-borne coatings to more friendly to the environment coatings as waterborne, powder, high solids or low VOC coatings, and radiation curable coatings. The main four major types of compliant coatings that exist in market place are water-borne coatings, high built coatings (low VOC's or solvent free), powder coatings and radiation curable coatings. Two areas consider both UV cure [11] and powder coatings [12] as VOC free. However UV curing of solvent free formulation is restricted with only few exceptions to flat surface with film thickness and pigmentation limitation. Also powder coating can not be applied to many surfaces

1.2.1. Water-borne Coatings

Recently, much research has focused on replacing solvent-based paints with water-based paints. The advantages of waterborne paints include being nonpolluting, easy to handle, quick drying, economic, and environmentally friendly [13,14].

The difference between waterborne and solvent-borne paints is due to the difference between solvents and water. Density, surface tension and thermal conductivity of water are greater than most common used solvents in paint. The latent heat of evaporation for water is higher than common solvents; consequently thermodynamic driven evaporation of water is more slowly at room temperature. Higher Surface tension of water compared to common solvents which used in paints plays important rule in film formation [15].

Different emulsion polymers and copolymers are available for different water-borne formulations. Water-borne two-component polyurethane coatings that met or exceeded the performance levels of typical solvent-based systems were prepared and evaluated. A key advantage is the use of isocyanate cross-linkers normally employed in solvent-borne systems without any modification or chemical change. Blending of the isocyanate cross-linkers offered a wide formulation latitude in the resultant water-borne two-component polyurethane coatings[16].

Pangelinan and Rhodes [17] demonstrated, by using scanning electron microscope (SEM), that the barrier properties do not exist because of the highly porous microstructures of many water-based coatings. These porous micro structures of the water-borne coating adversely affect their barrier properties. However, such deficiency can probably be addressed by improving the cohesive resistance of the coatings.

1.2.1.1. Water- borne paint formulations

Water-borne paint is not simply solvent –borne paint in which solvent has been replaced by water. Water-borne paints are more complex and difficult to formulate than solvent – borne coatings [15]

The technique of preparation of water-borne paints is different than solvent- borne paints. Generally, emulsion paints formulations are prepared using two common stages. The first stage is grinding of pigments and extenders in mill base or high speed homogenizer. The second stage includes low speed mixing of emulsion resin with the grinding mill base ingredients with other required additives and adjustment of paint viscosity by thickener addition [18, 19].

Choosing optimal surfactant additives, adding solvent coalescing and/ or other related additives according to the required application are important to improve water-borne paint formulations. High performance water-borne paints based on different types of emulsion polymers and suitable extenders are prepared by adjusting CPVC and choosing optimum concentrations of additives. In addition the prepared formulations can resist corrosion of mild steel by adding corrosion inhibitor as extra additive with studied and optimum concentration [20].

1.2.1.1.1. Effect of surfactants

It was observed that the film integration of styrene/butadiene increased with decreasing the length of ethylene oxide segment [21, 22]. Early studies observed surfactant absorption into latices [23]. Other study showed that non ionic surfactant could act as plasticizer for acrylic latex [24]. Recent studies showed that surfactants play important rules in water sensitivity of latex films [25]. Studies of optimum surfactant concentration showed that the best film surface corresponded with surfactant concentration providing full surface coverage of the latex [26, 27].

1.2.1.1.2. Effect of co-solvents

Co-solvents are commonly used to provide adequate flow, film formation, wetting, and drying behavior of the paint. They can affect basic properties like pot life, gloss andultimate barrier protection. Combination of water soluble solvents like glycol ether and water insoluble solvents as benzyl alcohol, are required for optimum performance properties [28].

1.2.1.2. Acceptable pigments for water-borne paints

In developing paint formulations, the type of pigmentation packages that are acceptable for water-borne systems are taken into consideration. For white gloss enamels, titanium dioxide is used as the main pigment and talc is often used as an extender pigment. Due to environmental concerns the use of lead and chromate based corrosion inhibiting pigments is no longer acceptable [29]. In an extensive study conducted by Jackson, M.A. [30], anti-corrosive pigments that function well in waterborne epoxy primer are those, which are relatively inert and don't release large amount of water soluble salts which adversely affect the cationic character of the water miscible amine functional curing agent.

In choosing the extender pigments, good results are often obtained if a combination of particle size, shape, low oil and water absorption are employed, which may improve barrier protection. Typical pigments used include calcium metasilicate, barium sulfate, talc, micaceous iron oxide, and mica [31].

1.2.1.3. Film formation of water-borne paints

In one component solvent –borne coating, a polymer dissolves completely in a solvent and a dry film is formed by evaporation of the solvent. In water-borne coating, not all the polymer particles dissolve in water. Additives must be added to keep the solid polymer dispersed. Solid particles in water-borne paints must spread out with the evaporation of water where coalescence is thermodynamically favored over individual polymer spheres to minimize the total surface area and consequently decrease its free energy. This process are described in three steps [15]:

1. Colloid concentration by evaporation the bulk of the water in newly applied paint, consequently particles move and slide until the are densely backed and close together without change in shape.
2. Coalescence by the action of water remaining between particles, where water tries to reduce its surface at both water-air and water- particle interface. As this happens, deformation of all sphere solid polymer particles occurs and dodecahedral honeycomb structure was formed instead.
3. Further coalescence and macromolecules interdiffusion can occur under certain conditions by diffusion of polymer chain across the particle boundaries, to form more homogeneous film to improve mechanical strength and water resistance of the film [15, 32, 33].

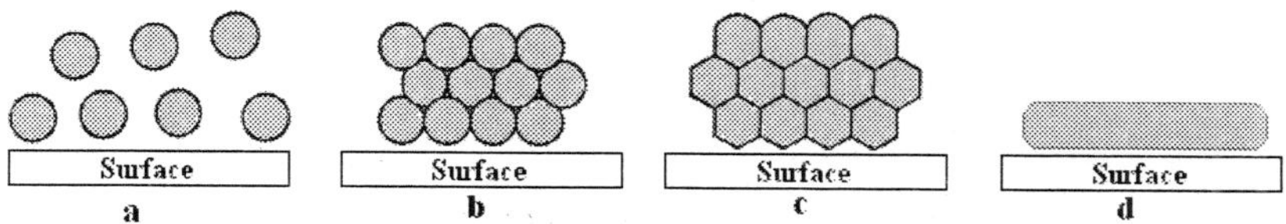

Figure 3. Latex coalescence and film formation of water-borne paints: a) wet film, b) water evaporation and colloid concentration, c) coalescence and particle deformation and d) interdiffusion of polymer chains

Figure (3) represent a schematic representation of latex coalescence and film formation of water-borne paints.

1.2.2. High solid coatings

High solids have been developed to comply with the regulations and legislations of the environment. High solids refer to coatings with more than 80% non-volatiles by volume. In practice, paints with 70 and 60% volume solids are usually included in high solids category [34]. According to European and American paints, high solid paints are those with solid content more than 50-55%. In other markets of protective coatings industry it is assumed that high solids coatings have a minimum volume solids content of 65% [35].

The physical drying coatings, such as the vinyls and chlorinated rubbers have gradually been replaced by chemical drying two component coating systems. The epoxy and polyurethane based coatings are currently representing the major part of high solid coatings in the protective coatings market. Zinc based primers in combination with high solids epoxy coatings and polyurethane topcoats have been the state of the art in corrosion protection and weathering resistance for many years [36].

Research on high solids alkyd polymers has been receiving increasing attention in the enamel industry to reduce VOC emissions [37, 38]. The major challenge in the coating industry is, however, to synthesize high solid alkyds and obtain the improved film performance with high gloss, gloss retention and long-term color stability [34].

Two principles are used to achieve high solids, which are developed in North America [35]:

1.   Low molecular mass and high cross-linking formulation.
2.   Usage of reactive diluents with high dissolving capacity.

The easiest way of reducing the molecular mass is by increasing oil length of the alkyd resin. Unfortunately, increase in oil length can lead to unfavorable problems as:1) low-molecular-weight resin with unsatisfactory properties, slow drying, sagging and inferior pigment stabilization [39]. 2) Reduction of film compactness as a result of loose packing of molecules, which in turn leads to a reduction of the intermolecular forces. 3) Decreasing the number of hydrogen bond forming groups (except castor oil).

On the other hand, Using of reactive diluents accompanied with some disadvantages as [34]: 1) decreasing the viscosity of a coating, 2) instability during storage and 3) early embrittlement of the film on ageing.

More recent way to achieve high solids is creation of highly branched structure in the polymer backbone by use of multifunctional monomers or building blocks [40]. Highly branched architecture helps to achieve resins with viscosity lower than its linear counterparts at a comparable molecular weight [41]. Several articles have been recently published where in hyperbranched macromolecules as well as dendrimers were explored as one of the precursors to make high solid alkyds [42-45]. Researchers have reported about air-drying dipentaerythritol (DPE) modified high solids alkyd resins based on castor and linseed oil. However, due to characteristic yellowing tendency, linseed oil finds limited use in premium coating applications[46].

On the other hand, high performance high solids epoxy siloxane hybrid binder coatings have been developed. These coatings comply with the most stringent regulations in both

Europe and USA. With a volume solids content of 90% and a solvent content of 120 g/l, the epoxy siloxane hybrid coating has a significant lower VOC than all current or planned regulatory requirements in US or Europe [36].

1.2.3. Powder coatings

Powder coatings have grown in popularity as environmentally friendly coatings because of the absence of solvents. They are defined as the technology to deliver a dry powdered coating particle to a substrate, and then soften it to allow for a homogeneous film to be formed. Powder coating is heated to its melt temperature where a phase change occurs, causing it to adhere to the substrate and fuse to form continuous coat. Different application techniques from electrostatic spraying to fluidized bed systems have been developed to increase efficiency and the polymers have been modified to overcome the high melt flow temperatures so other than metal substrates can be powder coated. Automotive bodies, housing materials, wires and cables are the most famous applications of this technique. Polyethylene and epoxy resins are the predominant types of paints used [47]. The most important factors for successful handling of powder technology are the fluidizability of the powder, the dispersing system, the application equipment and materials recovery [35].

Powder coatings have evolved continuously for more than half a century. The main advantages of powder coatings are: [35, 48]

1. Absence of solvents so there is no emission of VOC.
2. Very high utilization rate,>95% due to recycling.
3. High film build in one coat.
4. There is no hazardous material for storage transportation and waste.
5. Very high performance with competitive low cost.

Disadvantages of powder coatings can be summarized as follows: [35, 48, 49]

1. High film thicknesses, 70- 100um are required for good flow.
2. Color matching and color changing are difficult compared with liquid coatings.
3. Coating sharp edges and corners is difficult and needs special techniques.
4. High baking temperature.

1.2.3.1. Manufacturing technology

The manufacture process includes [48]:

1. Dry blending all ingredients.
2. Passing through an extruder at elevated temperature to mill the pigments, melt the resins and achieve a homogeneous mix.
3. Cooling the extrudate rapidly.
4. Milling the flakes of extrudate to final size in a multi-stage process.
5. Removing undersize and oversize particles.

There are some limitations and disadvantages of this process:

1. Pigment dispersants and any other Liquid additives should be converted to solid form firstly.
2. The used technique in extruder is constant (shearing time equals the dwell time) and this may not succeed with all types of pigments
3. Extruder temperature must be followed up to ensure that extruder temperatures are low enough to avoid premature cross-linking.

1.2.3.2. Developments of powder coatings

Many trials have been advanced to avoid process problems and put limits to some disadvantages of powder coatings. Radiation curable powder coatings using UV and near infrared (NIR) radiation can avoid premature cross-linking where, melting and cross-linking processes are complete in few seconds. Unfavorable film thickness of > 70um is due to high particle sizes of approximately 30- 40 um in the powder coatings to ensure safe handling. To attain both reduction of film thickness and attractive surfaces, finer powder particle size must be used. Unfortunately finer particle size raises problems with stabilization, handling and safety. Several projects as standing wave atomization, powder production with super critical CO_2 and powder slurry have been initiated to solve this problem [35, 50].

1.2.4. Radiation curable coatings

Radiation cured coatings have replaced many conventional low solids and solvent borne coatings. They are eco-friendly where, they permit low emission formulations and they rationalize the consumption of resources because coatings curable by ultra violet (UV) radiation or electron beam (EB) require very little energy for film formation and cross-linking [35, 51,]. Radiation curing uses electron beam (EB), ultraviolet (UV) light, or visible light to polymerize a reactive and usually solvent-free coating material [52].

Radiation curable coatings react through unsaturated sites on oligomers which are diluted in the liquid state with monomers or solvents or combined with reactive photoinitiator and other required additives in minute amounts (surfactants, stabilizers, etc.) according to the relevant application Accordingly, Various formulations including acrylated prepolymers based on a polyurethane or a bis-phenol A core were blended with mono- and difunctional reactive diluents, isobornyl acrylate and hexanediol diacrylate, respectively. The influence of photo-initiator content and UV-dose were examined in experiments combining the different factors. The approach allowed formulating on rational bases a reactive blend yielding a coating exhibiting sufficient elasticity without exhibiting tackiness by blending the two types of prepolymers [53].

In presence of radiation source double bonds as active sites are capable of reacting to form cross-linked polymers. Photoinitiators are activated by UV radiation and as a result a radical polymerization process occurs. The rate of polymerization depends on the intensity of the radiation used. Pigmented system required higher energy in put. Depending on the formulation of the UV coating, release of water or solvents at 70-90 °C is useful to produce attractive surfaces [35].

Radiation curable prepolymers (oligomers) are generally low- to medium- molecular weight mono- or multi- functional unsaturated materials, and they provide the primary properties of the cured film such as high abrasion resistance, high tensile strength, elongation, good solvent resistance, and acceptable levels of hardness and flexibility. Unsaturated polyesters, acrylated prepolymer (unsaturated acrylics, urethane acrylates, polyester acrylates, polyether acrylates and epoxy acrylates) and polyene/ thiol systems are different types of oligomer or prepolymer.The oligomer used in the formulation plays an important role in determining the final properties of the finish. Resins used in conventional solvent-based coatings can be chemically modified for use in radiation-cured systems by introducing acrylate functionality [54].

Monomers are used in UV/EB curable systems to provide final film properties and viscosity control of the resin. They are also important in determining the speed of cure, crosslink density, and final surface properties of the cured film. Acrylic monomers are widely used in UV-curable systems due to their high reactivity, moderate cost and low volatility. However, there is a growing interest in preparing in industrial applications of cationic processes as, novel siliconcontaining monomers bearing epoxycyclohexane groups. They possess reactivities much higher than that of typical epoxides and show rates of photopolymerization, which resemble those of acrylate monomers [55]. The resulting polymers possess excellent adhesion, chemical resistance and mechanical properties. For these reasons, there is a growing interest in industrial applications of cationic processes [56].

UV/EB curing technologies are rapidly growing fields, due to the many environmental and performance benefits for chemists, manufacturers, and consumers.

1.2.4.1. UV curing coatings

The UV-curing technique is based on the polymerization of a multifunctional system induced by an incident UV radiation to obtain a three-dimensional network. The reaction allows transforming at room temperature in a fraction of a second a liquid system into a solid having rubbery or glassy properties. It has been widely applied in many industrial fields for the manufacturing, the decoration and the protection of different materials, including wood.

The main advantages of the technique are [57]:

1. The energy needed for the process is low.
2. No solvent is used.
3. The polymerization is very fast and highly efficient.
4. The cure is selectively limited to the irradiated area.
5. Rapid production rate, lower process costs, high chemical stability, high dimensional stability.

However, some limitation of using UV curing technique are found as in high thickness coating films where UV radiation penetration decreases through coating thickness [52].

1.2.4.2. EB curing coatings

Electron beam curing (EBC) is used to achieve a high-quality surface coating. The low-energy EBC technique can be applied in practice to cure coatings on rigid non-porous substrates. Its famous curing applications are for top coatings of furniture, laminated panels and for the thin coil coatings on flexible substrates (as in food packaging). Major advantages of the EB curing processes are solvent free system, less energy consummative, a much higher production rate, and a processing ability at ambient temperature [52]. Applications of EB curing on metal substrates are rather limited due to insufficient adhesion and flexibility shown by EB cured coatings when applied on metal parts. However, adhesion and flexibility can be improved by various ways, in particular via a proper choice of the chemical pretreatment of the substrate [58].

REFERENCES

[1]   Lambourne, R; Strivens, TA. *Paint and Surface Coatings: Coatings Theory and Practice*, 2nd, Ch 1: Paint composition and applications a general introduction by R Lambourne, William Andrew Publishing, USA, 1990

[2]   Wicks, ZW; Jones, FN; Pappas, SP. *Organic Coatings: Science and Technology*, 2nd Ed., ch1: What are coatings?; John Wiley & Sons, NJ, 2006, 4.

[3]   Elsner, CI; Cavalcanti, E; Ferraz, O; Di Sarli, AR. Evaluation of the surface treatment effect on the anticorrosive performance of paint systems on steel. *Progress in Organic Coatings,* 2003,48, 50-62.

[4]   Hare, CH. *Corrosion and the preparation of metals for painting*, Unit 26 Federation Series; on Coatings Technology, Federation of Societies for Coatings Technology, Philadelphia, 1978.

[5]   Allen, KW. *Strength and structures aspect of adhesion*, University Press of London, 1965, 1, 14

[6]   Brown, RFG; Carr, C; Taylor, ME. Effect of pigment volume concentration and latex particle size on pigment distribution. *Progress in Organic Coatings*, 1997, 30, 185-194

[7]   Gowri, S; Balakrishnan, K. The effect of the PVC/CPVC ratio on the corrosion resistance properties of organic coatings. *Progress in Organic Coatings,* 1994, 23, 363-377

[8]   Stieg, FB. The influence of PVC on paint properties. *Progress in Organic Coatings,* 1973, 1, 351-373.

[9]   A. Kalendov, A; Vesely, D. Study of the anticorrosive efficiency of zincite and periclase-based core–shell pigments in organic coatings. *Progress in Organic Coatings*, 2009, 64 5–19.

[10]   Almeida, E; Santos, D; Fragata, F; Rincon, O; Morcillo, M. Alternative environmentally friendly coatings for mild steel and electrogalvanized steel to be exposed to atmosphere. *Materials and corrosion*, 2001, 52, 904-919.

[11]   Hoyle, CE; Kintle, JF. Eds., Radiation curing of polymeric materials; *ASC Symposium Series 417*, American Chemical Society: Washington, DC, 1990.

[12]   Jilek, JH. *Powder Coatings; Federation of Societies for Coating Technology: Blue Bell*, PA, 1991.

[13]   Mohamed, HA; Badran, BM; Aglan, HA. Water-borne methylamine adduct as corrosion inhibitor for surface coatings. *Journal of Applied Polymer Science*, 2001, 80, 286-269.

[14]   Badran, BM; Mohamed, HA; Aglan, HA. Effect of different polymers on the efficiency of methylamine adduct as corrosion inhibitors for surface coatings, *Journal of Applied Polymer Science*, 2002, 85, 879- 885.

[15]   Forsgren, A. *Corrosion control through organic coatings*, Ch3: Waterborne Coatings, Taylor and Francis Group, USA, 2006.

[16] Buckley, William O. / Groeneveld, Marianne K. Crosslinker blends in two-component waterborne polyurethanes. *European Coating Journal*, 1997, 10, 942-947.

[17] Pangelinan, AB; Rhodes, PR. *Annual Meeting of the Societies for Coatings Technology;* Federation of Societies for Coatings Technology, Atlanta, GA, Oct. 27-29, 1993.

[18] Gotsis, C; Sofokleous, K. Comparison of grinding kinetics between a typical ball mill and a ball mill fitted with a breaker plate. *Metallurgical and Materials Transactions* B, 1998, 29 17-24.

[19] Reck, E; Wilford, J. New wetting and compatibility techniques for TiO_2: Application of new techniques provides a means to select pigments for improved formulations. *European Coatings Journal*, 1998, 1 (2) 28- 33.

[20] Mohamed, HA. New waterborne paints with different binders and corrosion inhibition application, *Corrosion Technology and Research.*, 2010, 7(1), 85-89

[21] Bradford, EB; Vanderhoff, JW., Morphological changes in latex films. *J Macromol Chem.*, 1966, 1, 333-360.

[22] Bradford, EB; Vanderhoff, JW. Additional studies of Morphological changes in latex films. *J Macromol .Sci. Phy.*, 1972, B6 (4), 671-694.

[23] Vijayendran, BR. *Polymer colloids II*, Fitch, RM, Ed, Plenum: 1980, 209

[24] Eckersley, ST; Rudin, A. The effect of plasticization and pH on film formation of acrylic latexes. *J Appl. Polym Sci.*, 1993, 48, 1369- 1381.

[25] Lauren, NB; Fellows, CM; Gilbert, RG. Effect of surfactant systems on the water sensitivity of latex films, *Prog. Org. Coatings*, 2004, 92(3), 1813-1823.

[26] Dobler, F; Pith, J; Lambla, M; Holl, Y. Coalescence mechanisms of polymer colloids. 1. Coalescence under the influence of particle-water interfacial tension, *J Colloid Interface Sci.*,1992, 152,1-11.

[27] Elliot, PT; Wetzel, W; Xing, L; Edward, L; Glass, J; Particle, Coalescence, ACS, Series 663 Technology for Waterborne Coatings J. Edward Glass Editor, 1997, The American Chemical Society,

[28] Wallker, FH; Cook, MI; Dubowik, DA. Proc. XXII Waterborne, High *Solids and Powder Coatings Symposium*, 119, 1995.

[29] Walker, FH; Cook, MI. Two component waterborne epoxy coatings, *ACS Symposium Series 663 Technology for Waterborne Coatings* J. Edward Glass Editor, 1997, The American Chemical Society.

[30] Jackson, MA. An Evaluation of Anti-Corrosive Pigments. J. *Protective Coatings & Linings.*, 1990, 37, 54-64.

[31] Galgocci, EC; Weinmann, DJ. *Proc. XXII Waterborne, High Solids and Powder Coatings Symposium,* 1995, 119.

[32] Gauthier, C; et al., *ACS Symposium Series 648*, Film formation in water-borne coatings, In: T; Provder, MA; Winnik, MW; Urban, American Chemical Society, Washington, 1996.

[33] Gilicinski, AG; Hegedus, CR. New applications in studies of waterborne coatings by atomic force microscopy, *Prog. Org. Coat.,*1997, 32, 81-88

[34] Haseebuddin, S; Parmar, R; Waghoo, G; Ghosh, SK. Study of hexafunctional polyol in high solids air-drying alkyd:Improved film performance, *Progress in Organic Coatings,* 2009, 64, 446–453

[35] Goldschmidt, A; Streiberger, H. Basf handbook on basic of coating technology, 2nd Ed,

Environmental compatible paints and coatings, Vincentz Network, Hannover, Germany, 2007.

[36] Keijman, JM. *High Solids Coatings: Experience in Europe and USA,* http://ppgamercoatus.ppgpmc.com, Cited 30 May 2010.

[37] Lindeboom, J. Air-drying high solids alkyd pants for decorative coatings, *Prog. Org. Coat.,* 1998, 34, 147–151.

[38] Bakker, PJ;. Boekee, D. Alkyd resins having a low dynamic viscosity for use in high-solids coatings, *US Patent ,* 1999, 5, 959,067.

[39] Zabel, KH; Klaasen, RP; Muizebelt, WJ; Gracey, BP; Hallett, C; Brooks, CD. Design and incorporation of reactive diluents for air-drying high solids alkyd paints *Prog. Org. Coat.,* 1999, 35, 255–264.

[40] Hult, A; Malmstrom, E; Johansson, M. *Polymeric Materials Encyclopedia,* vol. 5, CRC Press, New York, 1996, 3172–3177.

[41] Voit, BI; Turner, SR. *Polymeric Materials Encyclopedia,* vol. 5, CRC Press, New York, 1996, 3177–3185.

[42] Rolf, ATM. van Benthem. Novel hyperbranched resins for coating applications, *Prog. Org. Coat.,* 2000, 40, 203–214.

[43] Manczyk, K; Szewczyk, P. Highly branched high solids alkyd resins, *Prog. Org. Coat.,* 2002, 44, 99–109.

[44] B. Pettersson, K. Sorensen, *Proceedings of the 21st Waterborne, Higher Solids and Powder Coatings Symposium,* 1994, 753–764.

[45] Tomalia, DA; Frechet, JMJ. *Dendrimers and other dendritic polymers,* John Wiley & Sons Inc., New York, 2002, 1-208

[46] Karakaya, C; Gunduz, G; Aras, L; Mecidoglu, IA. Synthesis of oil based hyperbranched resins and their modification with melamine-formaldehyde resin, *Prog. Org. Coat.,* 2007, 59, 265–273.

[47] Schweitzer, PA. *Paint and Coatings: Application and Corrosion Resistance,* p 80, 81, Taylor& Francis Group, USA, 2006.

[48] Morrison, S. *Take a powder,* http:// specialchem4coatings.com, Cited, 19, Dec, 2007.

[49] Hester, CI; Niholson, RL; Cassidy, MA. *Powder Coating Technology,* 1990, Noyes Data Corporation, USA.

[50] Woltering, J; Kreis, W; Steritberger, HJ. Powderclearcoat for Conventional Application Equipment, *Proceeding XVII Surcar,* 1997.

[51] Hoyle, CE; Kinstle, JF. *Radiation Curing of Polymeric Materials,* American Chemical Society, ACS Washington, DC,1990, p.1.

[52] http://www.specialchem4coatings.com/tc/radiation-curing/index.aspx, as retrived on 1 June, 2010.

[53] Pichavant, L; Coqueret, X. Optimization of a UV-curable acrylate-based protective coating by experimental design, *Progress in Organic Coatings,* 2008, 63, 55–62.

[54] Pourreau, DB; Smyth, S. New low-viscosity acrylic-urethane prepolymers and their acrylated oligomers for moisture and UV-curable coatings, *J Coatings Tech,* April 2005, 2(15).

[55] Decker, C. *Prog. Polym. Sci.,* 1996, 21, 593–651.

[56] Lazauskaite, R; Grazulevicius, JV. Synthesis and cationic photopolymerization of electroactive monomers containing functional groups *Polym. Adv. Technol.,* 2005, 16, 571–581.

[57] Fouassier, JP; Rabek, JF. Radiation Curing in Polymer *Science and Technology*, 1993, 1–4, Elsevier, London.

[58] Molenaar, F; Buijsen, P; Smit, CN. *Progress in Organic Coatings*, 1993, 22, 393-399.

In: Paints: Types, Components and Applications
Editor: Stephanie M. Sarrica

ISBN: 978-1-61761-813-0
© 2011 Nova Science Publishers, Inc.

Chapter 7

NEW DEVELOPMENTS IN PAINT AND COATINGS TECHNOLOGY

***Saeed Farrokhpay**[*]*
Ian Wark Research Institute, University of South Australia,
Mawson Lakes, Australia

INTRODUCTION

Paint and coatings are almost as old as human itself. Over 35,000 years ago, when man was living in caves, he decorated his cave walls with drawings using natural materials such as clays, chalks and animal fats. Now, in the 21st century, the paint and coating global market value is as high as $86 billion [1]. It is clear that drive provided by the emergence of new technologies, the diversity of coatings and their impact on the society will be dramatically increased in coming years, an impact even much greater than what has been observed over the last 100 years. While the first production of paint goes back to the 19th century, a lot of development such as introducing synthetic binders, coil coatings, powder coatings, high solid paints and industrial water based coatings have been seen since then [2]. In fact, paint and coatings industry is growing day by day around the globe. Today, coatings are used to protect metals and buildings from corrosion, in addition to their traditional role for decoration. For highlighting the importance of protective coating, it should be noted that annual global cost of corrosion is about $300 billion which accounts for almost 4% of the worlds GNP [3]. Therefore, all major paint and coating companies are investing huge amounts on their research and development sector to formulate products compatible and suitable for today's aggressive environment. The current trends and challenges in paint and coatings technology are briefly reviewed in this chapter.

---

[*] Corresponding author : Tel: + 61883025791, Email: Saeed.Farrokhpay@unisa.edu.au

CURRENT CHALENGES

Environmentally Friendly Coatings

Environmentally friendly products are advanced area in the development of new paints and coatings. The main goals which have been achieved during last few decades are replacement of toxic anti-corrosive pigments with less toxic materials, and the progressive elimination of solvents in paint formulation [4]. It is predicted that finding less hazardous replacement for raw materials in paint formulation, including solvents, lead and other toxic metal based ingredients, will continue to dominant research and development in coming years. The environmentally friendly technologies value in USA was $13.8 billion in 2005, accounting for about 70% of the total US paint and coatings market [5]. In fact, increased environmental regulations which seem to be a challenge for the industry, have been a major driver for progress and innovation in paint and coatings. Some new development in this area are waterborne metallic paint for automotives, chromate free primer for aircraft structures [1], and water based two packs epoxy/amine primers for high quality industrial applications [6]. Statutory regulations have forced paint users to increase the use of more environmental friendly coating systems, which is why water based coatings are particularly attractive and important. However, achieving high gloss waterborne enamels, with similar properties to conventional solvent based coatings which have been used for many year, is still a challenge. To achieve the optimum visual and economical benefits of a pigment, the dispersion process is critical; the degree to which the pigment particles are dispersed is indeed a major factor in determining the quality and stability of the finished paint products [7]. The level of pigment aggregation is observed to be higher in water based paints than in similar conventional solvent based paints [8]. Polymeric dispersants are often used to assist dispersion; a comprehensive review of polymeric dispersant stabilization of TiO_2 pigment particles has been recently published [9]. The adsorption mechanisms of polymeric dispersants of varying functionalities on TiO_2 pigment particles and the effect of polymer adsorption on the dispersion of the pigment particles have been investigated previously [10-11]. Recent work has shown that the pigment distribution in dry film and pigment stabilization in wet paint are major factors in obtaining high gloss water based paint film [12-13].

Using Nanotechnology in Paints and Coatings

Nanotechnology, as the name implies, includes the science of nano scale materials (1 nm = 10^{-9} m). Nano particles scatter visible light much less effectively than their larger counterparts and thereby enable the production of a new range of material with novel properties. For example, transparent titanium dioxide is a common benefit from this technology. Size of nano particles is smaller than wave length of visible light. The surface to volume ratio of these particles is very high, meaning that their surface properties dominate their bulk properties. Nano particles improve many of desired properties such as chemical, heat resistance and opacity. Nanotechnology is being used in many applications in modern paint and coating industries, of which a summary is presented in Table 1 [14].

Table 1 shows that, in addition to those properties commonly associated with the application of nanotechnology in paint and coatings, there are many "other" (as high as 45%) of un-reported properties which can be obtained using this new technology. This "other" component may represent, for example, properties such as thermal stability, high performance thin film layers, chemical resistance, electrical conductivity combined with higher transparency, optical clarity, or replacement of hazardous products by more inert nano materials [14]. It is estimated that the global market value for nanotechnology in coatings and adhesives is nearly $3.7 billion in 2010 and is expected to increase to about $19.2 billion in 2015 [15]. Some commercially available nano particles and their application in surface coatings are presented in Table 2.

Table 1. Using nanotechnology in paints and coatings

Application	% of use
Scratch resistance	18%
Self-cleaning properties	12%
Enhanced durability	7%
Water repellency	6%
Antimicrobial finish	6%
UV/ yellowing resistance	6%
Other	45%

**Table 2. Some commercially available nano particles
and their application in surface coatings [31].**

Nano particles	Application
SiO_2	Scratch resistance, corrosion resistance
TiO_2 (rutile)	UV absorber, scratch resistance, optical effect
TiO_2 (anatase)	Self cleaning, anti fogging, anti bacterial
Al_2O_3	Scratch resistance, corrosion protection
AlO(OH)	Scratch resistance, corrosion protection
ZnO	UV absorber
Fe_2O_3	Medical applications
Ag colloids	Antibacterial
Metal colloids (Pd, Pt, Au, Ru, Cu)	UV absorber
Carbon nanotubes	Conductive coatings
ATO (Antimony Tin Oxide) & ITO (Indium Tin Oxide)	IR/UV absorber

The appearance and usefulness of nano particles brings many advantages and opportunities to paint and coatings. In fact, coating industry is among the first to tap the potential of nanotechnology. Nano materials have unique properties due to their small particle size, therefore their incorporation into conventional coating formulations improves the coatings properties and also provides a series of new functionalities for the coatings [16]. For example, incorporating nano particles such as nano size TiO_2, Fe_2O_3, ZnO, SiO_2, Al_2O_3 and

$CaCO_3$ into coatings formulation can improves the mechanical, rheological, anticorrosion, and light resistance properties of the coatings [17-21]. Many of nano particles such as nano ZnO and TiO_2 are non-toxic in nature, adding another advantage to coating industry. Nano particles smaller than 100 nm are able to re-inforce polymer matrix without disturbing the desired properties of the coating, most importantly, its transparency. For example, nano ZnO and TiO_2 are mostly used as UV blocking agents [22-23], whereas, nano SiO_2 and Al_2O_3 are used to improve scratch and abrasion resistance of the coating [24-25]. These coatings possess good mechanical properties such as hardness, mar and abrasion resistance, and also provide good colour and gloss retention after long term exposure to UV radiations, chemicals, detergent and solvent [26]. It has also been found that although the micro size pigments provide more saturated hues, intense colours can be achieved using ceramic pigments with particle size less than 50 nm, despite their small particle size [27].

One of the most interesting groups of nano particles is TiO_2 which is used for a wide range of applications [28]. TiO_2 powder is used as pigment in paint, plastic, enamel, paper and cosmetics, with a total production of about 6 million tons per year [13]. Early attempts to commercialize nano TiO_2 were not successful because of the high degree of agglomeration of the powder and difficulties in re-dispersing them in coatings [29]. Although the stability of TiO_2 pigment dispersions have been thoroughly studied [12-13], there is not much reported for nano size products. However, a recent review of the challenge of dispersion of nano particles in polymeric systems showed that regardless of chemistry or shape of nano particles, larger aggregated and poorly dispersed material are always present [30].

Scratch and abrasion resistance

Scratch resistance of coating can be improved by using conventional micron size inorganic fillers, but they often cause matt or semi-matt appearance to coatings, by scattering visible light. However, by using nano particles, scattering of light can be significantly reduced. Nano particles with particle size around 40 to 60 nm are effective fillers. Nano particles such as ZrO_2, AlOOH or SiO_2 can be used in UV curable lacquers to improve abrasion resistance without affecting the transparency of clear coats [3].

Silica particles and clays are among the most widely studied inorganic fillers for improving the scratch/abrasion resistance of transparent coatings [32]. The refractive index of these particles is 1.46 and 1.54 for fumed silica and bentonite clay, respectively, which are very close to those of most resin-based coatings [32]. However, high concentrations of silica particles are generally required to obtain a significant improvement in the scratch and abrasion resistance [32]. Consequently, these high loadings can lead to other problems associated with viscosity, thixotropy and film formation [32].

On the other hand, the refractive index of alumina particles larger than 100 nm is high, i.e. 1.72, which causes significant light scattering, result in hazy appearance in most clear coatings [32]. Therefore, the use of alumina particles in transparent coatings is very limited, in spite of the fact that alumina is significantly harder than silica [32]. Currently, only coatings with a high refractive index, such as melamine-formaldehyde resins, can be used with micron size alumina particles to improve scratch resistance and maintaining transparency [32].

UV resistance

Photochemical degradation caused by UV rays is common mode of failures of most of the coating systems. It causes the oxidation and decomposition of polymer films along with inorganic or organic pigments. Organic UV stabilizers also undergo deterioration after certain periods. Using nano particles such as TiO_2 or ZnO improve UV resistance property by not only absorbing but also reflecting those harmful rays. Addition of organic UV absorbers or nano TiO_2 to such coatings is very common [22, 33-34]. Therefore, reinforcing organic coatings by incorporation of inorganic nano particles has been widely investigated [16, 22, 34-36]. However, there are some limitations in using nano particles in clear coatings. For example, addition of nano TiO_2 to a clear coating may adversely affect its transparency [34]. Nano silica is not only an excellent UV absorber but it also provides a more transparent coating among other inorganic nano particles [25, 37]. The addition of nano silica particles in a coating can also increase the durability of the coating [38-40]. This has been attributed to its high UV absorbance which prevent degradation of the organic polymeric coating [36]. However, there is little experimental evidence to support such a claim, especially for nano silica.

Fire resistant property

Most of the flame retardant coatings like ammonium polyphosphate and melamine are not effective during fire [3]. Mechanical and chemical properties of flame retardant coating can be improved by incorporating nano materials such as nano magnesium aluminium-layered double hydroxide (LDH) to different flame retardant coating system. Nano LDH absorbs the heat and send out water and carbon dioxide when burns and hence lowers the temperature of the substrate [3]. It was shown that coating with 1.5% of nano LDH can significantly increase the fire resistant time (the time period in which back temperature of the test plate reaches to 300 °C) [3].

Anti-corrosive properties

Anticorrosive properties of coatings with optimum level of nano particles show better results than conventional coating [3]. Corrosion resistance of a coating is influenced by pigment-binder (P/B) ratio. It is an important factor by which properties of coating can be determined and it is also related to transportation of harmful corrosive species in electrolyte through the coating systems. Because of greater surface activity of nano particles, they can absorb more resin compared to conventional pigments and thus, reduce the free space between the pigment particles. Therefore, incorporation of nano particles increases the density of the coating and reduces the transport path of corrosive species enhancing the protective performance [3]. However, only optimum amount of nano particles should be added to the coating system for obtaining maximum anti-corrosive benefits. Excess amount of nano particles should be avoided as the amount of resin may not be enough to wet all nano particles, which may result in formation of discontinuous film, and in turn, leading to rise to defects in the coating systems [3].

Challenges in using nano particles

The main problem in using nano particles in paint and coatings is their dispersion stability, although some improvement have been achieved by applying chemical or electrochemical modifications to their surfaces [41]. For example, the surface of nano silica contain OH functionalities that cause inherent hydrophilicity [40]. Furthermore, stable binder is required to inhibit photocatalytic activities of nano TiO_2. And finally, extensive use of nano particles may cause different type of environmental problems, such as new type of toxic materials and other environmental hazards. Ultrafine particles can catalyse chemical reactions inside body which might be dangerous [3]. However, the level of hazard risk of nano particles can be considerably controlled by surface modification of nano particles and also better fixation of nano particles in nano composites [42].

Smart Coatings

The most interesting area of development in paint and coatings over the next few decades will take place in relation to the increased functionality of coatings, and the development of so-called 'smart coatings'. These coatings can react to external stimuli in an intelligent way and can be categorized in different ways including the functional ingredients of the coatings, their application, and also their fabrication methods. It is predicted that these new coatings will open up a range of new functions and play a major role in the future of paint and coatings industry. This means, changing the color in our living room as the temperature changes, using temperature sensitive window in cars which can switch from heat reflection during summer to heat adsorption during winter, sensitivity to electricity and gas allowing them to change color on command and warn of electricity malfunctions or gas leaks, and the possibility for TV screens that can be painted onto walls using electrical conducting coatings [1]. Other examples are coatings with adsorption properties to remove unpleasant smells, coatings that can make both machine and soldiers invisible to the naked eyes, and coatings that changes surface composition to better inhibit corrosion [43].

The development of coatings that are capable of repairing themselves is a real possibility; once bump or scratch is detected, for example, such a coating would release repair components from built-in micro or nano size containers to cover the damage [1]. Self-healing coatings automatically repair themselves via different systems including encapsulation [44-45], reversible chemistry [46], micro-vascular network [47], nano particle and monomer phase separation [48-49], and hollow fibres [50]. However, most of these systems, have major chemical and mechanical limitations which reduce their use in paint and coatings [51]. A recent review provides an insight into the expanding area of research in smart materials with self-healing properties, discussing both chemical (reversible and polymeric) and non-chemical (irreversible and micro-vascular) systems [52].

SUMMARY

The current challenge in paint and coatings industry is to apply sophisticated technology such as nanotechnology to achieve environmentally friendly and durable coatings. Nanotechnology is being used in many applications in modern paint and coatings. It is predicted that environmentally friendly applications such as waterborne systems are continuing to be areas where advances will be made. Smart coatings, which are currently at the research stage, are definitely growing.

It is clear that driven by the emergence of new technologies and developments such as nanotechnology and smart coatings, the diversity of coatings and their impact on the human society will increase in coming years.

REFERENCES

[1] Akzo Nobel, *The Global Coatings Report*. 2006, Akzo Nobel n.v.: The Netherlands.

[2] Johansson, K. Growing greener. *Eur Coat J* 2006, 12, 16-17.

[3] Khanna, AS. Nanotechnology in high performance paint coatings. *Asian J Exp Sci* 2008, 12(2), 25-32.

[4] Deyá, C; Romagnoli, R; del Amo, B. A new pigment for smart anticorrosive coatings. *J Coat Technol Res* 2007, 4(2), 167-175.

[5]   Challener, C. Environmentally friendly paints and coatings. *J Coat Technol* 2006, 3(7), 28-34.

[6] Gummeson, JJ; Gerlitz, M. Advanced in waterborne 2K epoxy/amine primers for heavy duty applications In: *International Waterborne, High-Solids and Powder Coatings Symposium*, Los Angles, 2004.

[7] Bouvy, A. Polymeric surfactants in polymerisation and coatings. *Eur Coat J* 1996, 11, 822-826.

[8] Clayton, J. Pigment/dispersant interactions in water based coatings. *Surf Coat Int* 1997, 9, 414-420.

[9] Farrokhpay, S. A review of polymeric dispersant stabilisation of titania pigment. *Adv Colloid Interface Sci* 2009, 151 (1-2), 24-32.

[10] Farrokhpay, S; Morris, GE; Fornasiero, D; Self, P. Role of polymeric dispersant functional groups in the dispersion behaviour of titania pigment particles. *Prog Colloid Polym Sci* 2004, 128, 216-220.

[11] Farrokhpay, S; Morris, GE; Fornasiero, D; Self, P. Effects of chemical functional groups on the polymer adsorption behaviour onto titania pigment particles. *J Colloid Interface Sci* 2004, 274(1), 33-40.

[12] Farrokhpay, S; Morris, GE; Fornasiero, D; Self, P. Titania pigment particles dispersion in water-based paint films. *J Coat Technol Res* 2006, 3(4), 275-283.

[13] Farrokhpay, S; Morris, GE; Fornasiero, D; Self, P. Stabilisation of titania pigment particles with anionic polymeric dispersants. *Powder Technol* 2010, 202(1-3), 143-150.

[14] Morrison, S. *Small is beautiful: the appeal of nanotechnology*. 2006. viewed online 30/3/2010 *http://www.specialchem4coatings.com*.

[15] Boehm, F, *Nano Technology in Coating and Adhesive Applications: Global Markets*; BBC Research: Wellesley, MA, 2010.

[16] Xu, T; Xie, CS. Tetrapod-like nano-particle ZnO/acrylic resin composite and its multi-function property. *Prog Org Coat* 2003, 46(4), 297-301.

[17] Dhoke, SK; Mangal Sinha, TJ; Khanna, AS. Effect of nano-Al_2O_3 particles on the corrosion behavior of alkyd based waterborne coatings. *J Coat Technol Res* 2009, 6(3), 353-368.

[18] Lowry, MS; Hubble, DR; Wressell, AL; Vratsanos, MS; Pepe, FR; Hegedus, CR. Assessment of UV-permeability in nano-ZnO filled coatings via high throughput experimentation. *J Coat Technol Res* 2008, 5(2), 233-239.

[19] Sung, L; Comer, J; Forster, AM; Hu, H; Floryancic, B; Brickweg, L; Fernando, RH. Scratch behavior of nano-alumina/polyurethane coatings. *J Coat Technol Res* 2008, 5(4), 419-430.

[20] Zhou, SX; Wu, LM. Preparation technology and product development of nanocomposite coatings. *Mater Rev* 2002, 16, 41-43.

[21] Jalili, MM; Moradian, S; Dastmalchian, H; Karbasi, A. Investigating the variations in properties of 2-pack polyurethane clear coat through separate incorporation of hydrophilic and hydrophobic nano-silica. *Prog Org Coat* 2007, 59, 81-87.

[22] Allen, NS; Edge, M; Ortega, A; Sandoval, G; Liauw, CM; Verran, J; Stratton, J; McIntyre, RB. Degradation and stabilization of polymers and coatings: nano versus pigmentary titania particles. *Polym Degrad Stab* 2004, 85(3), 927-946.

[23] Ammala, A; Hill, AJ; Meakin, P; Pas, SJ; Turney, TW. Degradation studies of polyolefins incorporating transparent nanoparticulate zinc oxide UV stabilizer. *J Nanopart Research* 2002, 4, 167-174.

[24] Cayton, RH; Brotzman, RWJ. Nanocomposite coatings-applications and properties. *Mater Res Soc Symp Proc* 2002, 317-322.

[25] Zhou, S; Wu, L; Sun, J; Shen, W. The change of the properties of acrylic-based polyurethane via addition of nano-silica. *Prog Org Coat* 2002, 45(1), 33-42.

[26] Dhoke, SK; Mangal Sinha, TJ; Dutta, P; Khanna, AS. Formulation and performance study of low molecular weight, alkyd based waterborne anticorrosive coating on mild steel. *Prog Org Coat* 2008, 62(2), 183-192.

[27] Cavalcantea, PMT; Dondib, M; Guarinib, G; Raimondob, M; Baldic, G. Colour performance of ceramic nano-pigments. *Dyes and Pigments* 2009, 80(2), 226-232

[28] Kubacka, A; Serrano, C; Ferrer, M; Lünsdorf, H; Bielecki, P; Cerrada, ML; Fernández-García, M. High-performance dual-action polymer-TiO$_2$ nanocomposite films via melting processing. *Nano Letters* 2007, 7(8), 2529-2534.

[29] Fernando, RH. Nanocomposite and nanostructured coatings: recent advancements In: *Nanotechnology Application in Coatings* American Chemical Society: Washington, 2009; pp 2-21

[30] Schaefer, DW; Justice, RS. How nano are nanocomposites? *Macromolecules* 2007, 40(24), 8501-8517.

[31] Sepeur, S, *Nanotechnolgy: Technical Basics and Applications*; Vincentz: Hannover, 2008.

[32] Cayton, RH; Sawitowski, T. The Impact of nano-materials on coating technologies In: *Nanotech 2005*, 2005.

[33] Zhang, Y; Zhou, GE; Zhang, YH; Li, L; Yao, LZ; Mo, CM. Preparation and optical absorption of dispersions of nano-TiO$_2$/MMA (methylmethacrylate) and nano-TiO$_2$/PMMA (polymethylmethacrylate). *Mater Res Bull* 1999, 34(5), 701-709.

[34] Allen, NS; Edge, M; Ortega, A; Liauw, CM; Stratton, J; McIntyre, RB. Behaviour of nanoparticle (ultrafine) titanium dioxide pigments and stabilisers on the photooxidative stability of water based acrylic and isocyanate based acrylic coatings. *Polym Degrad Stab* 2002, 78, 467-478.

[35] Hosseinpour, D; Guthrie, JT; Berg, JC; Stolarski, VL. The effect of interfacial interaction contribution to the mechanical properties of automotive topcoats. *Prog Org Coat* 2005, 54(3), 182-187.

[36] Jalili, MM; Moradian, S. Deterministic performance parameters for an automotive polyurethane clearcoat loaded with hydrophilic or hydrophobic nano-silica. *Prog Org Coat* 2009, 66, 359–366.

[37] Sawitowska, T. The use of nano additives in plastic coating and composite In: *International Congress of Nanotechnology (ICNT)*, San Francisco, 2005.

[38] Xiong, M; Wu, L; Zhou, S; You, B. Preparation and characterization of acrylic latex/nano-SiO$_2$ composites. *Polym Int* 2002, 51(8), 693-698.

[39] Bauer, F; Gläsell, HJ; Decker, U; Ernst, H; Freyer, A; Hartmann, E; Sauerland, V; Mehnert, R. Trialkoxysilane grafting onto nanoparticles for the preparation of clear coat polyacrylate systems with excellent scratch performance. *Prog Org Coat* 2003, 47, 147-153.

[40] Barna, E; Bommer, B; Kürsteiner, J; Vital, A; Trzebiatowski, OV; Koch, W; Schmid, B; Graulea, T. Innovative, scratch proof nanocomposites for clear coatings. *Compos A* 2005, 36, 473-480.

[41] Zhang, W; Li, L; Yao, S; Zheng, G. Corrosion protection properties of lacquer coatings on steel modified by carbon black nanoparticles in NaCl solution. *Corros Sci* 2007, 49(2), 654-661.

[42] Reijnders, L. The release of TiO$_2$ and SiO$_2$ nanoparticles from nanocomposites. *Polym Degrad Stab* 2009, 94, 873-876.

[43] Schuman, TP. Smart corrosion inhibition strategies: substrate, coating, and inhibitors. *J Coat Technol* 2007, 4(2), 60-69.

[44] Brown, EN; White, SR; Sottos, NR. Microcapsule induced toughening in a self-healing polymer composite. *J Mater Sci* 2004, 30(5), 1703-1710.

[45] Caruso, MM; Delafuente, DA; Ho, V; Sottos, NR; Moore, JS; White, SR. Solvent-promoted self-healing epoxy materials. *Macromolecules* 2007, 40(25), 8830-8832.

[46] Chen, X; Wudl, F; Mal, AK; Shen, H; Nutt, SR. New thermally remendable highly cross-linked polymeric materials. *Macromolecules* 2003, 36, 1802-1807.

[47] Toohey, KS; Sottos, NR; Lewis, JA; Moore, JS; White, SR. Self-healing materials with microvascular networks. *Nat Mater* 2007, 6, 581-585.

[48] Cho, SH; Andersson, HM; White, SR; Sottos, NR; Braun, PV. Polydimethylsiloxane-based self-healing materials *Adv Mater* 2006, 18, 997-1000.

[49] Lee, JY; Buxton, GA; Balazs, AC. Using nanoparticles to create self-healing composites. *J Chem Phys* 2004, 121(11), 5531-5540.

[50] Pang, JWC; Bond, IP. A hollow fibre reinforced polymer composite encompassing self-healing and enhanced damage visibility. *Compos Sci Technol* 2005, 65, 1791-1799.

[51] Cho, SH; White, SR; Braun, PV. Self-healing polymer coatings. *Adv Mater* 2009, 21, 645-649.

[52] Syrett , JA; Becer, CR; Haddleton, DM. Self-healing and self-mendable polymers. *Polym Chem* 2010, 1, 978-987.

In: Paints: Types, Components and Applications
Editor: Stephanie M. Sarrica

ISBN: 978-1-61761-813-0
© 2011 Nova Science Publishers, Inc.

Chapter 8

INTUMESCENT POWDER COATINGS

Heinrich Horacek
A-4048 Puchenau, Am Wiesenrain 1, Austria

ABSTRACT

New regulations for hazardous air pollutants drive operations to compliant coatings. Powder coatings are the most popular choice. Also the sizes of furnaces for powder coatings have increased in the past. Today large parts up to 7 m length and 5 tons weight can be treated. These recent developments open the possibility to manufacture building parts made of steel and protected by intumescent coatings like columns and pipes continuously: in a first step the specimens are cast or moulded. In a second step the hot parts are powder coated. Intumescent additives may be added to many classical formulations of powder coatings but the amount of additives is limited and cross liking binders allow low or no intumescence. Therefore special interest deserve polyurethanes comprising organophosphorus polyesters which exert intrinsic intumescence without further additives. Another limitation has to be seen in the requirement that the curing temperature has to be lower than the temperature of intumescence. UV curing intumescent powder coatings avoid this limitation and combine the advantages of UV curing and powder coating. As powder coating is not restricted to steel a large field of activities is opened.

1. INTRODUCTION [1, 2, 3, 4]

The National Emission Standards for Hazardous Air Pollutions address low level of volatile organic compounds, reduction or elimination of certain hazardous alkyl phenols. They drive operations to compliant coatings as powder, electro, autophoretic, UV, high solid and water borne. Recently powder coating has been the most popular choice. Electro, autophoretic and UV continue to increase but on a smaller scale due to limitations involving substrate type, multiple colours and cost. High solids are available for a wide variety of applications including acrylics, alkyds, urethanes and epoxies. By definition water borne paints use water as solvent in solutions or as dispersant in dispersions. They are sensitive to dirt and oil which tend to create blisters and lead to poor adhesion and accelerated corrosion, nevertheless high solid and water borne coatings are replacing solvent based paints at an increasing rate. Metals generally require a minimum of 5-6 stages and separate the cleaning and conversion stages. Plastics generally use a minimum of 6-8 stages. The most typical types of applications are dip, flowcoat and spray. Electrocoating and autophoretic are dip processes by design, although dipping can be used to apply waterborne and powder coatings but is not practicable for high solids. Spray is the most common method of paint application, including air spray, air assisted airless, airless, HVLP(high volume low pressure) and elestrostatic. HVLP benefits the environment by reducing the amount of bounce back of paint off the part resulting in an improved transfer efficiency and lower paint usage.

Air assisted and airless guns operate at higher pressures 15- 60 bar and 35- 300 bar respectively but like HVLP they achieve a better transfer efficiency.

Electrostatic applications where the paint is negatively charged and attracted to a grounded part work well with water borne because of the conductivity of water. However a voltage block or isolation is needed to prevent operator injury. For the application of liquid and powder electrostatic guns and rotary atomizers are used both commonly. The Environmental Protection Agency defines air dried paints as those that cure below 90°C with baked finishes cured at higher temperatures. As Figure 1 illustrates, liquid coatings cure at an initially slower rate then powder coatings in order to allow the water and solvent to evaporate before the coating "skins" over ,otherwise "solvent popping" occurs. On the other hand powder ovens should heat the part quickly to liquefy the coating and permit it to flow. The exhaust rate for powder is usually based upon the colour of the powder and whether products of combustion interfere with the colour. Today powder ovens are available even for large parts of 7m length and of weights up to 5 tons.

Advantages of powder coating:

1.   100% solid content
2.   no VOC
3.   no losses in fluidised bed or electrostatic powder surface coating EPS
4.   short time of coating without time consuming drying, about 10 minutes

Disadvantages of powder coating:

1.   off site in expansive and size limited furnaces
2.   high costs for heating and tempering
3.   limited film thickness of 2,5 mm

New developments try to overcome some of the disadvantages for instance by radiation curing where the energy requirements are much lower than for thermal baking. As result the substantial shrinkage which occurs in a very short time in the UV curing of acrylate systems using radical photoinitiators the adhesion to smooth surfaces such as metals is generally reduced by the stress of the film. In this respect cationic systems show fewer limitations.

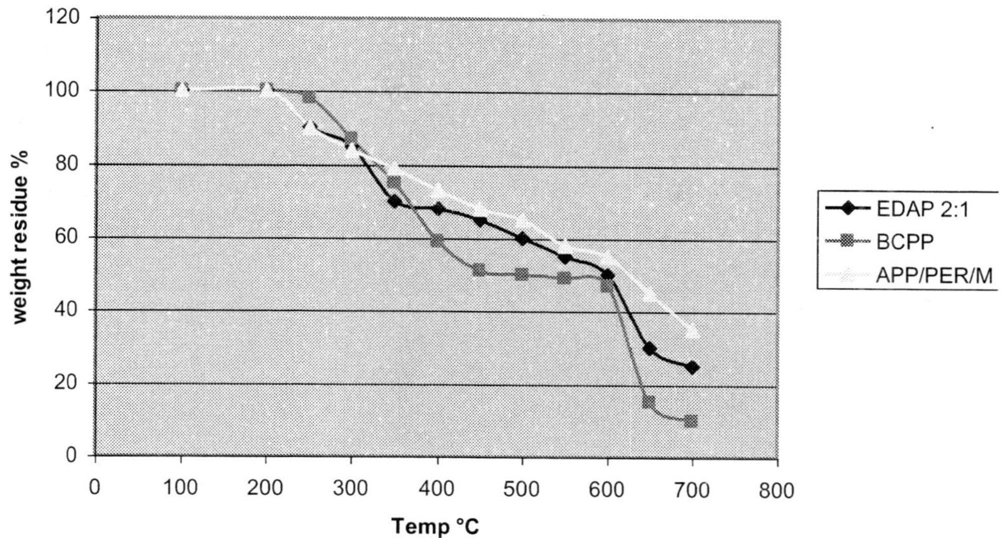

Figure 1. TGA of intumescent additives, N_2, 5K/min

Table 1. Thermoplastic powders used in fluidised bed and EPS

	Vinyls	Polyamides	Polyester	PE	PP	Cellulosics
Polymers	PPVC	PA11,	12	PBuT		
Primer required	yes	yes	no	yes	yes	yes
Melting point °C	130-150	186	160-170	120-130	165-170	160-170
Preheating/postcure	290-230	310-250	300-250	230-200	250-220	280-230
Specific gravity g/cm^3	1,2-1,35	1,0-1,15	1,3-1,4	0,9-1,0	0,9-1,0	1,15-1,35
Shore D	30-55	70-80	75-85	30-50	40-60	65-75

Table 2. Thermosetting powders used in EPS

	Epoxy	PUR	Polyester	Hybride	Acrylic
Polymers			unsaturated	Epoxy/Acrylates	
Melting point °C	120-200	160-220	160-220	140-210	120-200
Curing time (min)/temp.°C	1-30/240-135	10/200	10/200	8/190	10/200
Hardness	H-4H	H-2H	H-2H	H-2H	H-2H
Storage temp. °C	30	30	30	30	30

When intumescent paints are focused, solvent based and water borne paints are available. They are applied by spraying. Also off site applications are gaining market shares. For electrostatic powder surface coating (EPS) thermoplastic as well as thermoset are used [3,4]

In principle all powders applied in fluidised bed can be sprayed electrostatically when ground to small particle size.

In order to combine the advantages of powder coating and UV curing UV curing powder coatings were developed. Also this technology was checked for its use in intumescent coatings.[5,6]

Radiation curing coatings could be designed with curing rates of seconds. The energy requirements were much lower than for thermal backing. As result of the substantial shrinkage which occur in a very short time in the UV curing of acrylates using radical photo initiators, adhesion to smooth surfaces such as metals were generally reduced by the stress in the film. The shrinkage was caused by the gain of density during polymerisation and by the transition of liquid to solid state.

Intumescent powder coatings had to expand in the presence of heat and fire which was only possible for thermoplasts or for slightly crosslinked thermosets . Under this aspect plastisized Polyvinylchloride PPVC was chosen for char formation and low combustion. Polypropylene was selected because it is degraded to well flowing low molecular products by the addition of peroxides. From the group of thermosets Polyurethane PUR was the material of choice because it allowed formulations with low or no cross liking in the reaction with twofunctional intumescent polyesters as well as the addition of monofunctional Bicyclopentaerythritol phosphate an intumescent additives. But high amounts of intumescent additives reduced the flowability of the formulations and their concentrations were therefore limited.

This was the reason why three different intrinsic intumescent polyesters were

synthesized :
pentaerythritol phosphate diethylene glycol polyester
pentaerythritol phosphite diethylene glycol polyester
N,N dihydroxyethylmethylamine diethylphosphonate fumar acid polyester.

2. EXPERIMENTAL

2.1. Ingredients

In Table 3 all ingredients used for intumescent powder coating formulations and their providers are listed.

As intumescent polyesters were not available on the market they had to be synthesized.

2.2. Synthesis of intumescent polyesters

2.2.1. Synthesis of pentaerythritol diphosphite diethylene glycol polyester (1) and of pentaerythritol diphosphate diethylene glycol polyester (2)

A 2000ml three necked round bottom flask was equipped with a mechanical stirrer, reflux condenser, thermometer, addition funnel and dry N2 inert.

In case (1) a solution of 0,5 mol (132,5g) dichloropentaer ythritoldiphosphite [7] in 500g toluene and in case (2) a solution of 0,5 mol (148,5g) dichloropentaerathritoldiphosphate [8] in 500g toluene were added over a period of 1 h to a solution of 0,53m(56,2g) diethylene glycol (Hoechst) and 101(1mol) triethylamine (BASF) in 800 g toluene. Both solutions were heated at reflux for 5 h. The precipitated triethyl amine hydrochloride was filtered. The filtrates were concentrated by heating to a final temperature of 160°C/5mm. The solid final products were characterized:

Case(1) Melting point Tm = 90°C, elementary analysis: 37,2%C, 5,8%H, 18,6%P
Acid number (DIN 53240) :2, hydroxyl number (DIN 53402): 112
Case(2) Melting point 80°C, elementary analysis 34,5%C, 5,5%H, 16,2P
Acid number: 150, hydroxyl number: 1050

The high hydroxyl number indicated hydrolysis of polyester (2) in the presence of watery bases during titration when the hydroxyl number was determined.

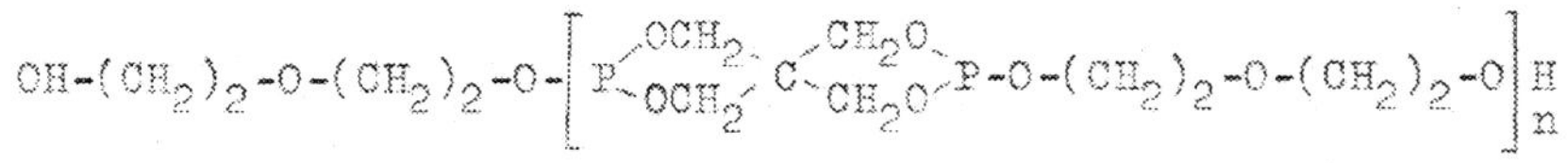

$$\text{OH-(CH}_2)_2\text{-O-(CH}_2)_2\text{-O-}\left[\text{P}\begin{smallmatrix}\text{OCH}_2\\\text{OCH}_2\end{smallmatrix}\text{C}\begin{smallmatrix}\text{CH}_2\text{O}\\\text{CH}_2\text{O}\end{smallmatrix}\text{P-O-(CH}_2)_2\text{-O-(CH}_2)_2\text{-O}\right]_n\text{H} \qquad (1)$$

$$\text{OH-(CH}_2)_2\text{-O-(CH}_2)_2\text{-O-}\left[\overset{\text{O}}{\text{P}}\begin{smallmatrix}\text{OCH}_2\\\text{OCH}_2\end{smallmatrix}\text{C}\begin{smallmatrix}\text{CH}_2\text{O}\\\text{CH}_2\text{O}\end{smallmatrix}\overset{\text{O}}{\text{P}}\text{-O-(CH}_2)\text{-O-(CH}_2)_2\text{-O}\right]_n\text{H} \qquad (2)$$

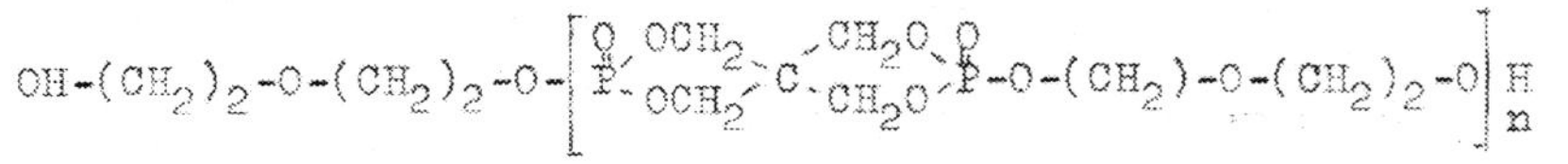

2.2.2. *Synthesis of N,N dihydroxyethylmethylamine diethylphosphonate fumar acid polyester*

The same equipment was used as already mentioned. The 2000ml flask was loaded with 1,1 mol(280g) dihydroxyethylmethylamine diethylphosphonate (Levagard 4090 N, Bayer Chemicals). Through a powder funnel 1mol (98,1g) maleic anhydide (DSM fine Chemicals) was added. The temperature was raised to 60°C. When the solution was clear, the temperature was raised until reflux occurred.. The mixture was stirred at reflux for 6 h and 19 g water were removed. Toluene was distilled off under vacuum. After cooling the low molecular impurities were removed by solving in benzene and precipitating in petrol ether. The solid residue was characterized:

Melting temperature Tm = 95°C, elementary analysis: 45,7%C, 7,0%H, 4,4%N, 9,8%P
Acid number: 18, hydroxyl number: 90

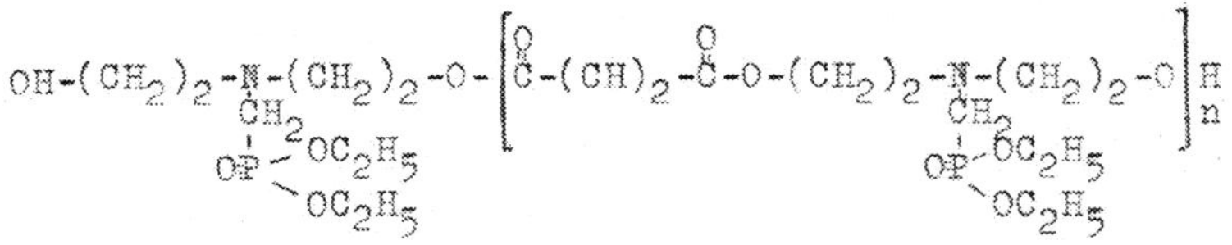

$$\text{OH-(CH}_2)_2\text{-N-(CH}_2)_2\text{-O-}\left[\overset{\text{O}}{\text{C}}\text{-(CH)}_2\overset{\text{O}}{\text{-C}}\text{-O-(CH}_2)_2\text{-N-(CH}_2)_2\text{-O}\right]_n\text{H}$$

2.3. Characterization of Intumescent Ingredients

The intumescent ingredients were characterized by thermal gravimetric analysis (TGA) on a Mettler Toledo TMA/SDTA with TGA/SDTA 851 Modul. The samples were placed in aluminium oxide crucibles of 900 µl volume (ME 511119,960) with 12 mm diameter covered by punctured lids. The TGA measurements took place under N2 with 80 ml/min at a heating rate of 5K/min. The TMA measurements were performed in aluminium oxide crucibles with 7 mm diameter and 4,6 mm height covered by 6mm diameter lids in air at a heating rate of 50K/min.

In Figure 1 the weight losses of BCPP and a mixture of 3m APP, 1m PER and 1m Mel were determined in dependence of temperature. The thermal stability decreased in the order APP/PER/M, EDAP 2:1 and BCPP. BCPP had a melting point of 220°C. All samples exerted intumescence about 300°C.

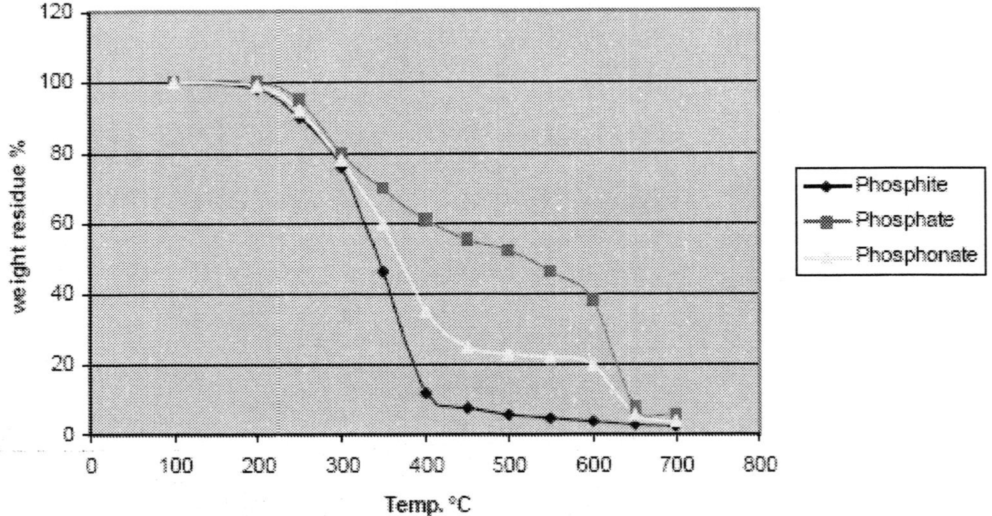

Figure 2. TGA of phosphorusorganic polyesters, N_2, 5K/min

In the same way the polyesters were investigated by thermal gravimetric analysis (TGA) in Figure 2. All polyesters degraded above 200°C. At 300°C intumescence was observed. The thermal stability of the polyesters decreased in the order phosphate- polyester most stable, phosphonate- polyester medium and phosphite polyester the lowest stability.

TMA measurements indicated that intumescence started at 250- 300°C in every case but the foams were not stable enough for exact data of expansion factors.

Table 3. Formulations of intumescent thermoplastic powder coatings

Nr. 1 (PPVC)	Nr.2 (PP)
70 pbw PPVC	70pbw PP
	0,2pbw Perroxide
15pbw APP	25pbw EDAP
5pbw PER	
5pbwM	
4pbw TiO2	4pbw TiO2
1 pbw Ca stearate	0,8pbw Ca stearate
Sum: 100	100

2.4. Manufacturing Process of Intumescent Powder Coatings

In Table 3 and 4 the formulations for intumescent powder coatings are summarized

All formulations of Table 3 and 4 were extensively mixed in a Loedige mixer. The blends were extruded on a twin screw extruder Leistritz LSM 30.34GL9R at 180°C, 50 rpm and a through put of 10 kg/h followed by water bath and pelletizer. All mixtures were ground under cooling on a Baumeister pin mill. The powders were classified by seaving. The particle size distributions were measured on a MasterSizer XSB.OD(Malvern Instrument). Only powders with particles between 30 and 50μm were used.

Table 4. Formulations of intumescent thermoset powder coatings

Nr. 3 (PUR)	Nr. 4 (UV cured)	Nr.5 (Phosphite)	Nr.6 (Phosphonate)
38pbwRocute194	29pbw Roskydal LS2266	59pbw polyphosphite(1)	62pbw polyphosphonate(2)
41pbw Crelan LS2147	23,5pbw Rocute 560	36pbw Crelan LS2147	33pbw Crelan LS2147
0,1pbw Sn-octoate	17,5pbw DER 663 UE		
0,2pbw Diazabicyclooctane			
18pbw BCPP	15pbw APP		
	5pbw Mel		
	5pbwPER		
2pbw TiO2	3pbw TiO2	4pbw TiO2	4pbw TiO2
0,7pbw Montan wax	1pbw Irgacure 819	1pbw Hostalub	1pbw Hostalub
	1pbw Darocure 1173		
Sum: 100	100	100	100

2.5. Spraying and Curing of Intumescent Powder Coatings

The electrostatic powder spraying was carried out on a Wagner powder spraying device with a PEM-C3 spray gun, a steering device EPG 2007 and a powder injector PJ 2020 PRS. The powders were whirled up and were blown through a high voltage field of 50 kV. The electrically loaded powders were sprayed on preheated and grounded closed steel columns. The sinter or curing processes were exerted at 200°C in 30 minutes.

In Figure 3 the melt viscosities at low shear forces 230 sec-1 were measured in dependence of the temperature in a Rheograph 2000 for the thermoplastic formulations PPVC(Nr.1) and PP (Nr.2) according DIN 54811.

More complicated is the curing of thermosets PUR(Nr.3), Phosphite(Nr.5) and Phosphonate(Nr.6) in dependence of temperature and time given in Figure 4. At time 0 the product temperature was 100°C and the heating rate was 5K/min. When 200°C was reached after 20 minutes further heating was stopped.

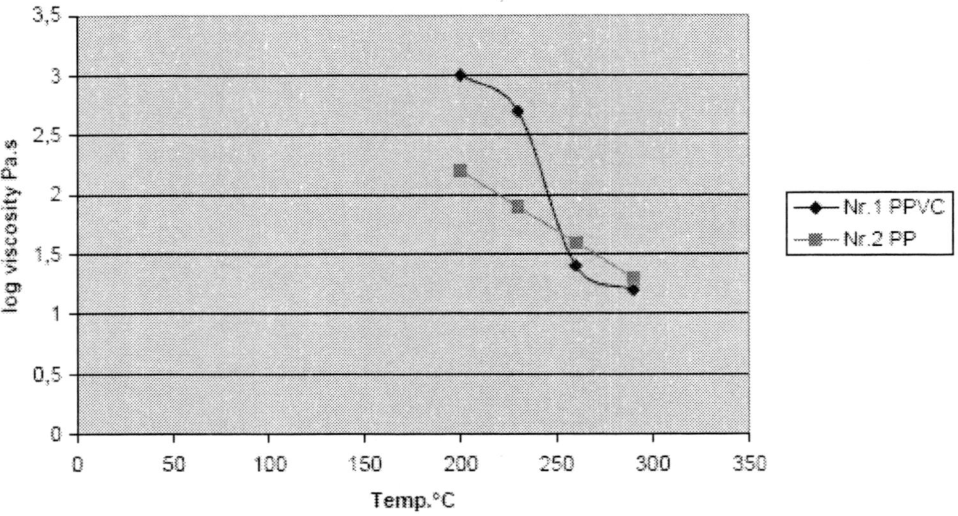

Figure 3. Viscosity as a function of temperature for thermoplastics

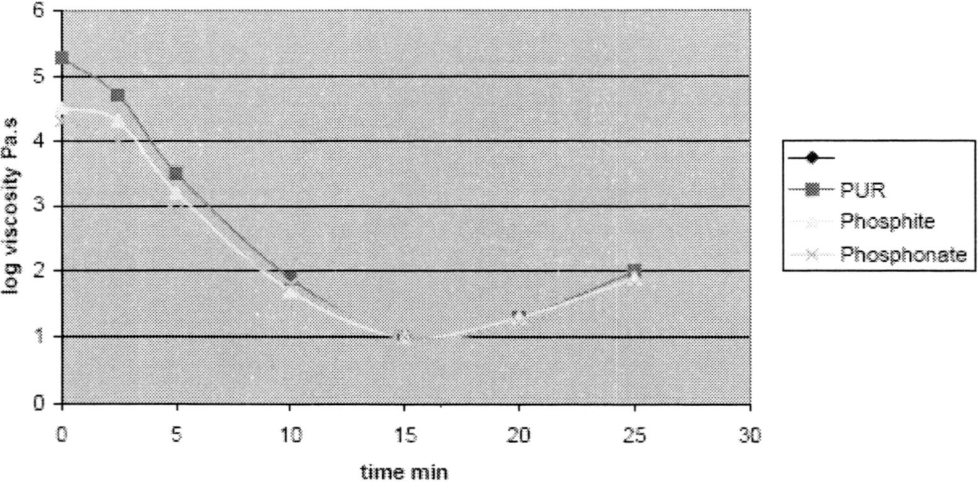

Figure 4. Viscosity as a function of curing time, heating rate 5K/min starting from 100°C

In comparison with the viscosities of thermoplastic formulations the viscosity minima of thermosets were much lower.

The UV curing thermoset powder (Nr.4) was heated to 200°C and cured under a mercury gallium lamp with 1.700-2.900 mJ/cm2 energy input and at a distance of 30 cm within 2 minutes. It showed the highest shrinkage of 0,2 cm3/g in comparison with thermoplastics of about 0,08 cm3/g. Due to the low particle volume concentrations pvc the surfaces of the intrinsic powder coatings Nr.5 and 6 were smooth and exerted more gloss. For better comparison a final thickness of 750 μm was attempted for all coatings Nr.1-6.

2.6. Fire Performance

The coated closed steel columns with a horizontal perimeter to area factor 270 m^{-1} equipped with several thermocouples for continuous measurements of temperature were placed in a furnace heated according to the ISO curve. In Figure 5 the temperatures of the uncoated column as well as of the columns protected by the six paints were recorded in dependence of time. The measure of efficiency was the time lag until 500°C were reached. Under this aspect following ranking was achieved in Table 5:

3. RESULTS AND SUMMARY

Intumescent powder coatings exerted restrictions in concentrations of intumescent additives. Smooth surfaces were only achieved at low additive concentrations. Nevertheless PPVC with low amounts of intumescent ingredients showed the best performance. In consequence intrinsic intumescent powder coatings had no such restrictions but the most appropriate polyphosphate ester hydrolysized. The coating based on polyphosphite ester was not as efficient as that based on polyphosphonate ester.

The mode of application is still in question because fluidised bed allows higher thickness up to 2,5mm and is the ideal method for continuous coating of two dimensional articles of continuous length.

Until now intumescent powder coatings are not available on the market. In literature only few patents [5,6,9,10] treat the subject. The present results and the many advantages of powder coatings should encourage further work. As powder coating is not restricted to steel protection a large field of activities is opened.

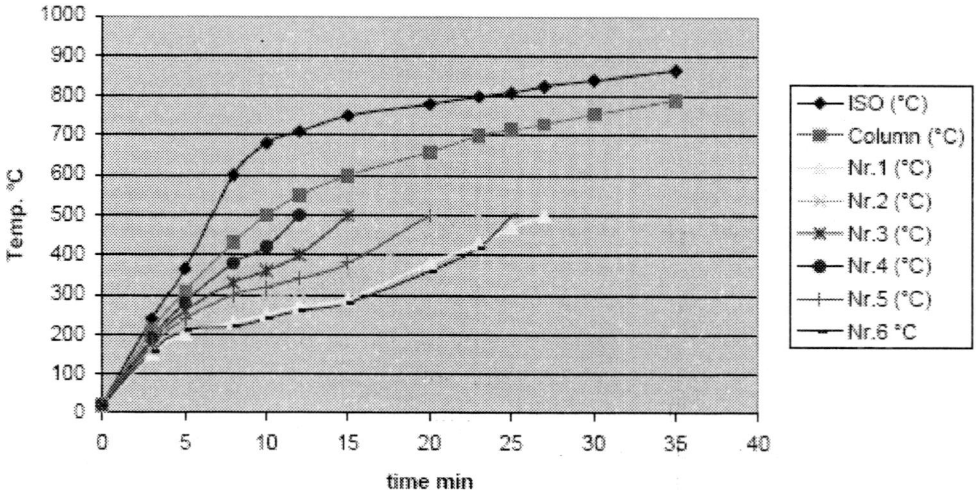

Figure 5. Fire test of intumescent powder coatings Nr. 1-6, dft= 750µm, Hp/A= 0270m^{-1}

Table 5. Time until 500°C are reached and expansion factor

Nr.	Time (min)	expansion factor
	(Thickness after expansion/750µm) 15min at 500°C	
1	28	20
3	27	17
6	25	15
2	24	18
5	22	10
6	15	3

REFERENCES

[1]   Herr, C. *metal finishing Nov.*, 2007, 40-43

[2]   Enviral and Meeh, *Pulverbeschichtungen,* mo Jg.59, 2005, 12-14

[3]   Kirk Othmer 4th ed. Vol.3 615 *Coatings*

[4]   Kirk Othmer 3rd ed. Vol. 19.1 *Powder Coatings.*

[5]   WO 03066749 A1, Leigh and Co. Radiation curable intumescent coatings, 2003.

[6]   EP 1 659 157 A1, *Horacek,* 2005.

[7]   Lucas, HJ; Mitchell, Jr. FW; Scully, CN. *J.Am.Soc.*, 1950, 72, 5491-5495.

[8]   Rätz, R; Sweeting, O. *J. Org. Chem.*, 1963, 1608-1912.

[9]   WO 02077110 A1, 2002, Chance and Hunt Lim. Ferro Lim., Fire Retardant intumescent coating.

[10] WO 02096996 A1, 2002, Leigh and Co. *Coating composition.*

In: Paints: Types, Components and Applications
Editor: Stephanie M. Sarrica

ISBN: 978-1-61761-813-0
© 2011 Nova Science Publishers, Inc.

Chapter 9

THE PREPARATION OF AG-NANOPARTICLE-EMBEDDED PAINTS AND THEIR ANTIMICROBIAL ACTIVITY

Renat R. Khaydarov[1], Rashid A. Khaydarov[1], Olga Gapurova[1], Svetlana Evgrafova[2] and Seung Y. Cho[3]

[1]Institute of Nuclear Physics, Tashkent, Uzbekistan
[2]V.N. Sukachev Institute of Forest SB RAS, Krasnoyarsk, Russia
[3]Yonsei University, Wonju, South Korea

ABSTRACT

Over the last decades silver has been engineered into nanoparticles, structures from 1 to 100 nm in size. At present day silver nanoparticles are widely used as antibacterial/antifungal agents in a diverse range of consumer products. This paper deals with the authors' research in the field of preparation and studying antimicrobial properties of silver nanoparticle-embedded paints. The silver nanoparticles have been synthesized using the technique based on using cellulose fibers as reductant. For conducting microbiological tests the commercially available water paint was mixed with 200 ppm silver nanoparticles solutions in various ratios of 20:1, 50:1 and 100:1 in order to impart antimicrobial properties to the paint. To evaluate the antibacterial and fungicidal properties of Ag nanoparticles in paints we have used *Escherichia coli*, *Salmonella typhimurium*, *Aspergillus niger*, *Staphylococcus aureus*, *Pseudomonas aeruginosa*, *Candida albicans* cultures. The tests conducted have demonstrated that synthesized silver nanoparticles added to water paints show a pronounced antibacterial/antifungal effect, despite the fact that they tend to be agglomerated into clusters. It has been shown that larger concentrations of silver nanoparticles have a greater antibacterial/antifungal efficacy in Ag-nanoparticle-embedded paints.

Keywords: silver, paint, nanoparticles, bacteria, fungi

INTRODUCTION

Silver first received regulatory approval for use as an antimicrobial agent in the early 20th century (Chopra, 2007). Over the last decades silver has been engineered into nanoparticles, defined as structures up to 100 nm in size, have been extensively investigated and have found applications in many areas of science and technology (Lewis 1993, Murphy et al. 2005, Li et al. 2005) due to their unique size-dependent optical, electrical and magnetic properties. At present day, most of silver nanoparticles applications are connected with their usage as antibacterial/antifungal agents (Buzea et al. 2007). For instance, there is a growing interest in preparation of bactericidal cotton fibers containing silver nanoparticles for textile industry (Lee and Jeong 2004, Lee and Jeong 2005). There is also a promising trend to impregnate commercially available paints with silver nanoparticles (Revina et al. 2001, Khaydarov et. al. 2009a). Microbes are known to come into contact with walls in a number of ways: by deposition of dust and fine aerosols, by human skin contact, and by splashes from liquids. The nanosilver-based wall paint can potentially prevent the formation of mould inside buildings and, by conjecture, the growth of algae on outside walls.

We have reported recently (Khaydarov et. al. 2010) on a novel method of continuous fabrication of aqueous dispersions of silver nanoparticles using cellulose fibres. Our assays conducted on *Escherichia coli*, *Staphylococcus aureus* and *Bacillus subtilis* cultures returned low values of minimum inhibitory concentrations, that proves the promising potential of the synthesized colloidal silver solutions for bactericidal applications.

The purpose of this paper was to check the efficacy of synthesized nanosilver as antimicrobial agent against a range of microbes on the surface of water paints.

MATERIALS AND METHODS

The morphology of silver nanopartilces on the surface of paint samples was observed by field emission scanning electron microscopy (FE-SEM; JSM-6700F, JEOL, Japan). The size and shape of the nanoparticles in solution were determined by transmission electron microscopy (TEM) (LEO-912-OMEGA, Carl Zeiss, Germany). The concentration of silver nanoparticles in solutions was determined by neutron activation analysis (NAA) (Soete et al. 1972, Khaydarov et. al. 2008). Samples were irradiated in the nuclear reactor of the Institute of Nuclear Physics (Tashkent, Uzbekistan). The product of nuclear reaction $^{109}Ag(n,\gamma)^{110m}Ag$ has the half-life $T_{1/2}$ = 253 days (Khaydarov et. al. 2004). The silver concentration was determined through measurements of the intensity of gamma radiation with the energy of 0.657 MeV and 0.884 MeV emitted by ^{110m}Ag. A Ge(Li) detector with a resolution of about 1.9 keV at 1.33 MeV and a 6144-channel analyser were used for recording gamma-ray quanta.

The silver nanoparticles used for embedding paints have been synthesized via the cellulose fiber - based technique we have described previously (Khaydarov et al. 2010). The obtained nanosilver particles suspended in water solution were spherical with diameter of 13 ± 5 nm (cf. Figure 1).

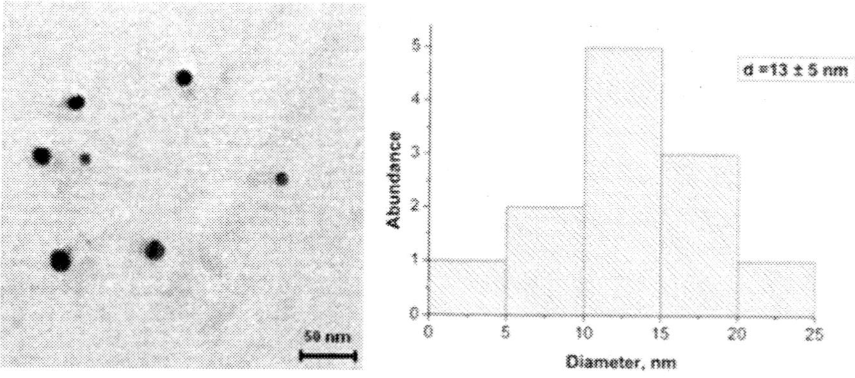

Figure 1. TEM micrograph (scale of 50 nm) of silver nanoparticle solutions prepared from aqueous solution (800 mg/l) of AgNO3 at the temperature of 25°C obtained after five days of exposure using the technique described in (Khaydarov et. al. 2010)

Common household acrylic paint widely used for renovating and decorating purposes has been used in our experiments. Silver nanoparticle-containing acrylic paint was obtained by diluting the initial paint sample with the Ag colloidal solution in such a way as to get the desired value of nanosilver concentration within the desirable testing range. For the laboratory antibacterial tests a 22 mm x 22 mm pasteboard was covered with silver nanoparticle-containing acrylic paint. To evaluate the antibacterial and fungicidal properties of silver particles *Escherichia coli* was used as a representative Gram-negative bacterium; *Staphylococcus aureus* and *Bacillus subtilis* were used as Gram-positive bacteria; *Aspergillus niger, Aureobasidium pullulans* and *Penicillium phoeniceum* were used to represent cosmopolitan saprotrophic fungi. Within the laboratory tests in order to evaluate the antibacterial and fungicidal properties of Ag nanoparticles added to an acrylic paint, samples with Ag nanoparticle content as well as control samples were immersed in a thin layer of beef-extract agar. 1 mL of suspension with approximately 10^5 CFU/mL density of the microorganisms to be tested was distributed uniformly on agar surface and incubated at 28 °C (CFU = colony forming units). Antimicrobial activity was evaluated according to the presence or absence of microbial growth just above the sample after a 24-h incubation for bacteria and a 72-h incubation for fungi. All microbiological tests were performed in triplicate.

For conducting field tests the commercially available acrylic paint was thoroughly mixed with 200 ppm silver nanoparticle solutions in various ratios of 20:1, 50:1 and 100:1 in order to get 10 ppm, 4 ppm and 2 ppm Ag-nanoparticle-embedded paint. In field tests we have used building bricks as the surface of putting paints on. The bricks were painted at the temperature of 20°C by applying 2 layers consecutively by the brush over the period of 7 days. The antimicrobial effect of Ag-nanoparticle-embedded paint has been estimated after 1, 7 and 30 days of painting. The bactericidal effect of the paint within field tests was studied using *Escherichia Coli ATCC 25922, Klebsiella pneumoniae ATTCC 13833, Staphylococcus aureus Wood-45* and *Pseudomonas aeruginosa 508* cultures. The antimicrobial influence of the paints has been checked as follows. Suspension of strains (0,1 mL), prepared in sterile physiological solution in the concentration of 10^7 CFU/mL was coated on the surface of painted bricks. The incubation periods were chosen to be 1, 3, 6 and 24 hours at room temperature. The sowing was performed on plates with nutrient medium.

RESULTS AND DISCUSSION

Laboratory Tests

We have used SEM measurements in order to observe silver nanoparticles on the surface of the modified paint. As one can see from Figure 2, most of initial silver nanoparticles agglomerated into clusters up to 300 nm in size because of attractive interaction forces between them.

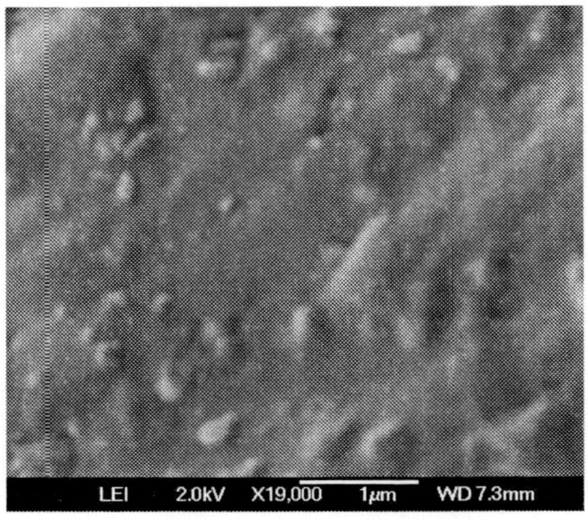

Figure 2. Sample of Ag-nanoparticle embedded water paint (Note the spherical structures corresponding to nanoparticles)

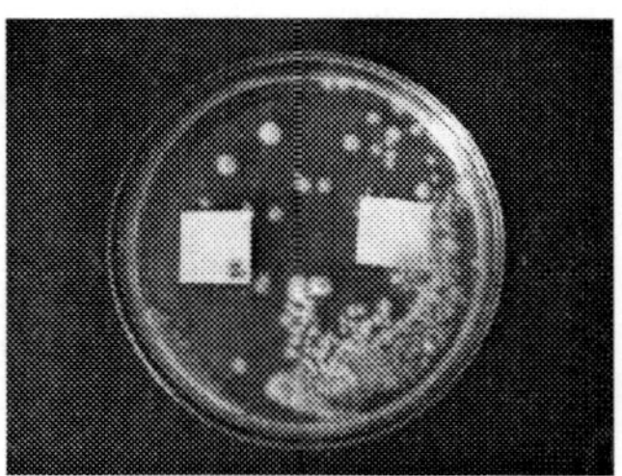 

Figure 3. Growth of *P. phoeniceum* (left) and *A. pullulans* (right) cultures on pasteboard samples modified by silver nanoparticles. (Note the white spots corresponding to microbial colonies). Sample #1 is a control sample, i.e. it was covered with non-modified paint; sample #2 was covered with the paint modified by silver nanoparticles of 0.8 $\mu g/cm^2$ density

The photographs in Figure 3 show growth of *P. phoeniceum* and *A. pullulans* cultures on pasteboard samples modified by silver nanoparticles. The absence of visible growth of the microbes on the samples #2 confirms the antibacterial/antifungal effect of Ag-nanoparticle-embedded paints. The analogous experiments conducted by us (Khaydarov et. al 2009b) with *A. niger* and *S. aureus* cultures have also demonstrated a pronounced antifungal/antibacterial efficacy of the water paint modified by silver nanoparticles of 0.8 $\mu g/cm^2$ density.

Field Tests

Field tests have been conducted using building bricks as the surface of putting paints on (see Table 1).

Table 1. The antimicrobial effect of Ag-nanoparticle-embedded paints on different microbial strains

Microbe	Paint samples	Amount of bacteria on the surface of the paint, CFU			
		1	3	6	24
Escherichia Coli ATCC 25922	Control sample	150	10	1	0
	2 ppm nanosilver	120	10	0	0
	4 ppm nanosilver	100	0	0	0
	10 pm nanosilver	20	0	0	0
Klebsiella pneumoniae ATTCC 13833	Control sample	300	40	10	1
	2 ppm nanosilver	250	30	5	0
	4 ppm nanosilver	130	30	12	0
	10 pm nanosilver	50	5	0	0
Staphylococcus aureus Wood-45	Control sample	400	100	50	10
	2 ppm nanosilver	320	90	20	0
	4 ppm nanosilver	200	70	10	0
	10 pm nanosilver	50	10	1	0
Pseudomonas aeruginosa 508	Control sample	300	120	50	1
	2 ppm nanosilver	270	100	30	0
	4 ppm nanosilver	200	80	30	0
	10 pm nanosilver	80	30	5	0

As seen from the table 1, all tested Ag-nanoparticle-embedded paint samples demonstrated a bactericidal effect on the surface of bricks. Increase of the nanosilver concentration from 2 ppm up to 10 ppm leads to more pronounced antimicrobial effect. The revealed antimicrobial efficiency of the paint samples without silver nanoparticles is connected with the bactericidal influence of some components of the paint. We suppose that with the course of time paint-volatile components will be evaporated that will reduce the antimicrobial efficiency of the control samples of the paint without nanosilver.

It would be fair to say that the mechanism of the bactericidal effect of the paint impregnated with silver nanoparticles is not well understood as yet. Lok and co-authors (Lok et. al 2007) have recently reported that "Nanosilver represents a special physicochemical system which confers their antimicrobial activities via Ag^+". By contrast, according to (Morones et al. 2005) the bactericidal effect of silver nanoparticles on micro-organisms is connected not merely with the release of silver ions in solution. Following their report, silver nanoparticles can also be attached to the surface of the cell membrane and disturb its proper function drastically. They are also able to penetrate inside the bacteria and cause further damage by possibly interacting with sulfur- and phosphorus-containing compounds such as DNA.

The lack of knowledge on the mechanism of the bactericidal effects of silver nanoparticles (Kim et. al. 2007) in combination with the growth of applications of nanosilver in various branches of industry over the last years has caused concerns that silver nanoparticles may have a toxic effect on human health. According to recent studies (Soto 2005, Braydich-Stolle et al. 2005, Hussain et al. 2005, Grodzik and Sawosz 2006) silver nanoparticles can exhibit toxicity to animal and human cells. In this connection we have observed an outside wall of a house painted with nanosilver-modified paint. The NAA tests conducted over a period of eight months showed that no significant loss of silver nanoparticles in the wall paint occurred. Thus the results indicate that silver particles impregnated to the paint do not present a *direct* health hazard, as owing to sufficient bonding of silver nanoparticles to the material the potential for excessive exposure to Ag ions or nanoparticles will be negligible.

CONCLUSION

Silver nanoparticles have been considered to be a commercially viable addition for use in the paint industry. The tests conducted demonstrate that Ag-nanoparticle-embedded acrylic paints show a pronounced antibacterial/antifungal effect, despite the fact that the silver nanoparticles tend to be agglomerated into clusters. The demonstrated sufficient bonding of silver nanoparticles to the paint indicates that the potential for excessive exposure to Ag ions or nanoparticles is negligible.

REFERENCES

Braydich-Stolle, L., Hussain S., Schlager J. & Hofmann M. C. (2005). In vitro cytotoxicity of nanoparticles in mammalian germline stem cells, *Toxicological Sciences, 88(2)*, 412-419.

Buzea, C., Pacheco, II. & Robbie, K. (2007). Nanomaterials and nanoparticles: Sources and toxicity. *Biointerphases, 2(4)*, MR17-MR71

Chopra, I. (2007). The increasing use of silver-based products as antimicrobial agents: a useful development or a cause for concern, *Journal of Antimicrobial Chemotherapy, 59*, 587–590.

Grodzik, M. & Sawosz, E. (2006). The influence of silver nanoparticles on chicken embryo development and bursa of Fabricius morphology, *Journal of Animal and Feed Sciences, 15(Suppl 1)*, 111-114.

Hussain, S. M., Hess, K. L., Gearhart, J. M., Geiss, K. T. & Schlager, J. J. (2005). *In vitro*

Khaydarov R.R, Khaydarov R.A., Gapurova O., Estrin Y., Evgrafova S., Scheper T. & Cho S.Y. (2009) "Antimicrobial effects of silver nanoparticles synthesized by an electrochemical method" in Book *"Nanostructured Materials for Advanced Technological Applications"*, Springer, Netherlands, 215-218

Khaydarov, R A., Khaydarov, R R., Gapurova, O. & Estrin, Y. (2010). A novel method of continuous fabrication of aqueous dispersions of silver nanoparticles. *Int. J. Nanoparticles, 3*, 77-91.

Khaydarov, R A., Khaydarov, R R., Gapurova, O., Estrin, Y. & Scheper, T. (2008).

Electrochemical method of synthesis of silver nanoparticles. *J Nanopart Res.*, Doi:10.1007/s11051-008-9513-x

Khaydarov, R. A., Khaydarov, R. R., Estrin, Y., Evgrafova, S., Cho, S., Scheper, T. & Endres, C. (2009). "Silver nanoparticles: Environmental and human health impacts", in Book: *"Nanomaterials: Risk and Benefits"*, Springer, Netherlands, 287-299.

Khaydarov, R. A., Khaydarov, R. R., Olsen, R. L. & Rogers, S. E. (2004). Water disinfection using electrolytically generated silver, copper and gold ions, *Journal of Water Supply RT-Aqua*, *53*, 567-572.

Kim J. S., et al. (2007). Antimicrobial effects of silver nanoparticles, *Nanomedicine: Nanotechnology, Biology, and Medicine*, *3*, 95- 101

Lee, H. J. & Jeong, S H. (2004). Bacteriostasis of nanosized colloidal silver on polyester nonwovens. *Textile Research Journal*, *74*, 442-447.

Lee, H. J. & Jeong, S. H. (2005). Bacteriostasis and skin innoxiousness of nanosize silver colloids on textile fabrics. *Textile Research Journal*, *75*, 551-556.

Lewis, L N. (1993). Chemical catalysis by colloids and clusters, *Chem Rev.*, *93*, 2693-2730 .

Li, Y., Wu, X. & Ong, B. S. (2005). Facile synthesis of silver nanoparticles useful for fabrication of high-conductivity elements for printed electronics. *J Am Chem Soc.*, *127*, 3266-3267.

Lok, C. N. et al., (2007). *J Biol Inorg Chem.*, *12*, 527-534.

Morones, J. R., Elechiguerra, J. L. & Camacho, A. et al (2005). The bactericidal effect of silver nanoparticles. *Nanotechnology*, *16*, 2346-2353.

Murphy, C. J., Sau, T. K. & Gole, A. M. et al (2005). Anisotropic metal nanoparticles: synthesis, assembly, and optical applications. *J Phys Chem B*, *109*, 13857-13870.

Revina, A. A. et al. (2001). *Himicheskaya Promyshlennost'*, *4*, 28-32.

Soto, K. F. et al., (2005). *Journal of Nanoparticle Research*, *7*, 145-169.

In: Paints: Types, Components and Applications
Editor: Stephanie M. Sarrica

ISBN: 978-1-61761-813-0
© 2011 Nova Science Publishers, Inc.

Chapter 10

CERAMIC SURFACE PAINTINGS AND PIGMENTS FROM THE AGUADA CULTURE (ARGENTINA): XRD AND SEM-EDX ARCHAEOMETRIC STUDIES

S. R. Bertolino,[*] V. Galván Josa[†]and G. Castellano[‡]
Universidad Nacional de Córdoba - CONICET, Argentina

Abstract

Scanning electron microscopy (SEM) combined with energy-dispersive spectrometry (EDX) and X-ray diffraction (XRD) are nondestructive analytical techniques well established in materials characterization and, recently, they have become more popular in archaeometric investigations, since their simultaneous application allows mineralogical, chemical and topographic analysis. There is still a lack of detailed information regarding their suitability for studying samples such as thin paint layers on pottery or micro-granular phases, where standard procedures are not always appropriate.

This chapter focuses on the use of X-ray and SEM techniques for the mineralogical and chemical characterization of pigments, and surface treatments and paintings in pottery from Ambato and Portezuelo styles of the Aguada Culture (Catamarca, Argentina, ca. 600-1000 AC). An image-treatment software was developed to solve the difficulties for paint discrimination, which implements a new methodology to process backscattered electron images. This software brings to evidence small mean atomic number contrasts among paints and the paste in ceramics with a minor detail loss. Quantitative mineral compositions were obtained by Rietveld refinement of XRD patterns. In Tricolor Ambato pottery, reddish paint resembles the paste due to the presence of hematite and Fe-clays; sometimes the white paint contains Pb-rich instead of Ca-rich phases. Black paint has scarce Mn-minerals but often, like in the case of Black Incised type sherds, no particular phase is identified as a color source, suggesting possible organic pigments ("carbon black") or resulting from the firing technique.

The polychrome paints in sherds of Aguada Portezuelo style were made over a white Ca-rich base and they contain Fe-Mn (black), Fe-Mn-Ca (burgundy) and Fe-Ca (reddish). The white ones correspond to gehlenite, a firing product (possibly above 900-1000°C); but calcite and CaO also occur (above 900°C).

---

[*]E-mail address: bertolin@famaf.unc.edu.ar

[†]E-mail address: galvan@famaf.unc.edu.ar

[‡]E-mail address: gcas@famaf.unc.edu.ar

White and reddish pigments found at Piedras Blancas (Aguada Ambato) were also characterized. Due to their scarcity a new methodology was developed complementing XRD Rietveld refinements for mineral quantification with quantitative elemental analysis by SEM-EDX spectra, which proved to be consistent. A special sample holder for few milligrams was designed and constructed. In addition, surface charge accumulation effects were considered by determining the Duane-Hunt limit to assess the effective incident energy, which remarkably improved the sets of concentrations obtained.

The mineralogical and chemical differences found between Ambato and Portezuelo styles suggest that they are two distinctive entities not only regarding their designs, but also in what concerns the materials chosen and the technology used.

The results obtained by all means were consistent. The methodologies and characterization tools proposed here are suitable and can be recommended for routine analyses of different materials.

PACS 32.30.Rj, 368.55.Nq, 61.05.cp, 68.37.Hk, 82.80.Ej.

Keywords: SEM, BSE images, image processing, XRD, paint characterization, elemental analysis, Aguada culture.

1. Introduction

Different analytical techniques, originally developed in the field of materials science, are suitable to study art and archaeological objects. These techniques give art historians and archaeologists the opportunity of gaining information about the composition of such objects and, therefore, helping to answer questions regarding how, where, when or by whom such artifacts were made. Most of them are common tools of clay mineralogists and physicists, who can assist archaeologists providing information to typify ceramics, paints, pigments, raw materials, etc. Additionally, these significant data can help to understand the manufacturing processes and, if correlated with studies regarding technology of production, to provide clues in the interpretations of the social, political, economic and cultural context of the civilizations considered. The interdisciplinary study of ancient pottery by means of these techniques has lately proved to be of great help in understanding and solving some archaeological problems such as provenance of ceramic pieces and raw materials, and technologies used by potters [1, 2, 3, 4, 5, 6].

The analysis of paints and pigments is one of the most interesting aspects, also in the history of art, to understand deterioration processes and to create or enlarge a database. Such investigations are also valuable and in some cases indispensable for conservation, restoration and authentication projects [7].

X-ray analytical techniques, like X-Ray Diffraction (XRD), Scanning Electron Microscopy complemented with Energy Dispersive X-Ray spectrometers (SEM-EDX), X-Ray Fluorescence (XRF) and Total Reflection X-Ray Fluorescence (TXRF), are valuable tools for surface and material characterization which have widely been used for the analysis of archaeological pieces [8, 9, 10, 11, 12, 13]. The popularity of these techniques lies not only in their non-destructive and relatively low-cost character, but also in the advantage that their simultaneous application allows mineralogical (XRD) and chemical (XRF, TXRF and

SEM-EDX) analysis of pieces [14, 15]. The fact that only small amounts of archaeological samples are available has increasingly encouraged the application of these techniques in this area during the last decades [16, 17, 18, 19, 20].

X-ray diffraction applies to minerals and other crystalline phases identification. In order to achieve quantitative results, the Rietveld refinement method leads to crystal parameter determinations on single phase XRD patterns and also to mineral quantification on multi-phase XRD diagrams.

Laser Ablation-Inductively Coupled Plasma-Mass Spectrometry (LA-ICP-MS) [21, 22, 23, 24] and Neutron Activation Analysis (NAA) are techniques also used for chemical characterization, and permit a concentration limit of detection (LOD) of a few parts per billion (ppb=ng/g) in micrograms of sample [25, 26, 27, 28]; however, they present the disadvantage of high-cost analysis and limited availability of nuclear reactor centers in the case of neutron activation [21, 24, 29], disadvantages along with their destructive character which inhibit their frequent use. Also Mössbauer spectroscopy is commonly used for determining iron oxidation states in archaeological samples [30], an important issue when studying the color origin. Alternatively, Raman spectroscopy is a powerful complementary technique that gives information on molecular composition and crystal symmetry of organic and inorganic compounds, one of its main applications on the archaeology and art history being pigment and paint identification [31, 32]. Raman is applied directly on the material of interest (a picture, a ceramic piece, etc.) without sample preparation or any disturbance of the surface; with Raman Microspectroscopy spectra can be collected from a very small volume ($< 1\mu$m diameter), but in cases, several problems may arise depending on the excitation source energy [33].

In an electron microscope, energy dispersive spectrometry (EDX) and X-ray maps (XRM) apply to chemical composition determinations while secondary electron (SE) and backscattered electron (BSE) imaging are used to textural, topographic and chemical contrast analyses of ceramic pastes, surfaces, paints and pigments. Auger electrons are generated in regions close to the sample surface (a few tens of nanometers), and they can provide the chemical composition of that area [34, 35]. Although Auger spectroscopic analysis may be accurate, several experimental difficulties, in addition to the expensive character of the involved instruments, make it an infrequent technique.

Backscattered electron images may provide certain chemical-contrast information from the surface of the studied material, since the fraction of backscattered electrons increases with the mean atomic number Z of the irradiated sample [16]. However, when two regions on a sample have slight chemical contrast (i.e., similar mean atomic numbers), the 8-bit BSE gray level signals of those regions will look alike. In this case, simple human eye recognition in BSE images becomes difficult in view of its limited 30 gray-level discrimination power [36]. While this is a relevant question for surface characterization through BSE imaging, the problem has scarcely been faced in the literature, and only morphological developments have been done in this area [37, 38]. For these reasons, a methodology for image processing has been developed [39], in order to allow the maximum profit of the information contained in these images, with the capability of distinguishing between similar mean atomic number regions. This methodology will be described in section 4.1.3.

EDX spectral characteristic lines are built-up through ionizations that occur in all the interaction volume covered by the primary beam, whose typical dimensions may reach up to 5μm under normal operating conditions (20keV). So, EDX analysis is a very powerful tool for microchemical characterization of undisturbed samples. Nevertheless, it is not always suitable for surface paintings if the paint layer is thinner than the interaction volume. In that case, it may extract information not only from the painted surface but also from the substrate (the paste, in the case of ceramics). EDX spectra are very useful for the identification of major and minor elements present in each phase, though this is almost impossible with trace elements, since signal-to-noise ratio inhibits their detection. Similarly, by increasing the incident electron beam energy, the penetration depth increases, which allows to ionize deeper regions: important changes in the set of elements identified in the registered spectrum permit to infer if multiple paint layers have been applied. The quantification of element contents is possible if the surface analyzed is perfectly flat and polished, situation hard to meet on undisturbed samples of ceramic paintings or powders. In order to perform a quantitative analysis of the archaeological powder pigments studied here, a special sample holder for few milligrams was developed [40]; the characteristic line intensities were corrected for matrix and charge accumulation effects and the results were verified with estimations obtained from the quantitative mineral analysis carried out by Rietveld method procedures (see section 4.3.).

X-ray maps provide information about the distribution of selected element relative contents within an area. Like in the case of EDX spectra, if the paint layer is too thin, these images are not sensitive enough to show reliable abundance distributions of the selected elements. A new procedure was developed in a previous work to avoid loss of information [41]. As explained in section 4.4., the method is based on the acquisition of an RGB (red-green-blue) image from the XRMs, where elements with similar concentrations are grouped in one channel (R, G or B), and then they are all fused with the corresponding SE image. This method has the advantage that the chemical contrasts as well as the relative elemental distributions are shown in one image keeping the resolution of a SE image.

As mentioned above, a number of works have been devoted to the characterization of archaeological materials, although little has been done regarding the strengths and weaknesses of the X-ray analytical techniques in this area [42]. For this reason, the present chapter aims to give an important step forward in this direction discussing the suitability of these techniques and reporting new developments which contribute to the chemical and mineralogical quantitative characterization of the archaeological ceramic paints and pigments. These new methods are also appropriate to any other case where small quantities of sample are available and/or its surface cannot be disturbed.

This chapter focuses mainly on the application of X-ray and SEM techniques to characterize the mineralogical and chemical composition of pigments and the external surface treatments and paintings of pot sherds from Ambato and Portezuelo styles of the Aguada Culture, Ambato and Catamarca valleys respectively (Catamarca, Argentina). On the other hand, the measurement conditions and techniques suitability along with new developments on the mineralogical and chemical quantification and the treatment and processing of SEM-EDX images reported in previous works [39, 40, 41, 42], will also be discussed here, in an attempt to define the best methodology and to improve the quality of data obtained.

An exhaustive study of pigments and paints from the Aguada Culture has not been completed yet, only qualitative analysis being found in the literature [2, 5, 6, 43]. The characterization provided in this chapter will complement previous results in this area [5,

6, 42, 39], and may contribute with the elaboration of a database of archaeological pigment and paint composition used in ancient America.

2.   The Aguada Culture

The Aguada Culture [44] is one of the most spectacular of northwestern Argentina because of its iconographic complexity (also called the "draconian style") and technological quality. According to different studies [45, 46, 47], this culture emerged as a consequence of a social and ideological change that occurred in the fourth to twelfth centuries of the Christian era in the Ambato Valley [48], and progressively spread out regionally. It was a process of deepening and consolidation of social inequalities, which was based on the preeminence of a solar cult, long distance exchange of luxury and symbolic goods and the establishment of hereditary social hierarchies. Local ceramic styles developed within each cultural region (Ambato, central Catamarca and Hualfin valleys, and northern La Rioja). Ambato and Catamarca are the two contiguous valleys that, although not separated by any actual geographical barriers, show two very different and peculiar Aguada styles that are understood as an evidence of two autonomous political units. The typical Aguada Ambato ceramics are black incised but a tricolor one is also common, while the Aguada Portezuelo is painted with a polychromatic palette usually over a white background.

Archaeological sites locate between the Yungas (rainforests, east) and the pre-Puna (desertic to semi-desertic, northwest) areas; this was a strategic zone with access to valuable natural resources, favoring the settlement of human groups [49] since the Formative Period (600 BC to 300 AC). Piedras Blancas, within the Ambato Valley, belongs to the Regional Integration Period (300-1000 AC), where a well settled, more complex and unequal society developed [48, 50]. Portezuelo and South Tiro Federal (radiocarbon age ca 600-900 AC), in the Catamarca valley, corresponds to the Regional Integration Period as well [5].

The occurrence of ceramics of one style along with the other one (Ambato pieces are found at archaeological sites in Catamarca and other Valleys, while Portezuelo ones barely appear at some sites in the Ambato Valley), poses the question of the kind of relationships established between these two political units: Are we in front of a regional system of circulation of particular goods or both ceramic styles were locally produced in both valleys? Were they made in the same way and/or using the same materials? Neither their relationships and the materials used nor their technology has yet been well understood.

3.   Samples Analyzed

Pottery sherd samples were chosen from a collection for being representative or unique in their technological classes. Different samples were considered, all corresponding to the Aguada culture described above. Ambato style samples were collected from Piedras Blancas site, in the Ambato Valley, while the Aguada Portezuelo ones were found in Portezuelo and South Tiro Federal, sites within the Catamarca Valley, about 80 km south of Piedras Blancas.

The two typical Aguada Ambato ceramics analyzed are: the Aguada Ambato Black Incised pottery, which corresponds to samples labeled as B4, B7, B11, B27 and B52, and the Aguada Tricolor ware, involving samples B2, B22, B33, B39, B44, B49, B53, B54 and C1. The first group is a fine ware of very good quality, detailed engraved or incised decoration depicting animal, human and fantastic motives; it is found at many places outside the Ambato Valley, pointing toward its value as exchange good [51, 52]. The second one is a more roughly finished ware, of large vessels used for storage and beer processing, which can be plain or decorated with human and fantastic motives too, and painted in white, red and black [46, 53, 54]. These tricolor vessels are related to an older technological tradition, which acquired a new meaning within the Aguada context. Along with these two types, there are others with different finishing, from which samples B32, B36 and B37 were chosen.

The pigment samples consisted of aggregates of very fine particles found within containers, in different sectors at Piedras Blancas, only a few tens of milligrams being available for analysis. The analyzed samples have been labeled according to the following scheme: W1, W2, W3, W4, W5, W6 and W7 correspond to white pigments, whereas R1, R2, R3 and R4 are reddish pigments. Samples W2 and R1 were associated with a funerary trousseau of a child; children and animal sacrifices are known in Aguada culture as linked to the first stages of a house construction. The others were found related to domestic, working and garbage dump sectors. Sample W7 was spread from a vessel on a floor of a cooking and storage area; because of that it was suspected to be an organic ingredient for meals.

The Aguada Portezuelo style characterizes by a very fine ware, with quite complex and highly variable manufacturing and decorative techniques [43, 55, 56, 57]. One of the most outstanding characteristics is its noticeable polychromy. Decorative motives are painted in burgundy, reddish, black and yellow (a very atypical color in other Aguada ceramics of Northwest Argentina [57]) over a white slip and combining an elaborated background and figure interplay depicting motives either in negative or positive. In some cases, the colors have not been fixed by the firing and appear smooth and without brightness, also displaying pre- and post-firing paintings. Another decorative technique poorly studied for this ceramics is the existence of resistant negative painting [57]. The selected pot sherd samples, from the Central Catamarca Valley belong to "black over white" (Gui1, Gui2, GTF6) or "polychrome over white" (A11, A12, GP1, GTF5) type, their interiors are either black or brown (probably untreated).

4.   Characterization Techniques

Mineralogical qualitative identification was first accomplished by optical microscopy, so as to localize regions of interest and to establish the following steps for materials characterization. After this, mineralogical and chemical quantitative analyses were carried out on ceramics painted surfaces as well as on white and reddish pigments, combining mainly XRD, SEM and EDX. SE images, BSE images, XRMs and X-ray spectra were collected with different purposes, as described below (section 5.).

These techniques have different strengths and weaknesses when applied to the cases studied. Archaeological ceramic pieces and sherds have been buried for over 1000 years, suffered weathering and partial removal of the paints, so the paint layer thickness is not homogeneous and the surfaces are rough and not plane [58]. In addition, the amount of material that can be extracted is too small for conventional analyses (i.e. XRD). Nevertheless, considering all data together, they give a good characterization of both the paint and the paste.

4.1. Scanning Electron Microscopy

Images taken in an electron microscope may provide relevant information about the topography of the surface of interest by means of the detection of secondary electrons, i.e., electrons which originally belong to the sample and are ejected due to the interaction of the impinging electrons with it. On the other hand, incident (primary) electrons may be registered as well, after having elastically interacted with sample atoms; since the probability for this kind of interaction is strongly dependent on the mean atomic number irradiated, chemical contrast maps can be obtained through this backscattered electron signal, which is clearly distinguished from secondary electrons due to an important difference in their energies. Because of the fact that backscattered electrons have larger energies, their penetration allows detecting information rather spread in space, resolution being therefore poorer in this case. For this reason, topographical studies are accomplished by means of SE images, whereas chemical contrast is surveyed through BSE images.

4.1.1. Secondary electron imaging

The interactions of incident electrons with the sample include ionizations, and some of the electrons ejected from the target atoms may eventually emerge through the irradiated surface. The energies of these secondary electrons are much lower than that of primary electrons: mainly around 3 eV, exceptionally adding up to a few tens eV. For this reason, the emerging secondary electrons can only originate in regions quite close to the surface, so that SE image resolution is intimately related to the incident probe dimensions reaching the sample [35]. This good resolution is the reason for which sample topography, morphology and texture are mainly explored with this SE signal. By adequating the electron detector gain, low energy electrons are collected, typically energies lower than 50 eV.

4.1.2. Backscattered electron imaging

This signal may provide certain chemical-contrast information from an intermediate zone close to the surface of the interaction volume in the studied material, since the fraction of backscattered electrons increases with the mean atomic number Z of the irradiated sample [16]. BSE images therefore constitute an attractive alternative to this aim, since backscattered electrons allow the distinction of small variations ΔZ in the mean atomic number: in very low noise BSE images, a discrimination eventually up to $\Delta Z \approx 0.1$ may be achieved [35].

The results obtained so far, as well as those detailed in next sections, encourage the need of further manipulation of BSE images. Since, as explained previously, when two regions of a sample present slight chemical contrast (i.e., similar mean atomic numbers), 8-bit BSE signal gray levels will also be similar. In this case, simple human eye recognition in this kind of images becomes difficult in view of our limited 30 gray-level discrimination power [36]. Although this is a relevant question for surface characterization through BSE imaging, the problem has scarcely been faced in the literature, and only morphological developments have been done in this area [37, 38].

For these reasons, a methodology for image processing has been developed in a previous work [39], in order to allow the maximum profit of the information contained in these images, with the capability of distinguishing between similar mean atomic number regions. The methodology proposed in ref. [39] consists in the application of several filters, intended to subtract noise and fluctuations which are inherent to the measurement process [36, 59]. With this kind of procedure, pixels are not considered as isolated points, but global trends in the image are taken into account instead. After a smoothing process, in order to discriminate close gray levels, contrast is enhanced in the region of interest, which always implies certain detail loss. Finally, the program developed in ref. [39] permits an edge-enhancement procedure, allowing the identification of each pigment region.

4.1.3.   BSE image-processing routines

In order to make mean atomic number contrast more evident, circumscribe regions corresponding to each pigment, improve focus and reduce noise due to roughness, BSE images must be processed. Specific transformations have to be applied, modifying each pixel gray level according to its value and those of its neighbor pixels —not to its location. These functions are usually called filters [36], and its implementation is based on a convolution mechanism. Convolution filters can be represented by a rectangular $M \times N$ matrix, which is moved across the image so that its central element coincides with each image pixel. In each position, every pixel value is multiplied by the corresponding element of the convolution matrix. Finally, the central matrix element is replaced by this product sum. Some kind of test is necessary in order to determine which routines are more suitable for each case.

Different transformations have been implemented in the program developed in [39]. The processing routine followed for the application chosen here basically consists of three steps:

Smoothing.   These methods are pixel-to-pixel operations intended to reduce noise as well as contrast between neighboring pixels [36]. Since details are associated not with single pixels, but with the structure surrounding every location, it is important to bear in mind that any smoothing operation implies a loss of information. This means that no general criterion can be established for the best smoothing procedure. It must be noticed that certain filters may introduce undesirable noise, as is the case of the well-known Savitzky-Golay filter [60], which is inappropriate when dealing with sharp edges. The smoothing algorithms tested in the program developed involve low-pass filters [61].

Contrast enhancement.   The transformation implemented with the aim of exaggerating the contrast between regions which must be discriminated consists in stretching the gray-level scale in the range of interest [62, 63]. The function $T(i, j)$ which produces a new image covering most of the gray-level scale can be defined as:

$$T(i,j) = \begin{cases} b(i,j)\,v_1/u_1 & \text{if} \quad b(i,j) \leq u_1 \\[2mm] [b(i,j) - u_1]\,(v_2 - v_1)/(u_2 - u_1) + v_1 & \text{if} \quad u_1 \leq b(i,j) \leq u_2 \\[2mm] [b(i,j) - u_2]\,(255 - v_2)/(255 - u_2) + v_2 & \text{if} \quad b(i,j) \geq u_2 \end{cases}$$

where u_1 and u_2 are the minimum and maximum gray levels of the region of interest, whereas v_1 and v_2 are the minimum and maximum gray levels desired, and $b(i,j)$ is the smoothed image gray level in pixel (i,j). The choice of u and v values strongly depends on the particular image studied. For the determination of u_1 and u_2, the program permits the selection of rectangular windows of arbitrary size inside the image, in order to obtain the average gray level for each region.

The disadvantage of this step is the saturation of the image and partial loss of information outside the regions of interest. However, those slight differences in chemical contrast, originally hidden to human eye, are brought to evidence, which is the purpose of this particular step.

Edge enhancement. Transitions from a region to another with a different gray level may denote changes in composition. It is necessary to distinguish in the intensity fluctuations, those originated by surface imperfections or grain boundaries from those corresponding to compositional changes. Once the transformations detailed above have been applied, edges are poorly defined, since smoothing produces this blurring. Several algorithms have been developed for edge detection [36], most of them enhancing intensity changes, allowing the appropriate delimitation of each paint zone. The enhancement of edges corresponding to compositional variations was performed here by the application of Laplacian filters [64], which can be modified by the user in order to adequate it according to the particular case considered.

The implementation of these filters has been carried out by developing a program in MATLAB® environment. It presents a user-friendly graphic interface with drop-down menus and the possibility of working with several images simultaneously. It also allows studying details or specific image regions through a "zoom" tool. For phase analysis, the program allows to visualize line intensity profiles and to convert some gray levels to optional colors. At any stage, the program allows the use of several well-known transformations, as well as the implementation of any other filter designed by the user.

4.2. X-Ray Diffraction

Powder X-ray diffraction is a well known and powerful technique for the identification of mineral and other crystalline materials giving also information about the components' crystal structure, lattice parameters, symmetry, degree of crystallinity, stress, etc. However, quantification of crystalline components, particularly in the fine fraction, continues being limited by only the analysis of each species peaks set from the XRD patterns. This difficulty could be highly improved by treating the XRD patterns of multiple phase samples with the Rietveld method, after a careful identification of the components of the sample.

Hugo Rietveld developed a method to refine crystalline structures using neutron diffraction data [65, 66]. This refinement method quickly found applications in powder XRD, covering structural analysis [67], studies on crystal defects, phase transformations [68, 69], determination of amorphous content [70] and quantification of mineral mixtures [71, 72].

4.2.1.   XRD Rietveld refinement

The Rietveld refinement method consists in using a least squares approach to refine a theoretical XRD spectrum until it matches the measured profile [73]. The calculated diffraction pattern is based in a model that includes concentrations, structural parameters (spatial group, atom positions, thermic factors, etc.), microstructural parameters (concentration, crystal size, microdeformations) and instrumental parameters (instrumental contribution to the FWHM and peak shape, width of slits, size of the sample, penetration of the incident beam on the sample, etc.).

For the quantification of mineral phases, the DiffracPlus TOPAS® software [74] was used. To this aim, a Pearson VII profile function [75] was used for approximating the X-ray diffraction peaks; this model, based on a Gaussian function multiplied by a correction factor in the form of a series expansion in Hermite polynomials, was chosen since it takes into account the asymmetry of the peaks. The preferred orientation effects were corrected for the easily orientable phases (e.g. phyllosilicates and feldspars) through the March-Dollase correction model [76, 77]. The background was fitted using a sixth degree polynomial function. Parameters related to thermal fluctuations were not refined, since their influence will be negligible as compared to the uncertainties introduced by the refinement of occupancy factors, in view of the poor statistics associated to the minor phases. The values for these

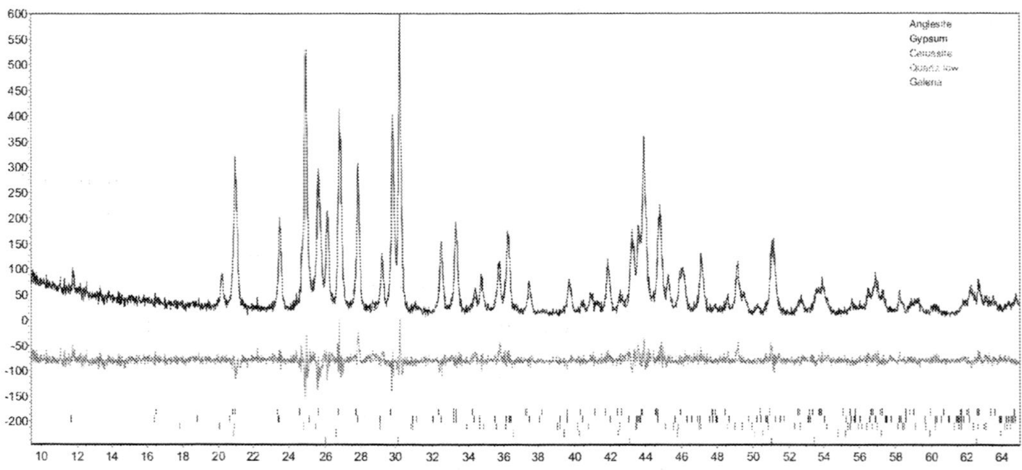

Figure 1. Rietveld refinement for the W2 XRD spectrum. The upper curves are the measured and calculated spectra respectively. The lower plot shows the difference between observed and calculated patterns.

parameters were taken from the Inorganic Crystal Structure Database (ICSD)[1] indexed files corresponding to each mineral. Figure 1 shows an example of the fit quality achieved in the refinement process for the sample W2.

In order to provide a figure of merit for the performance of each fit, the *Goodness-of-fit* is assessed through the weighted comparison of the predicted intensities $\tilde{y}_i$ with the experimental ones y_i:

$$GOF = \sqrt{\frac{1}{N-P} \sum_{i=1}^{N} \frac{(y_i - \tilde{y}_i)^2}{y_i}} \; ,$$

where i denotes each of the N channels in the considered XRD spectrum, and P is the number of free parameters in the fitting process [73]. The statistical quality of the fit is represented through the R_{e} parameter, defined as

$$R_{\mathrm{e}} = \sqrt{\frac{N-P}{\sum y_i}}$$

4.3.   Electron Probe Quantitation

In conventional electron probe quantitation, the characteristic intensity I_j emitted by each element of an unknown sample is recorded and then compared with the corresponding intensity I_j^o emitted from a standard of concentration C_j^o. As a first approximation, the intensity I_j may be taken as proportional to the mass concentration C_j of element j:

$$\frac{I_j}{I_j^o} \approx \frac{C_j}{C_j^o}$$

The comparison with the standard allows to eliminate geometrical and physical factors which are very difficult to determine [16]. The previous relationship has no general validity, and different matrix effects must be taken into account, usually referred to as "ZAF correction" [16, 35]: Z and A factors represent the generation, scattering and absorption effects, whereas the F factor involves secondary fluorescence enhancement; since these effects may differ from sample to standard, the ZAF correction is necessary in order to accurately relate the sample composition with the measured characteristic intensities. The magnitude of the Z, A and F correction factors strongly depends on the experimental conditions, mainly on the incident beam energy, X-ray take-off angle and differences in composition of the standards used as compared to the unknown samples. In some cases, it is possible to reduce their influence and even compensate effects, provided the experimental conditions are adequately chosen.

Matrix effects involve all elements in the sample, with a complex functional dependence on the concentrations. In the program MULTI [78] used here, an iterative method is implemented for assessing the complete set of concentrations C_j. This algorithm allows the quantification of an unknown sample with up to eighteen elements, which can be compared to several standards in any kind of combinations, ranging from one standard containing all the analytes to one standard for each analyte. Moreover, it is possible to perform analysis

---

[1] http://icsd.fiz-karlsruhe.de/icsd

of light elements if stoichiometric relationships are known. In addition, several ZAF correction models may be selected, from which the Gaussian model for the ZA correction [79] modified by Riveros and coworkers [80] along with Reed's fluorescence correction factor F [81] were chosen for the present analysis.

It is very important to take into account the different issues which may alter the X-ray spectrum acquired, and therefore distort the quantitative results obtained. If the sample is an insulator or the conductive coating is inadequate, surface charging effects can create a noticeable modification of the X-ray emission spectra, and the results of the quantifications will be incorrect. A way to evaluate charging effects is to determine the Duane-Hunt limit, which represents the maximum energy of the continuum X-ray spectrum produced by the incident beam impacting on the sample surface [82]. The conventional method to evaluate the Duane-Hunt limit value is to (linearly) fit the region before the cut-off energy. The energy values obtained this way were used as input for the quantification procedure with the program MULTI [78].

4.4. X-Ray Imaging

X-ray images provide information about the relative abundance of selected elements within a sampled area. This feature makes X-ray maps particularly useful for producing a spatial distribution map for specific elements.

XRMs differ from SE images in two fundamental issues: Firstly, spatial resolution for the case of SE images is around the incident electron beam diameter (~ 200Å), whereas for XRMs this resolution is determined by the interaction volume diameter, which approximately reaches 5μm in normal experimental conditions. Secondly, the acquisition time required for registering an XRM with reasonable statistical uncertainties, is quite larger than that corresponding to an SE or BSE image. For example, the probability for a primary electron to produce an X-ray photon within the solid angle subtended by the detector is around 10000 times smaller than the probability of collecting a backscattered or secondary electron. For this reason, if a 1024×768 SE or BSE image is registered in a few seconds, a 256×256 XRM takes more than 3 hours [83]. If a wavelength-dispersive spectrometer is used, this acquisition time is considerably incremented.

A methodology for pseudo compositional maps has been developed in a previous work [41]. This involves enlarged XRMs fused with an SE image, which allows obtaining distributions for elements or groups of elements preserving the spatial resolution provided by electron signals. The steps followed in this procedure are described in next section.

4.4.1. Fusion of XRM and SE images

XRMs are acquired in low resolution (256×256 pixels) in order not to exaggerate the time required for the data acquisition. With the aim of gathering all elemental distributions in a single picture, elements with similar spatial abundance are grouped in one color channel (R for red, G for green or B for blue). The criterion followed for the assignment of colors in the RGB image consists in identifying the elements responsible for some particular feature (in this case, the paint colors), and obtaining a correlation between the elemental distributions and the topography evidenced in the SE image. The correspondence between the

XRMs and the SE image is possible only if the number of pixels is the same in both. Under the assumption that the concentration distribution varies smoothly along the sample, spatial resolution may be altered by incrementing the number of pixels of these maps, and then performing a linear interpolation of the experimental data [36]. This apparent improvement in spatial resolution has several advantages, the main being the possibility of applying subsequent image processing routines with no substantial detail loss, whilst allowing an easier visualization, since compositional maps can be constructed in an RGB image format in which each color corresponds to the occurrence of an element or phase. The inconvenience of the poor statistics of the XRMs is overcome by the implementation of the smoothing and edge enhancement filters described above [83].

For the XRMs and SE image fusion process, the wavelet decomposition method [36] was used, enhancing those coefficients which preserve details in the SE micrograph. As shown in [41], in the final fused images the regions corresponding to each paint color are clearly evidenced, and no appreciable detail loss can be seen.

5. Sample Preparation and Experimental Determinations

For mineralogical determinations, samples prepared in random mounts were analyzed on a Philips X'Pert PRO PW 3040/60 diffractometer, with Cu-Kα X-ray radiation, Si monochromator, at 40kV and 30mA, step size of $0.02°$ 2θ and step scan at $\sim 1°$/minute for identification and $0.3°$/minute for mineral quantification. In most cases, since too small amounts were available, samples were mounted on an Si holder of low background where few milligrams can be analyzed. Quantification of mineral phases was accomplished by Rietveld refinement as implemented in the DiffracPlus TOPAS® commercial software. The LOD of this technique for the samples studied is estimated between 1 and 2%.

In order to attain SEM images and qualitative EDX spectra, natural fragments of Ambato and Portezuelo decorated sherd samples were coated with gold (11 nm thick) or carbon (25-30 nm thick) when Au interferes the signal of elements of interest, and analyzed with a LEO 1450VP scanning microscope, coupled with EDAX energy dispersive spectrometers, at the Laboratorio de Microscopía Electrónica y Microanálisis (LABMEM, UNSL). Measurement conditions were 15 to 20 keV, 0.5 nA, and particular experimental specifications are given below. In the case of qualitative analyses, all spectra were normalized to Si for a better comparison of relative intensities of present elements, because Si is a major component in all samples.

For quantitative analysis by SEM-EDX in the Ambato pigments characterized, the samples should be flat and the electron beam must impact normal to the surface. To achieve this purpose, all samples irradiated in the SEM were ground and compacted into a sample holder specially developed to this aim. The compaction was performed by applying a pressure of 20 ton/cm^2 during 20 minutes, by means of a Teflon plunger situated between the piston and the sample. After compaction, this plunger remains attached to the aluminum cylinder as a sample support, thus ensuring a homogeneous density. By this procedure, a small amount of sample (10 mg) is compacted covering an area of 0.5 cm^2, allowing several analyses in different regions of it. In Fig. 2 the main components of the sample holder designed are shown.

To acquire EDX spectra, the samples were not coated, due to the fact that, on one hand, carbon coating complicates the quantification of these elements, and on the other hand the presence of a gold coating (or other heavy metal layer) modifies the different absorption effects for each line, and implies the appearance of its characteristic peaks, which may overlap with some lines of interest. This is the case of gold M lines overlapping with sulphur K and lead M lines.

6.   Ceramic Surface Paintings

6.1.   SE and BSE Imaging

The Ambato and Portezuelo pot sherd samples were examined by these two imaging techniques. With the working conditions used, beam current of 300-400 pA and $\sim$ 25 to 40 sec of frame scan time, the minimum chemical contrast percent discriminable is lower than 4%, which will provide a good quality image.

Topographic (SE) and chemical contrast (BSE) images (Fig. 3) obtained from the surface of sherds clearly show the compositional differences of white (slip) and black Portezuelo paints. As explained above, SE images, though mostly topographic, in this case slightly reflect the chemical contrast. This is clearly seen on the BSE micrograph, where it is possible to distinguish not only the design contours but also the differences in mean atomic number of paints; black paint is therefore seen in lighter color because it has higher mean Z.

The SE images of Ambato Tricolor sherds reveal textural and grain-size differences resulting from the pigments used and the treatment given to their surfaces. Figure 4 shows the smooth and even finish of the reddish paint (1), while white paint (2) exhibits coarser grains and a rough surface; black paint (3) has intermediate features [6]. There seems to be a parallel trend of reddish material distribution suggesting that the surface could have been polished with a smoothing tool. In this case, it is evident that the black and white pigments

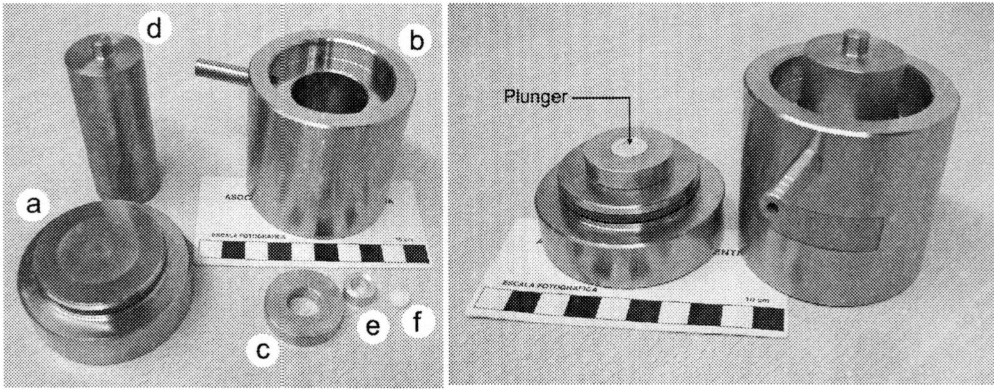

Figure 2. Sample holder designed for pigment analysis through EDX spectrum acquisition. a) supporting base, b) skirt, c) container ring, d) piston, e) sample supporting ring, f) plunger.

were painted over the reddish base, probably before firing this piece.

In general terms, BSE images of Aguada Ambato sherds show little or no differences in contrast among paints. Only in samples B7 and B53, BSE images show regions with slightly noticeable chemical contrast differences, which could help to identify different phases, as shown in Fig. 5. This motivated the search of new ways to enhancing the contrasts in these images as it is explained in the following section.

6.2. BSE Image-processing Applied to Ambato Ceramic Paints

The software developed to improve the chemical contrast was tested on Ambato pot sherd samples B7 and B53. Ultrasonic cleaning of these fragments was performed with the aim of removing contaminating powder remains. Samples were coated with carbon in order to ensure conductivity for irradiation in the SEM; this low atomic number coating was chosen because backscattered electrons are poorly attenuated [35].

In order to acquire adequate BSE images, a scanning time large enough was selected, with the aim of minimizing statistical fluctuations, using a 40 sec frame scan time. Examples of such images are displayed in Fig. 5. For both B53 and B7 samples, low contrast and little detail can be appreciated. As verified in section 6.3., the high quartz and feldspar contents both in paints and sherd of sample B53 inhibit to obtain a good chemical contrast in this image. Appropriate records were corroborated through linear profiles in BSE images collected at different acquisition times. Figure 6 shows an example of such profiles, in which white lines represent the approximate mean fluctuation values for reddish and black paints respectively; their different heights reflect the differences in mean atomic numbers. These fluctuations arise because the paint layer thickness is not homogeneous but rough and not plane, probably as a consequence of the burying of pieces for over 1000 years, being subjected to weathering and partial paint removal.

The software includes all the image transformations described above. These may be configured by means of different options, so that the best alternatives may be combined,

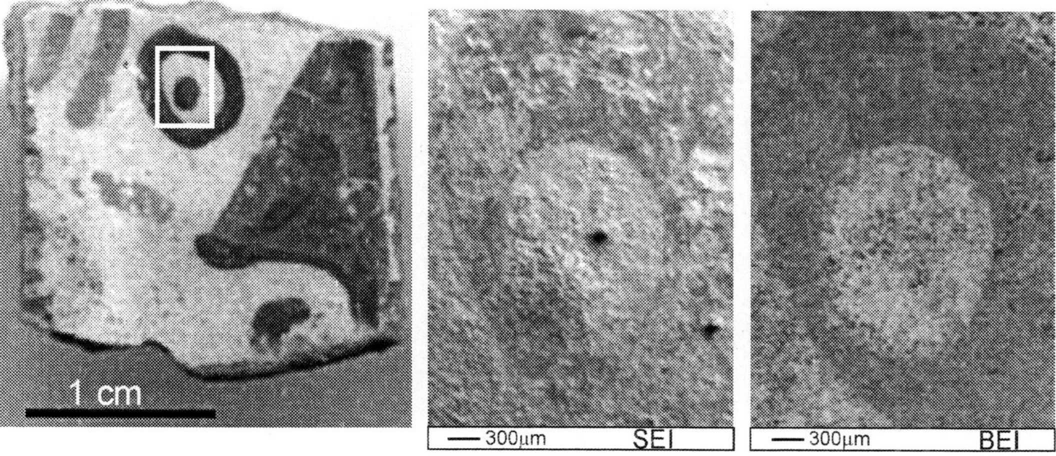

Figure 3. Comparison of SE and BSE images of sherd A11, Aguada Portezuelo; Au coating.

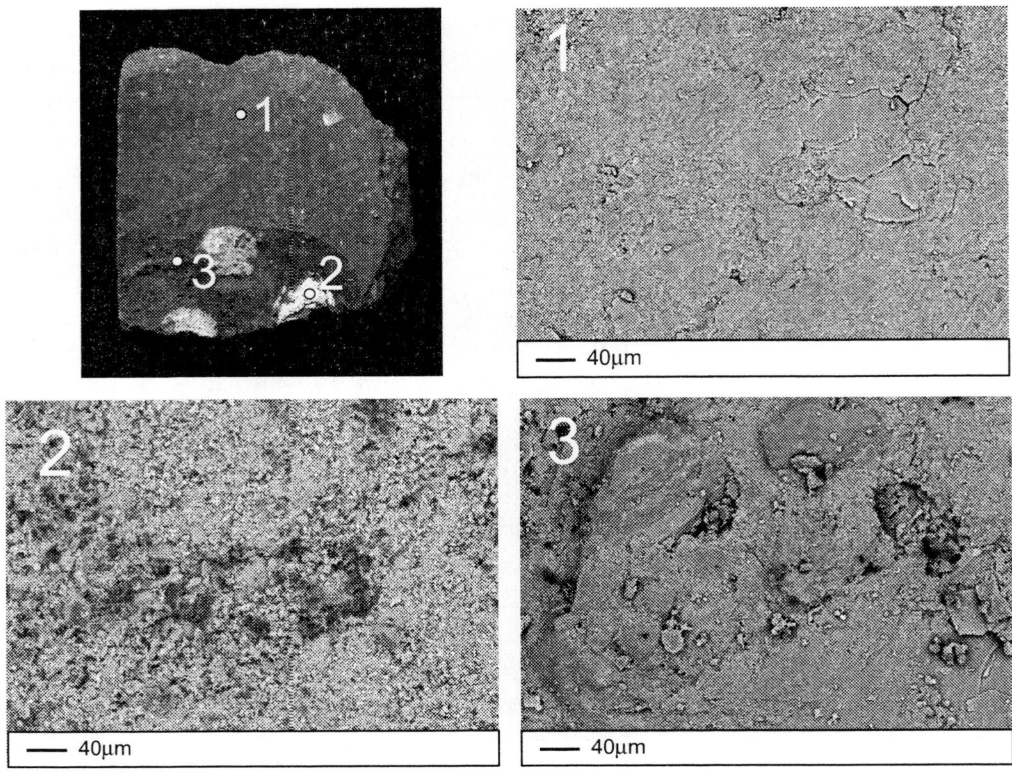

Figure 4. Ambato BSE micrographs, Au coating. Grain size and textures are different in 1-reddish (smooth), 2-white (grainy), and 3-black areas.

according to the specific situation under study.

For an adequate image interpretation, the first step which must be performed is noise minimization. Depending on its nature, several methods can be chosen for its reduction [36, 59]. In the case of the archaeological paints considered here, where noise originates in surface roughness, the optimal smoothing filter applied corresponds to a convolution matrix whose elements are all equal. Figure 7 shows tests performed with different matrix sizes, demonstrating that for this case the most appropriate is a 3×3 matrix size, since a stronger smoothing procedure leads to severe detail and grain boundary losses.

After smoothing, it is necessary to enhance contrast between those gray levels which are to be discriminated. The function $T(i, j)$ defined to this aim in section 4.1.3. is determined by the choice of the minimum and maximum gray levels of the region of interest, u_1 and u_2, and the desired minimum and maximum gray levels, v_1 and v_2. An example of this is shown in Fig. 8, where contrast is expected to be enhanced for the sample B7. In this case, a rectangular window containing pigments and naked sherd was selected, in order to obtain average gray levels inside it. The corresponding values resulted $u_1 = 175$ and $u_2 = 195$.

The last step in the implemented method consists in the enhancement of edges by means of the application of Laplacian filters, which have shown to be suitable for the images processed. Figures 9 and 10 show the unprocessed images and those after the full image processing completing the three steps for sample B7 and B53, respectively.

Figure 5. Photographs and BSE images of sherds B7 and B53.

After BSE image processing has been completed for sample B7, it can be seen that regions corresponding to each pigment have satisfactorily been separated. Contrast between the paste and white and black paints has been emphasized, and the final transformation has led to an adequate edge enhancement. Some details not visible in the original image can now be observed. Of course, the application of these transformations worsens the information outside the region of interest, but one can always return to the original image to start over with the process for each selected area.

The BSE images contain no information to identify the elements present in the sam-

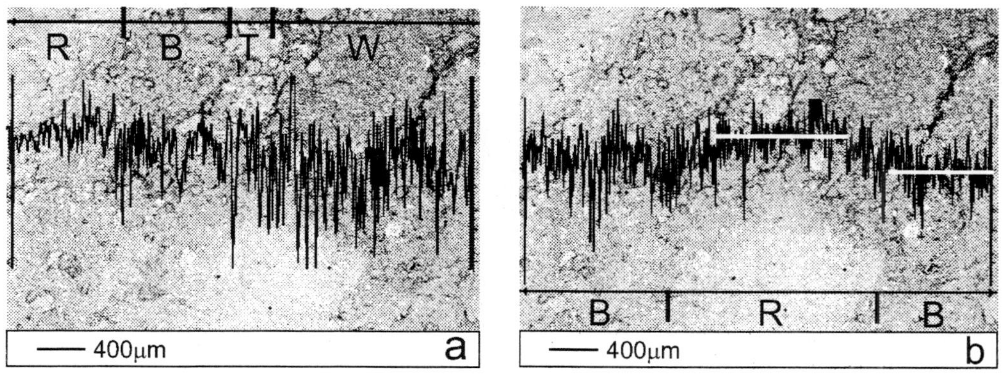

Figure 6. B53 scan line profiles.

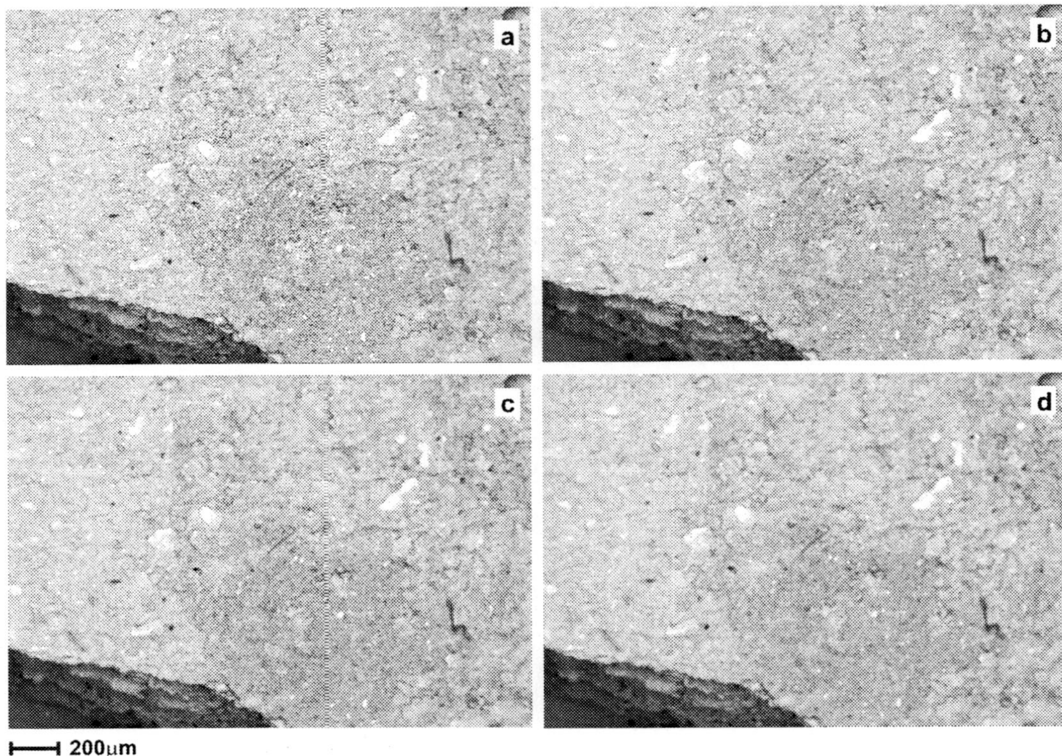

Figure 7. Application of smoothing filters of different sizes in sample B7: a) original image, b) 3×3, c) 4×4 and d) 5×5.

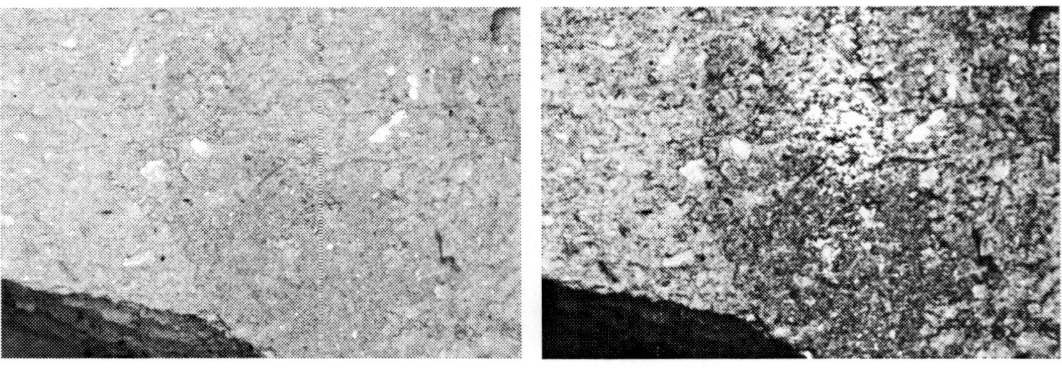

Figure 8. B7 smoothed image (left) and image obtained after contrast enhancement $((u_1, u_2) \rightarrow (v_1, v_2))$.

ple or their concentration levels. In order to discard the possibility that the low chemical contrast observed does not originate from external contaminants, mineral phases were quantified by an XRD Rietveld refinement procedure, as described below.

6.3.   X-Ray Diffraction

Aguada Portezuelo.   XRD was applied to most of the Portezuelo white paints, but it was possible only in a few black ones, due to the scarce amount of samples available. In all cases, the extracted materials also contained some grains from the paste, which is reflected in the bulk mineralogy mainly by the presence of quartz, feldspars, hematite, micas (as traces) and sometimes spinels (Fig. 11). Nevertheless, it was possible to identify the main components that cause the color.

Calcium minerals are the main components in all white paint samples from Portezuelo. Gehlenite is always present, along with subordinate analcime (Gui1) and/or calcite and sometimes traces of CaO. Sample GTF6 only contains analcime.

Gehlenite is an aluminosilicate that forms from mixtures of calcite with clay minerals (paste) at temperatures between 850 and 900 °C [84] or 1050°C [85]. Beyond that temperature it reacts and forms anorthite [56]. Buxeda i Garrigós and Cau Ontiveros described the reactions and new Ca-rich phases (CaO, gehlenite) produced in calcareous ceramics [84] and paid attention to the fact that they strongly depend on the firing atmosphere, the maximum firing temperature, calcite grain size, etc. During this reaction, if the dissociation of calcite was not completed (usually below 750 °C), the reaction sequence of gehlenite, calcite and CaO may coexist in the range of 700-900 °C. After firing, free metastable CaO may transform to portlandite (with atmospheric water) and this mineral to secondary calcite, with the addition of CO_2 from the atmosphere.

The mineralogy found, dominated by metastable Ca phases, suggests that the original white pigment used by potters was calcite (except GTF6). On this set of samples, it is also clear that at least the white pigment was painted before firing. The occurrence of solely gehlenite (GP1, GTF5) points out that the firing temperatures may have reached or overcome 900-1000°C; those that contain CaO and/or calcite seem to have been fired at temperatures possibly lower than 900°C. These temperatures are in agreement with those suggested by the paste characteristics. It is not possible to be certain on the origin of calcite detected by XRD. A secondary origin is more likely since CaO occurs in trace and one would expect higher CaO and less calcite contents if the reaction would have been stopped before complete dissociation of calcite. It should be considered that in samples with calcite,

Figure 9. Unprocessed (left) and optimized (right) BSE image for sample B7.

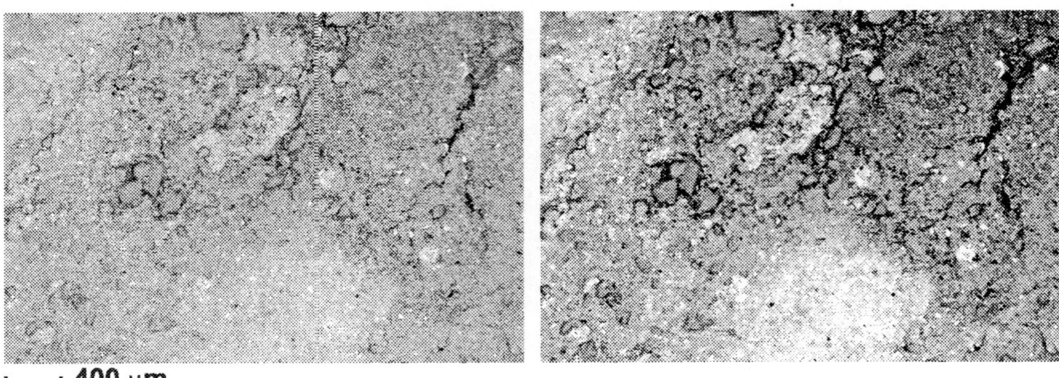

Figure 10. Unprocessed (left) and optimized (right) BSE image for sample B53.

micas are present in higher proportions than in the other ones; micas tend to disappear with increasing firing temperatures (at about 900°C its structure will be lost [86]), supporting the idea that these samples were fired at lower temperatures.

Regarding black paints, the XRD analysis allows to conclude they contain hematite, titanomagnetite (GTF5) and traces of hollandite and lithiophorite (Gui1).

There is no correlation in the mineralogy or chemical composition among the ceramic types analyzed: "black over white" or "polychrome over white", with either black or brown interior.

Aguada Ambato.   In white paintings from Ambato Tricolor, no Ca minerals were found by XRD analysis except in sample B49 and B2, as suggested from the results in BSE images and EDX spectra. As shown in Fig. 12, the main components are Pb minerals such as hatchite (B2, B49, B53, C1) and/or anglesite and traces of plumalsite (B2). The mineralogy of reddish paints is similar to that of the paste but with differences on quartz and feldspar content; this may suggest that the pieces were only polished (without addition of a slit) so that the coarse grains were removed; Fe-rich minerals present are hematite and traces of goethite. Black paints of Tricolor type are composed of Mn and Fe minerals such as hollandite, magnetite and hematite.

The black surfaces of Incised type (B4, B11) hardly show any mineral used as black pigment; it is likely that the color of these surfaces was given by organic ones (such as "carbon black" derived from the firing of wood or bones) or by the firing technique [42]. It is possible that the black surface treatment relates in composition to that of Portezuelo "graphitized" internal surfaces studied by De la Fuente [32], where carbon was identified by Raman spectrometry and interpreted as of biogenic origin.

In order to validate the results obtained for chemical contrast through BSE imaging (6.2.) and to discard the possibility that the low chemical contrast observed does not originate from external contaminants, mineral phases were quantified by a Rietveld refinement procedure. To this aim, a small amount of sample B53 was scraped from the surface; unfortunately, this was not possible to achieve in sample B7 due to the lack of sufficient material. By means of a quantification procedure, mean atomic numbers were determined for white

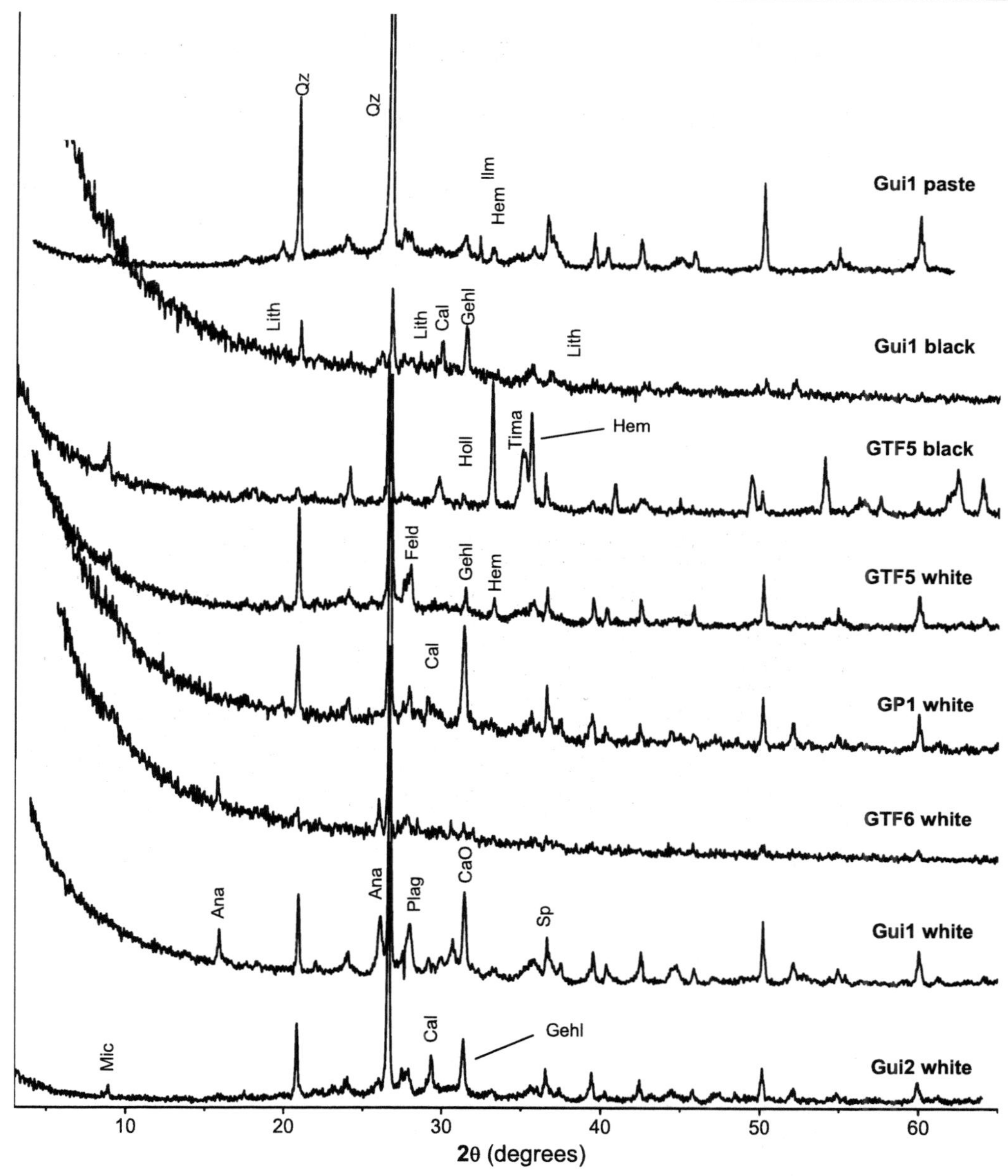

Figure 11. XRD patterns of paints from Aguada Portezuelo. Qz, quartz; Mi, micas; Ge, gehlenite; Cal, calcite; Feld, feldspars; Plag, plagioclase; Ana, analcime; Ilm, ilmenite; Hem, hematite; Sp, spinel; Tima, titanomagnetite; Holl, hollandite; Lith, lithiophorite

and black pigments, and for the ceramic body.

Table 1 shows XRD quantification for the white and black pigments and the paste for this sample. Structural formulas and ICSD numbers are also displayed. It can be seen that mean atomic number differences result lesser than 0.1. However, it must be mentioned that diffraction patterns may reveal certain information from remains of the ceramic body in the scraped paints, which might slightly mask greater differences in mean atomic numbers.

Although this may systematically influence the results of this analysis, it must be kept in mind that the uncertainties caused by the Rietved refinement are negligible, and these results are quite reliable. Regarding the mineralogical characteristics of each region, white pigment of sample B53 owes its color to the hatchite (a Pb-bearing mineral) abundance in the translucent quartz and feldspar matrices, the reddish paste color corresponds to the presence of hematite and Fe-rich clays after firing, whereas the black pigment is characterized by its hollandite content. These are cases in which pigment characteristic colors are

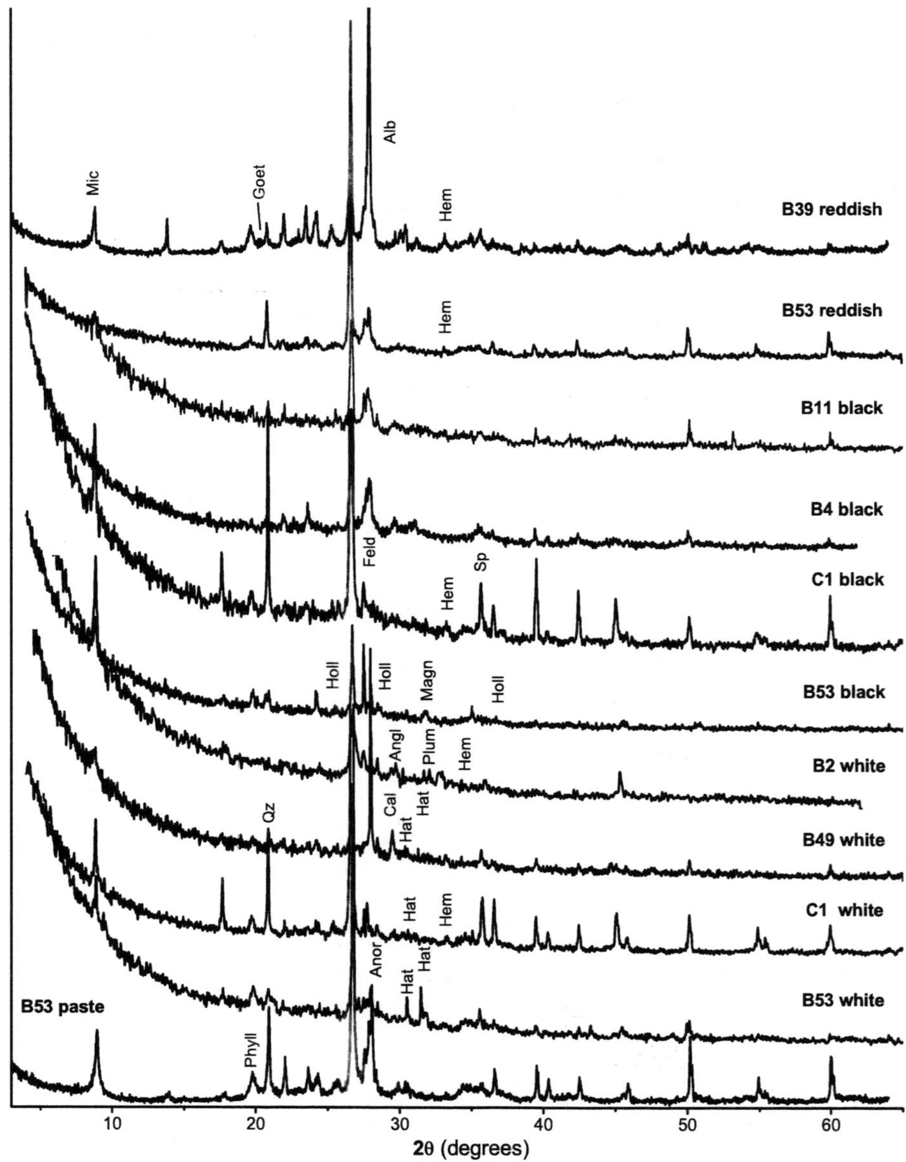

Figure 12. XRD patterns of paints from Aguada Ambato. Qz, quartz; Mi, micas; Feld, feldespars; Anor, anortite; Alb, albite; Hat, hatchite; Angl, anglesite; Plum, plumalsite; Hem, hematite; Magn, magnetite; Holl, hollandite; Phyll, phyllosilicates; Sp, spinel.

Table 1. XRD mineralogical characterization for B53 sherd and pigments.

Mineral	ICSD number	Structure	Mean Z	Sherd	Black	White
Quartz low	174	SiO_2	10.0	27.0	50.3	54.8
Hematite	56372	Fe_2O_3	15.2	≤ 0.5		
Muscovite	4368	$KAl_2(AlSi_3O_{10})(OH)_2$	8.7	3.5	1.8	6.7
Hollandite	60787	$Ba_{1.12}Al_{2.24}Ti_{5.76}O_{16}$	12.6		4.0	
Anorthite	86331	$Ca(Al_2Si_2O_8)$	10.6	69.0	44.0	42.0
Hatchite	27304	$PbTlAg(As_2S_5)$	45.0			≤ 0.5
Resulting mean Z in samples				10.40	10.37	10.34

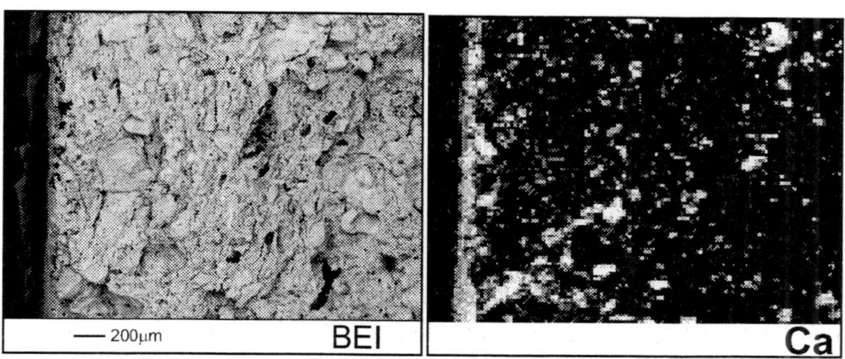

Figure 13. a) BSE image and Ca XRM of a cross-section of sherd Gui1, Aguada Portezuelo; Au coating. The paint layer is distinguished from the paste and becomes evident in the Ca mapping (light colors, right).

due to minority phases: for example, it is well known that even small proportions of lead mixed in a translucent medium result in a good white pigment [42], as verified in the EDX spectrum identification of a weak Pb-M signal, as well as XRD characterization. This small amount of lead hardly influences the BSE signal, though this slight increase can be brought to evidence through an adequate image-processing method, as shown above.

6.4. EDX Analysis

X-ray spectra acquire information on global composition from an interaction volume 3- to 5-μm deep [35]. The surfaces are rough and may have any tilt angle in the scanned area so that the elemental composition obtained is usually only qualitative. In some of the painted surfaces studied, the spectra may contain information from the paste, particularly if dealing with hard X-ray emission lines that may come from deeper zones of the specimen; on the contrary, soft X-ray emission lines come from shallower zones, which allows to better reflect the composition of the thin layer of paint [87].

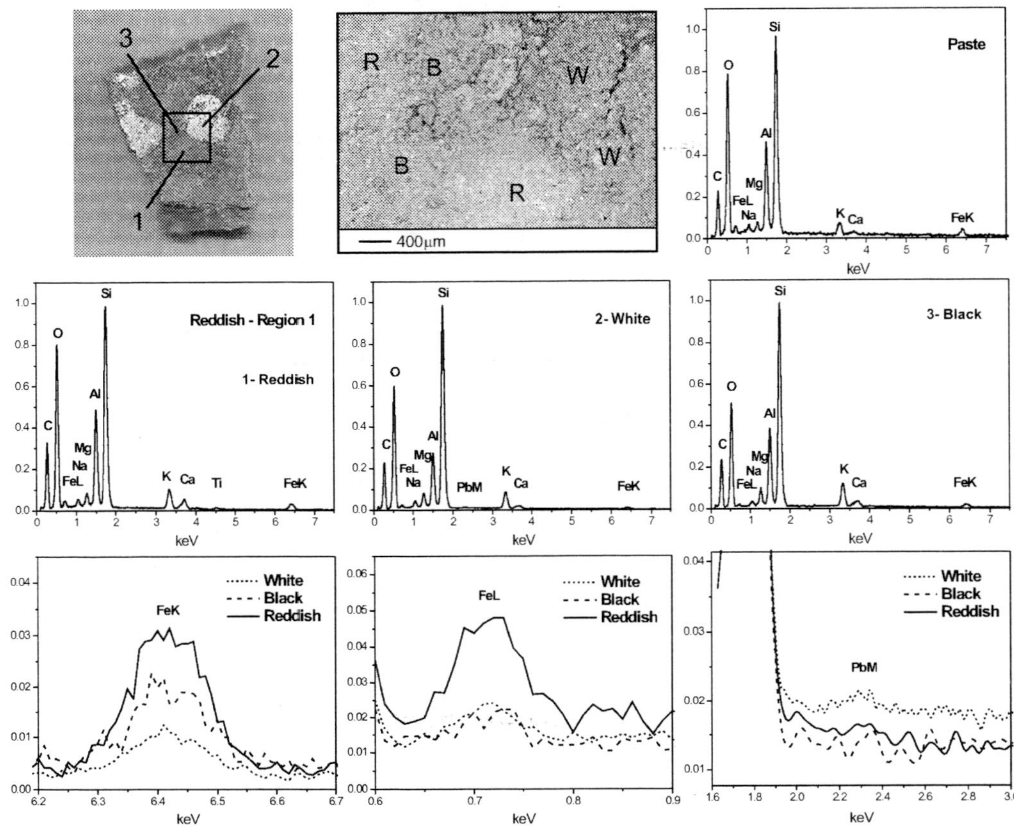

Figure 14. BSE Aguada Tricolor Ambato micrograph; C coating. Picture taken at 600 pA and 350 sec acquisition time. W, white; B, black; R, reddish; T, transition zone.

For the Portezuelo samples, X-ray spectra were collected from a 100 μm^2 area, in order to give qualitative information about the surface paints analyzed. In all samples, the white paint has a high Ca content; the black one, Fe and Mn (plus minor Ca); burgundy exhibits Fe, Mn and Ca, while reddish features Fe and Ca. Paints always contain the same major elements such as Si, Al, minor Mg, K, Na and traces of Ti, in proportions similar to that in the paste. This may be caused by the occurrence of gehlenite (an aluminosilicate Ca-mineral) and the irregular thickness and distribution of the paint layer as observed in Fig. 13. Ca is present in all paints; in the EDX spectra, peaks of typical elements for nonwhite paints (Mn, Fe) are generally weak. Colored paints are very thin layers and irregularly sparse on the surface; this might be reflected on the low intensity of EDX spectra which are also overlapped by the components (Ca) of the white paint used as a substratum.

In the case of Ambato painted sherds, the same experimental conditions were used. As can be seen in Fig. 14, EDX spectra are quite similar but permit to identify the colors. Unlike Portezuelo, in almost all sherds white paints are not Ca-rich; this element is actually slightly more abundant in the reddish ones. Weak Pb-M lines were detected in white surfaces; these lines are absorbed by lighter elements such as Si and Al that are the major components, so higher proportions of these elements could be suspected considering the identification of Pb minerals by XRD on this paint —as mentioned in section 6.3., small proportions of Pb mixed in a translucent medium result in a good white pigment. The composition of black paints is not clearly distinguishable as it was in Portezuelo sherds. Only Fe peaks change more visibly (Fig. 14). As expected, in reddish ones, Fe is more abundant than in the others but it is always present in considerable amounts. The Fe content in white paints may suggest contamination from the bottom paste. Differences are more pronounced when analyzing the Fe-L lines (soft X-ray); only superficial Fe-L photons will come out from the sample so they would reflect better compositional variations (Fig. 14). Like in Portezuelo analyses, Si, Al, minor Mg, K, Na and traces of Ti are always found, which in fact correspond to signal originated in the paste.

6.5. X-ray Maps

For the Portezuelo-style sample A12, Fig. 15 shows the SE image as well as the individual XRMs for the major elements (Mn, Fe, Si, Al, Ca); light elements such as O, C and Na exhibited a uniform spatial distribution. Four colors may be easily distinguished in the sherd surface photograph shown in Fig. 15: black, white, burgundy and red.

As shown in Fig. 16, point X-ray spectra were taken for each paint color region in

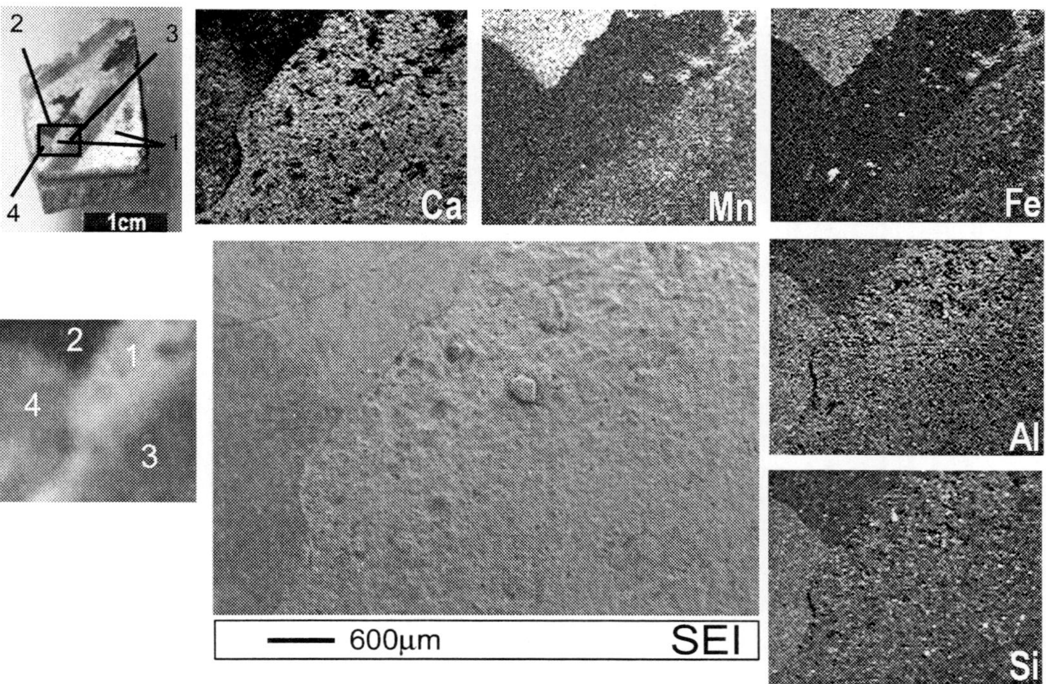

Figure 15. Micrographs for sample A12, decorated with white (1), black (2), burgundy (3) and reddish (4) paint; the pointed rectangle shows the irradiated zone. Ca, Mn, Fe, Al and Si XRMs from the irradiated zone; e) SE image from the studied region.

```
55 {
97 1274 m
97 1447 97 1274 97 1447 c
1046 1447 97 1447 1046 1447 c
1046 1307 1046 1447 1046 1307 c
952.67 1207.67 860.17 1075.67 768.5 911 c
676.83 746.33 606 577 556 403 c
520 280.33 497 146 487 0 c
302 0 487 0 302 0 c
304 115.33 326.67 254.67 370 418 c
413.33 581.33 475.5 738.83 556.5 890.5 c
637.5 1042.17 723.67 1170 815 1274 c
97 1274 815 1274 97 1274 c
closepath } bind def
/s56 {
362 795 m
287.33 822.33 232 861.33 196 912 c
160 962.67 142 1023.33 142 1094 c
142 1200.67 180.33 1290.33 257 1363 c
333.67 1435.67 435.67 1472 563 1472 c
691 1472 794 1434.83 872 1360.5 c
950 1286.17 989 1195.67 989 1089 c
989 1021 971.17 961.83 935.5 911.5 c
899.83 861.17 845.67 822.33 773 795 c
863 765.67 931.5 718.33 978.5 653 c
1025.5 587.67 1049 509.67 1049 419 c
1049 293.67 1004.67 188.33 916 103 c
827.33 17.67 710.67 -25 566 -25 c
421.33 -25 304.67 17.83 216 103.5 c
127.33 189.17 83 296 83 424 c
83 519.33 107.17 599.17 155.5 663.5 c
203.83 727.83 272.67 771.67 362 795 c
closepath 326 1100 m
326 1030.67 348.33 974 393 930 c
437.67 886 495.67 864 567 864 c
636.33 864 693.17 885.83 737.5 929.5 c
781.83 973.17 804 1026.67 804 1090 c
804 1156 781.17 1211.5 735.5 1256.5 c
689.83 1301.5 633 1324 565 1324 c
496.33 1324 439.33 1302 394 1258 c
348.67 1214 326 1161.33 326 1100 c
closepath 268 423 m
268 371.67 280.17 322 304.5 274 c
328.83 226 365 188.83 413 162.5 c
```

Figure 16. Point X-ray spectra for each color in sample A12. The presence of gold is due to the metallic covering.

order to identify the chemical elements for them: as discussed above, the black region is associated with important Mn and Fe peaks; an intense Ca peak and scarce Mn identifies the white paint; the intensity ratio between Fe and Ca is different from white to reddish; and the red regions differ from the burgundy because of the Mn presence in the latter. Bearing all these facts in mind and after comparing the individual XRMs, the image-fusion strategy followed in [41] was to group elements with similar spatial distributions (for example, Mn and Fe in Fig. 15), assigning to them one RGB channel of a new image. In the case of sample A12, this means that the new RGB resulting X-ray image was constructed assigning the red channel to the sum (pixel-by-pixel) of Al and Si XRMs, the green channel to Ca, and the blue one to the sum of Mn and Fe XRMs. The regions corresponding to each paint color in the resulting fused image [41] are clearly evidenced, with no appreciable detail loss.

7. Pigments

7.1. Scanning Electron Images

In the case of white and reddish Aguada pigments, previous to the acquisition of EDX and XRD spectra, secondary electron (SE) images of undisturbed samples were registered in

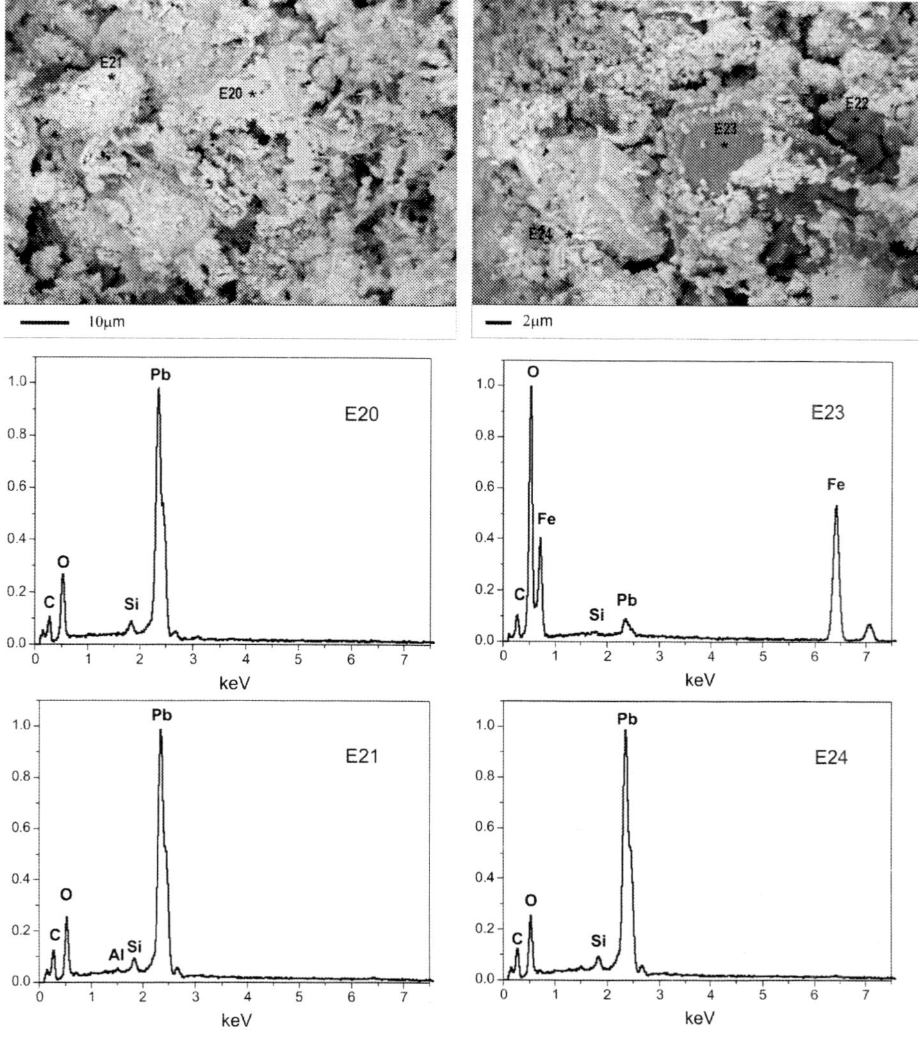

Figure 17. BSE images and X-ray spectra corresponding to the indicated points corresponding to the W2 sample.

order to analyze the particle size distribution and microstructure, and to identify or corroborate the occurrence of minor phases whose species are not easily identifiable by XRD (below the LOD).

BSE images were taken to distinguish regions with noticeable chemical-contrast differences, which might help to identify different phases. With the aid of EDX spectra taken from points with high magnification, a qualitative compositional estimate of small particles may be given, complementing the information provided by the BSE images acquired. Although these spectra are not representative of the overall sample, the identification of particles (or phases) present in very low proportion may be useful for the study of the provenance of raw materials.

The SEM images can also be helpful to infer about previous treatments. For example in Fig. 18, BSE images corresponding to the samples W3 evidence some fibrous structures (bottom left), but though having Ca this structures are not calcite, since in the EDX point spectra the carbon intensities are too low. These phases could not be identified by XRD since they are below the LOD associated to this technique, and because the EDX spectra show small peaks of other elements (Si, Al, S, Mg, P and K) it is difficult to interpret their nature. Gypsum as well as other asbestos minerals (tremolite and other Ca-anphyboles) may meet the shape and chemical composition; it is well known that when calcite is fired, CaO and CO_2, which may have fibrous shape, are produced; if so, it could be an indication that these pigments were previously cooked [42].

Regarding sample W7, the apparent defocalization may be due to the presence of organic materials such as fat as supporting medium for the pigment. The organic nature of the binder is supported by the high carbon concentration shown in the EDX spectra, bearing in mind that none of these pigment samples were carbon coated. The use of animal fat as a binder is common in pre-Inca cultures. Thus, in this case sample W7 corresponds to a prepared paint.

In the BSE images corresponding to the W1, W3, W7 white pigments and the R1 red-

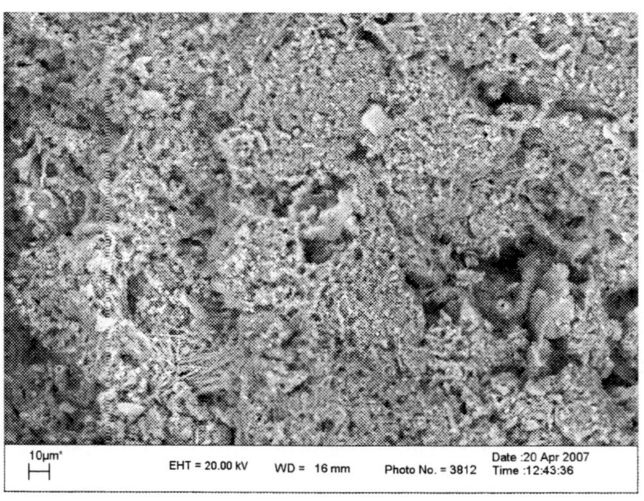

Figure 18. BSE micrograph of sample W3.

dish sample, some brilliant particles can be observed. Since the backscattered electron signal increases with the mean atomic number Z of the irradiated sample [16], this chemical-contrast is associated with the presence of small particles (about 2 μm diameter) containing Pb (see Fig. 19 and 20). In the R1 sample, particles of Ag (spectrum E1 of Fig. 20) can also be seen. These elements can not be detected through the EDX global spectrum or the XRD patterns, due to the fact that their concentration is smaller than the LOD ($< 0.1\%$). R2 contains a very fine grained mixture ($< 10\mu$m) of Fe-rich clay and iron oxide minerals.

7.2. X-Ray Diffraction

All pigment samples were also analyzed through XRD, mineral phases present in them being quantified by means of the Rietveld refinement procedure implemented in TOPAS$^\circledR$ software, as described above.

In the Rietveld refinement procedure, the effects induced by the preferred orientation inherent to the sample preparation method were considered. The major difficulties in the refinement process occur for the reddish pigments, because of the presence of iron phases (such as ferruginous clays) for which neither the concentration of the Fe^{+2} with respect to Fe^{+3}, nor their localization in the lattice (occupation factor) are adequately known. In addition, in some cases the major phase peaks are overlapped with the principal lines of minor phases, difficulting the deconvolution process and raising the LOD up to about 3%.

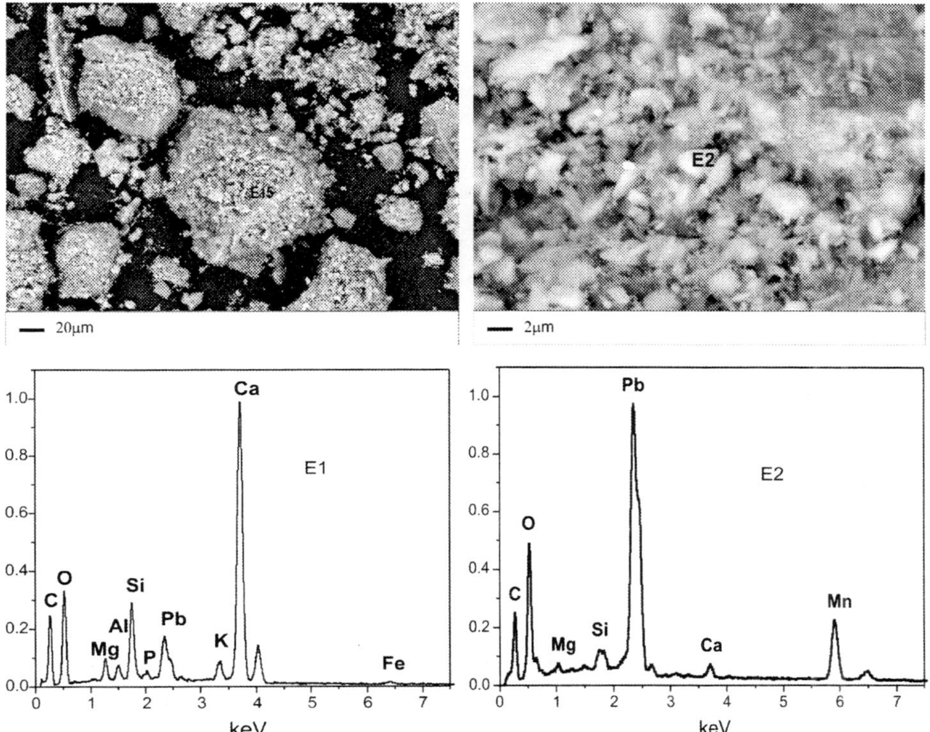

Figure 19. BSE images and X-ray spectra corresponding to the indicated points corresponding to the W7 sample.

Table 2. Mineralogical quantification obtained with XRD.

PHASE CONCENTRATIONS (wt %)

Mineral	ICSD code	Structure	WHITE							REDDISH			
			W1	W2	W3	W4	W5	W6	W7	R1	R2	R3	R4
Calcite	80869	$CaCO_3$	96.0		96.7	57.8	100		71.4		5.9		
Quartz	174	SiO_2	4.0	8.8	1.7	3.3			12.0	19.0	41.7	16.0	32.0
Albite	34917	$NaAlSi_3O_8$			1.6					33.3	17.6		54.1
Hematite	56372	Fe_2O_3							3.8	2.7	11.7	56.3	2.3
Muscovite	4368	$KAl_2(AlSi_3O_{10})(OH)_2$				7				45.0		8.2	11.6
Gypsum	409581	H_4CaO_6S		3.3				100			19.3		
Feldspar	88896	$K (Al Si_3O_8)$							12.4				
Anorthite	86331	$Ca (Al2 Si2 O8)$				31.9							
Litharge	15466	PbO							0.4				
Anlgesite	100625	$PbSO_4$		49.7									
Cerussite	36554	$PbCO_3$		27.7									
Galena	38293	PbS		10.5									
Tourmaline	156163	$NaFe_3Al_6Si_6O_18(BO_3)_3O_3F$										19.6	
Kaolinite	63192	$Al_2(Si_2O_5(OH)_4)$									3.9		
R_{wp} (%)			12.42	16.8	14.55	15.86	-	-	16.31	18.59	22.98	21.72	18.67
R_e (%)			9.54	13.71	11.65	10.32	-	-	9.52	12.34	13.47	9.78	7.50
GOF			1.30	1.23	1.25	1.53	-	-	1.71	1.51	1.71	2.22	2.49

The R factors [73] (numerical criteria of fit) obtained from the refinement results are listed in Table 2. The apparently high values for R_e (and hence for R_{wp}) evidence the difficulties mentioned above for the characterization of reddish pigments. Nevertheless, although the values obtained for R_{wp} seem to be too large, it must be pointed out that since a rather small amount of material is to be analyzed, statistics cannot be noticeably improved due to the experimental limitations this implies.

As shown in Table 2 and Fig. 21, white pigments W1, W3 and W5 are characterized by the content of more than 90% of calcite. This suggests that this mineral could be the color source. The W6 pigment is pure gypsum, whereas the pigments W4 an W7 contain, besides

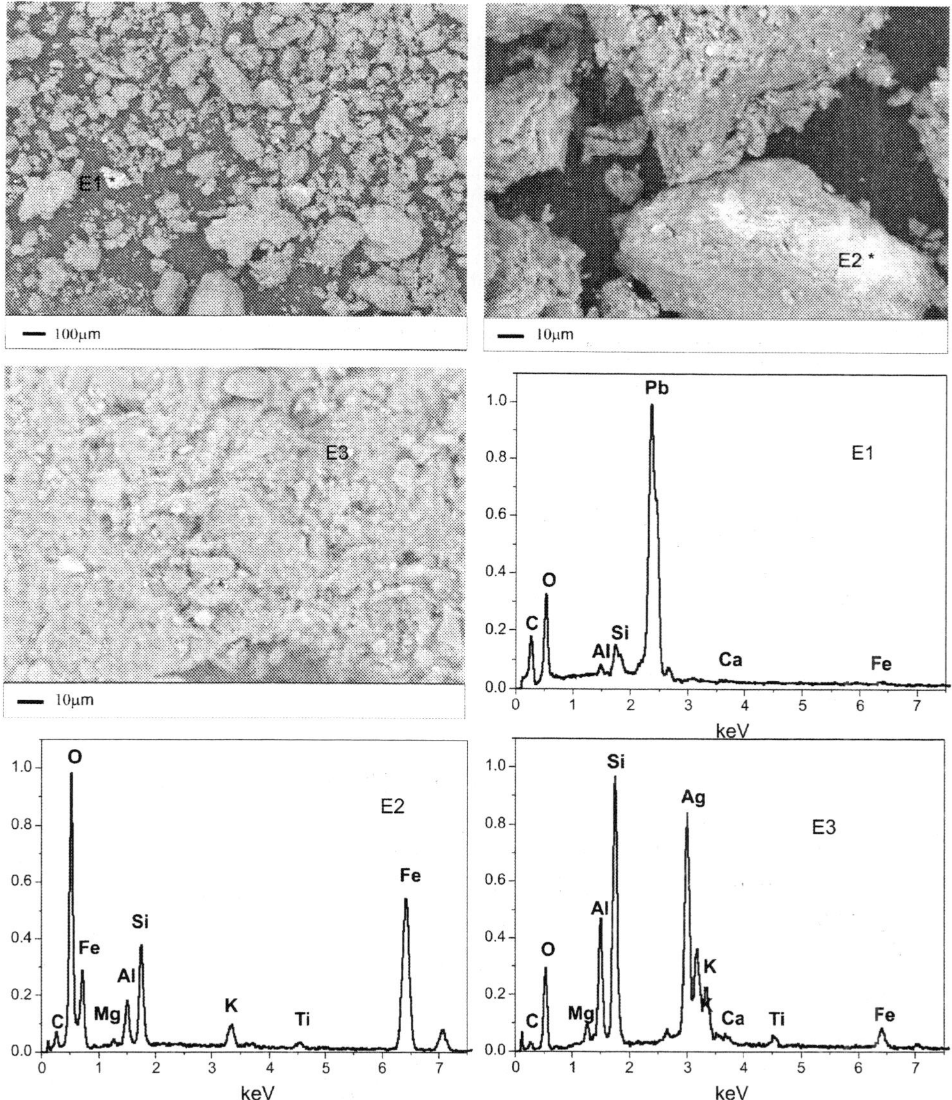

Figure 20. BSE images and X-ray spectra corresponding to the indicated points corresponding to the R1 sample.

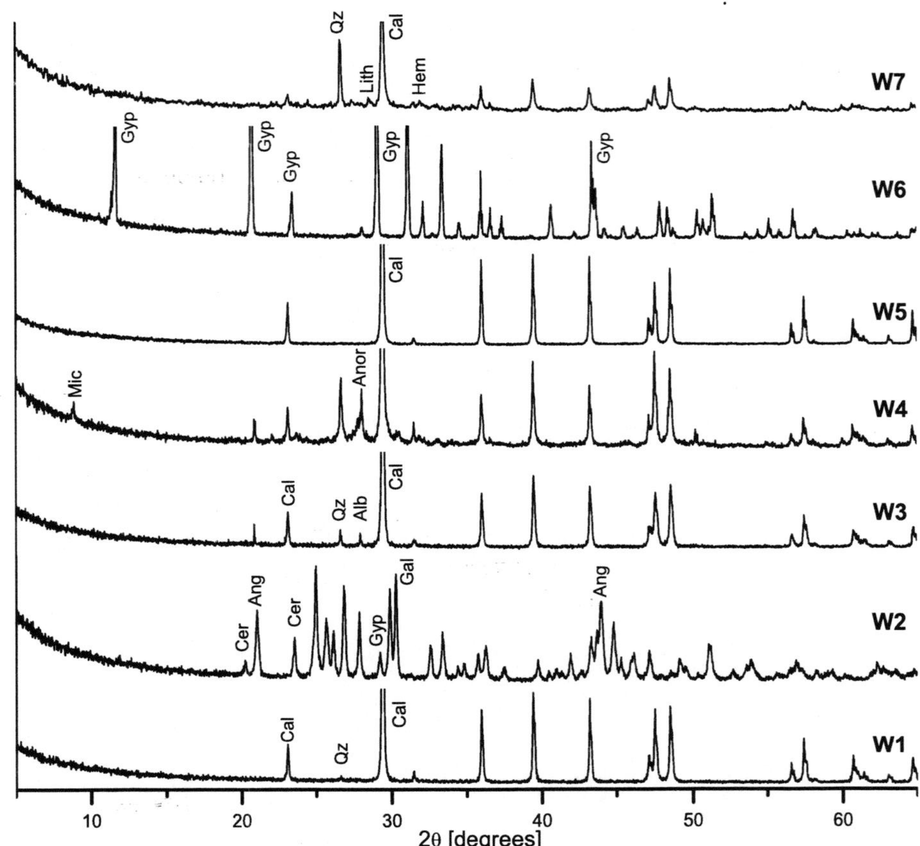

Figure 21. XRD patterns corresponding to the white pigments. The identified phases are also shown in the figure. Alb: Albite; Ang: anglesite; Anor: Anorthite; Cer: Cerussite; Gal: Galena; Gyp: Gypsum; Hem: Hematite; Lith: Litharge; Mic: Mica; Ort: Orthoclase; Qz: Quartz.

calcite (in a concentration lesser than 50%), quartz, micas and feldspars. These phases act as an embedding flux and could result in a glassy finish if the pigment is applied on the ceramic pot before cooking.

The W2 sample found in the mortuary context presents a very different mineralogical composition. It contains a high concentration of lead distributed in anglesite, cerussite and galena phases. Lead is also present in the sample W7 but in minor proportions, which would certainly improve the whiteness of the paint, as described in section 6.3. and Ref. [42]. The characteristic diffraction peaks corresponding to these phases can be seen in Fig. 1.

Regarding the reddish pigments, as displayed in Table 2 and Fig. 22, iron oxides are present mainly as hematite, in mixtures with different clays. The pigments R1 and R4 contain similar concentrations of hematite (2-3%), whereas R2 has a higher percentage of hematite (11%), and also contains calcite and gypsum. Finally, the R3 sample contains tourmaline (20%) and high amounts of hematite (56%).

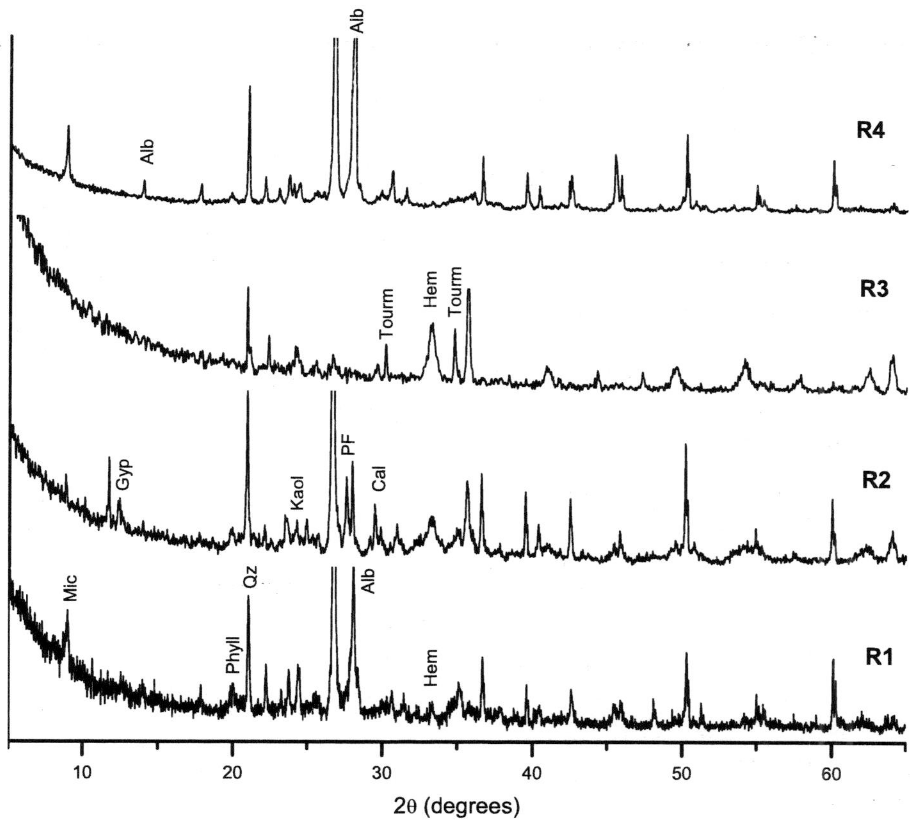

Figure 22. XRD patterns corresponding to the reddish pigments. The identified phases are also shown in the figure. Alb: Albite; Ang: Anglesite; Cal: Calcite; Cer: Cerussite; Gal: Galena; Gyp: Gypsum; Hem: Hematite; Kaol: Kaolinite; Mic: Mica; PF: Potassium feldspar; Phy: Phyllosilicates; Qz: Quartz; Tourm: Tourmaline.

7.3. EDX Analysis

The Sapphire Ultra-Thin Window X-ray detector used allows the detection of elements with atomic number greater than 5, with a resolution of 129.2 eV for the Mn-Kα line.

In all cases, the acquisition live time was 200 sec, nominal incident beam energy E_o=15 keV, 500 pA specimen current, 15 mm working distance and 33.42° take off angle. In order to obtain a representative analysis of each sample, the probe current was checked at the beginning and at the end of each measurement with a Faraday cup; this allowed to verify probe current variations remained within 1%.

The irradiation parameters used for the acquisition of spectra ensure a LOD of 500 ppm to 1000 ppm for most of the elements present in the samples studied. These considerations allow quantifying the major and minor elements with acceptable uncertainties.

To acquire EDX spectra, the samples were not coated, due to the fact that, on one hand, carbon coating complicates the quantification of these elements, and on the other hand the presence of a gold coating (or other heavy metal layer) modifies the different absorption effects for each line, and implies the appearance of its characteristic peaks, which may overlap with some lines of interest. This is the case of gold M lines overlapping with sulphur K and lead M lines.

As mentioned above, the Duane-Hunt limit value was fitted in order to determine the effective incident beam energies $E_o^{\star}$ which were used as input for the quantification routine by means of the program MULTI [78]. An example of such fitting procedure is shown in Fig. 23 for the reddish pigment R1. The fitted values for the effective incident beam energies corresponding to all the samples analyzed are displayed in Tables 3 and 4. It can be seen that this correction may become important (up to 20%).

The influence of the value taken for E_o on the concentrations obtained is exemplified in Table 5 for the R1 sample. Clearly the dependence of the ZAF matrix corrections on the incident energy is not equal for all elements, since relative concentrations may be strongly altered even for similar E_o values.

The quantification routines were assessed with the software MULTI [78], using SPI[®] commercial mineral standards. The global uncertainties, associated to the ZAF corrections and to the determination of characteristic intensities, are much lower than the errors due to charging effects and sample preparation.

Table 3 shows the results obtained for the quantification of white pigments through the X-ray characteristic spectra. Excepting for sample W2, white pigments are characterized by a high calcium concentration (beetwen 23 and 38%), which is in agreement with the el-

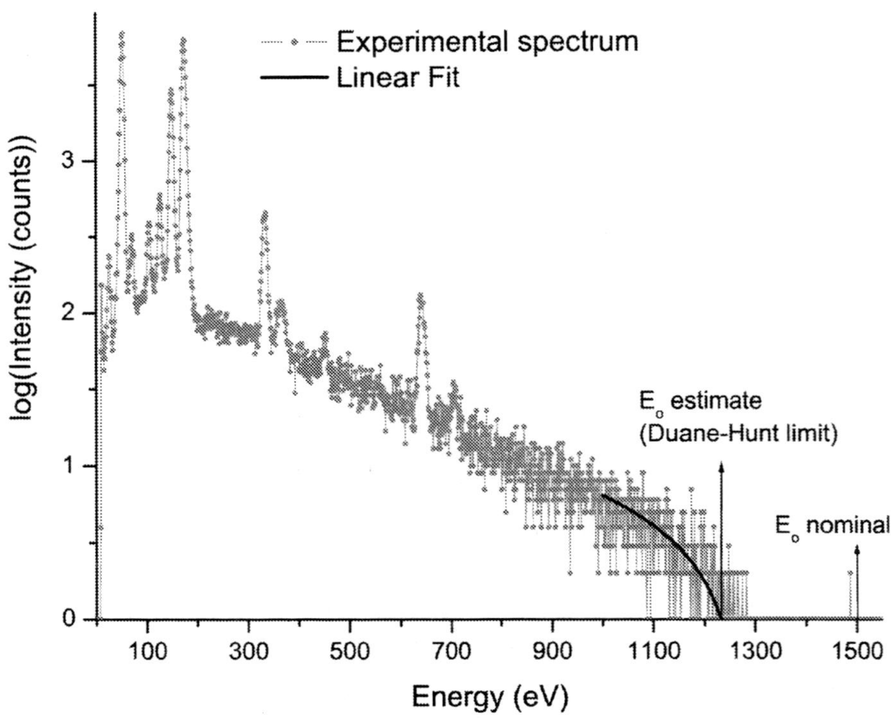

Figure 23. Determination of the effective incident energy value for R1 sample from the Duane-Hunt limit.

Table 3. Elemental concentrations of white pigments obtained with SEM-EDX analysis.

ELEMENTAL CONCENTRATIONS (wt %)

Sample	W1	W2	W3	W4	W5	W6	W7
$E_o^{\star}$ (keV)	14.6	14.6	15.0	14.0	14.8	14.8	13.1
Element							
C	11.4^a	1.2^a	10.9^a	10.0^a	13.9^a		9.9^a
O	47.6^a	20.6^a	46.3^a	51.3^a	43.7^a	60.9^a	44.1^a
Na		0.7	0.1		0.1		0.3
Mg	0.2		0.5	1.5	0.2		1.7
Al	0.3	0.9	0.4	2.4	0.4		1.8
Si	1.0	4.9	1.7	8.0	1.1		8.5
P			0.1				0.1
S		5.8	0.8			17.7	
K	0.3	0.6	0.5	1.5	0.4		3.8
Ca	38.2	1.6	36.1	23.1	37.2	21.3	26.6
Mn		1.0					
Fe	0.2	2.3	0.3	1.7	0.26		1.8
Pb		63.2^b					0.3^b

^{a}Assessed by stoichiometry
^{b}Calculated through Mα-line intensities

emental composition inferred from the mineralogical quantification. Some minor elements (e.g. Na or K) found in W3, W4 and W7 may be associated to feldspar phases. As expected from the XRD mineral identification, the W2 pigment, found in a funerary context, presents a very different elemental composition with respect to the other pigments studied: the major element is lead, calcium and sulphur being part of the gypsum quantified by Rietveld refinement.

The correct deconvolution of the spectra in the low energy region is very important because the detector can not separate the Pb-M lines from S-Kα line. However, the quantification of lead was performed taking into account the Pb-Mα line, since it is much more intense than the Pb-L lines for the electron beam energy chosen. To this aim, the EDX spectral processing was performed by taking into account radiative transition rates given by Perkins et al. [88]

In Table 4 the elemental composition of red pigments is shown. Samples R1 and R4 have similar compositions, nevertheless small differences in the Ca concentration can be seen. The Fe contents in both samples are consistent with the hematite phase obtained by XRD. According to the results obtained by SEM and XRD, the R2 sample bears a high Fe mass fraction (15.3%), and its Ca content is related to the presence of phases such as calcite, gypsum and albite. Finally, the R3 pigment presents the highest Fe concentration

Table 4. Elemental concentrations of reddish pigments obtained with SEM-EDX analysis.

ELEMENTAL CONCENTRATIONS (wt %)

Sample	R1	R2	R3	R4
$E_o^\star$ (keV)	12.3	13.7	13.5	12.9
Element				
C	2.9^a	0.7^a		
O	50.4^a	44.3^a	43^a	44.1^a
Na	1.1	0.6	0.4	5.0
Mg	1.7	0.8	0.6	0.8
Al	9.0	6.6	7	7.4
Si	23.8	21	22.8	27.7
S		2.7		
K	7	5	5.2	7.3
Ca	0.5	2	2	1.1
Fe	4.6	17.4	19	4.1

[a] Assessed by stoichiometry

Table 5. Mass concentrations obtained for sample R1 using the nominal incident energy (15 keV) and that determined through the Duane-Hunt limit (12.3 keV)

R1 SAMPLE (wt %)

Element	$E_o = 15$ keV	$E_o^\star = 12.3$ keV
C	3.06^a	2.87^a
O	46.02^a	50.41^a
Na	1.02	1.14
Mg	1.37	1.69
Al	7.12	9.03
Si	18.22	23.78
K	4.47	6.63
Ca	0.31	0.5
Ti	0.24	0.43
Fe	2.31	4.62
Total	84.16	101.1

[a] Assessed by stoichiometry

(41%), which is consistent with the amount of hematite and tourmaline detected by XRD.

In both sets of samples, EDX analyses produce compositions similar to those inferred through the XRD refinement procedure. In the case of reddish pigments, some discrepancies arise. It must be mentioned, however, that a wide variety of minerals bear similar crystallographic structures, and a number of ion substitutions could occur, which may imply important differences from the 'stoichiometric' formulas chosen for the phase quantification through XRD. In addition, the results provided by XRD are over-estimated, which is inherent to the phase concentration normalization accomplished by the Rietveld refinement method. Finally, a less important uncertainty is introduced by minor phases present in the sample which cannot be quantified by XRD and may add up to 3%.

8. Conclusion

The strengths and weaknesses of the non-destructive techniques SEM-EDX and XRD have been discussed when applied to the challenging characterization of archaeological ceramic surface paints and pigments of Aguada culture. These techniques are complementary, since individually they may lead to confusing interpretations, but combining them and considering all data together it was possible to attain a mineralogical, chemical and technological characterization of the materials studied.

Aguada Portezuelo paints (Au coated) are easily recognized and distinguished in their chemical contrast and mineralogical and chemical composition by all means used. For Ambato sherds instead, BSE images and XRMs do not show differences among paints even when working conditions are changed. This encouraged the development of an image-processing program in MATLAB® environment, which offers a user-friendly graphic interface with drop-down menus and the possibility of working with more than one image simultaneously. As corroborated through XRD characterization, the original BSE image for one of the samples (B53) presents a mean Z contrast lesser than 0.1, though it contains a small amount of Pb. Despite this conventional inconvenience, discrimination of zones with this low contrast was successfully accomplished after an adequate image treatment.

X-ray maps complement the information provided through BSE images. For the Portezuelo ceramic paints studied, XRM fusion with SE images using the wavelet decomposition, as implemented in [41], has shown to be a convenient alternative to the traditional segmentation method, since no significant detail losses are observed.

The chemical and mineral compositions of paintings from Portezuelo and Ambato are clearly different, particularly regarding the white ones. In the first case, the white base was painted before firing; the color is given mostly by Ca-rich metastable phases (gehlenite, CaO) and analcime (occasionally) that suggest variable firing temperatures (from $< 900°C$ to $1000°C$). These mineral differences do not correlate with Portezuelo ceramic types. In Ambato, the whites may sometimes be due to small contents of Pb-minerals. There is some heterogeneity within these groups too, but it does not affect the compositional characteristics identified on each style. It can be concluded that the ceramic pieces of Aguada Ambato and Portezuelo are two dissimilar entities that were made in very distinct ways; they not only differ on their designs but also on the chosen materials and the production technology applied by ancient potters.

Regarding pigments, the analyses showed that in the white ones the color is given by Ca bearing minerals with the exception of W2. W1, W3 and W5 contain calcite, W6 pigment pure gypsum, samples W4 and W7 calcite but with significant proportions of quartz, micas and feldspars. SE images of W7 revealed that it also contains a translucent and amorphous binder, probably animal fat, thus W7 corresponds to an elaborated paint. Reddish pigments due their color to hematite and Fe-rich clays except for R3 whose major contents are hematite, tourmaline and quartz.

The pigments found on a funerary context, W2 and R2, differ from the others: W2 (white) contains dominant Pb-bearing minerals (anglesite, cerussite and galena) and scarce gypsum, whereas R2 (reddish) characterizes by its kaolinite, K-feldspar and gypsum content (besides hematite) and the very fine grain size as well.

SEM-EDX chemical quantifications of pigments complemented the results obtained with XRD. Surface charge accumulation effects were taken into account by determining the Duane-Hunt limit in order to assess the effective incident energy, which allowed a remarkable improvement in the sets of concentrations obtained. XRD and EDX results are consistent.

In summary, the mineralogical and chemical differences found between Ambato and Portezuelo styles suggest that they are two distinctive entities not only on their designs but also on the materials chosen and the technology used. In addition to their intrinsic archaeological importance, these advances will help to develop a pigment and paint database for pre-incaic cultures from Northwestern Argentina.

The methodologies and characterization tools proposed here are suitable and can be recommended for routine analyses of different materials. Several developments and improvements may be faced, among which the inclusion of frequencial filters in both of the image-processing programs developed will be very simple, and would allow the subtraction of periodical fluctuations through Fourier analysis or similar techniques.

Acknowledgments

This work was financially supported by SeCyT (Universidad Nacional de Córdoba) and the FONCYT Agency to whom the authors are gratefully acknowledged. They are also thankful to F. Pasarelli and A. Laguens from the Museo de Antropología (UNC), and G. De La Fuente from the Escuela de Arqueología (Universidad Nacional de Catamarca) for providing samples. Special thanks are due to J. A. Riveros for his invaluable assistance.

References

[1] Druc, I. In *Archaeology and clays*; Druc, I.; Ed.; British Archaeological Report International Series 942; Oxford, 2001; pp. iii-vi.

[2] Bertolino, S. R.; Fabra, M. *Appl. Clay Sci.* 2003, *24*, 21-34.

[3] Garrison, E.G. *Techniques in Archaeological Geology*; Springer-Verlag: Berlin, 2003; 304p.

[4] Rice, P. M. *Pottery analysis*; The University of Chicago Press: Chicago, 2005; 559p.

[5] De La Fuente, G.; Kristcautzky, N.; Toselli, G.; Riveros, A. *Estudios Atacameños* 2005, *30*, 61-78.

[6] De La Fuente, G. *Glass Ceram. Conserv.* 2006, *14*, 3-6.

[7] Schreiner, M.; Frühmann, B.; Jembrih-Simbürger, D.; Linke, R. *Adv. X-ray Anal.* 2004, *47*, 1-17.

[8] Aloupi, E.; Karydas, A. G.; Paradellis, T. *X-Ray Spectrom.* 2000, *29*, 18-24.

[9] Cariati, F.; Fermo, P.; Gilardoni, S.; Galli, S.; Milazzo, M. *Spectrochim. Acta Part B At. Spectrosc.* 2003, *58*, 177-184.

[10] Fernández-Ruiz, R.; García-Heras, M. *Spectrochim. Acta Part. B At. Spectrosc.* 2007, *62*, 1123-1129.

[11] Fortina, C.; Memmi Turbanti, I.; Grassi, F. *Archaeom.* 2008, *50*, 30-47.

[12] Schleicher, L.; Miller, J.; Watkins-Kenney, S.; Carnes-McNaughton, L.; Wilde-Ramsing, M. *J. Archaeol. Sci.* 2008, *35*, 2824-2838.

[13] Barrios Neira, J.; Montealegre, L.; López L.; Romero, L. *Appl. Clay Sci.* 2009, *42*, 529-537.

[14] Appolonia, C. R.; Espinoza Quiñones, F. R.; Aragão P. H. A.; dos Santos, A. O.; da Silva, L. M.; Barbieri, P. F.; do Nascimento Filho, V. F.; Coimbra, M. M. *Radiat. Phys. Chem.* 2001, *61*, 711-712.

[15] Craig, N.; Speakman, R.; Popelka-Filcoff, R.; Glascock, M.; Robertson, J.; Shackley, M.; Aldenderfer, M. *J. Archaeol. Sci.* 2007, *34*, 2012-2024.

[16] Reed, S. *Electron Probe Microanalysis*, 2nd ed.; Cambridge University Press: Cambridge, 1993.

[17] Mirti, P. *X-Ray Spectrom.* 2000, *29*, 63-72.

[18] Sánchez Ramos, S.; Bosch Reig, F.; Gimeno Adelantado, J. V.; Yusá Marco, D. J.; Doménech Carbó, A. *Spectrochim. Acta Part B Atom. Spectrosc.* 2002 *57*, 689-700.

[19] Buxeda i Garrigós, J. *J. Archaeol. Sci.*, 1999, *26*, 295-313.

[20] Iordanidis A.; Garcia-Guines, J.; Karamitrou-Mentessidi, G. *Mater. Charact.* 2009, *60*, 292-302.

[21] Papadopoulou, D. N.; Zachariadis, G. A.; Anthemidis, A. N.; Tsirliganis, N. C.; Stratis, J. A. *Spectrochim. Acta Part. B Atom. Spectrosc. 59*, 1877-1884.

[22] Bao-Ping, L.; Jian-Xin, Z.; Greig, A.; Collerson, K.D.; Zhen-Xi, Z.; Yue-Xin, F. *Nucl. Instr. Meth. Phys. Res. B* 2005, *240*, 726-732.

[23] Cochrane, E.; Neff, H. *J. Archaeol. Sci.* 2006, *33*, 378-390.

[24] Pérez-Arantegui, J.; Resano, M; García-Ruiz, E.; Vanhaecke F.; Roldán, C.; Ferrero, J.; Coll, J. *Talanta* 2008, *74*, 1271-1280.

[25] Habicht-Mauche, J.; Glenn, S. T.; Milford, H.; Flegal, A. R. *J. Arch. Sci.* 2002, *29*, 1043-1053.

[26] Speakman, R. J.; Neff, H. *Am. Antiq.* 2002, *67*,137-144.

[27] Neff, H. *J. Arch. Sci.* 2003, *30*, 21-35.

[28] Vaughn, K. J.; Conlee, C. A.; Neff, H.; Schreiber, K. *J. Archaeol. Sci.* 2006, *33*, 681-689.

[29] Odell, H. *Lithic Analysis*, 3rd ed.; Springer: New York, 2006.

[30] Molera, J.; Pradell, T.; Vendrell-Saz, M. *Appl. Clay Sci.* 1998, *13*, 187-202.

[31] Smith, G. D.; Clark, R. J. H. *J. Archaeol. Sci.* 2004, *31*, 1137-1160.

[32] De La Fuente, G. A.; Pérez Martinez J. M. *Intersecciones en Antropología* 2008, *9*, 173-186.

[33] Edwards, H. G. M.; Chalmers, J. M. *Raman Spectroscopy in Archaeology and Art History*; Royal Society Of Chemistry: Cambridge, 2005.

[34] Polak, M.; Baram, J.; Pelleg, J. *Archaeom.* 1983, *25*, 59-67.

[35] Goldstein, J.; Newbury, D.; Echlin, P.; Joy, D.; Romig, A.; Lyman, C.; Fiori C.; Lifshin, E. *Scanning Electron Microscopy and X-Ray Microanalysis: A Text for Biologists, Materials Scientists, and Geologists* ; Plenum Press: New York, 1992.

[36] Russ, J. C. *The Image Processing Handbook*, 3rd. ed., CRC Press: Boca Raton, 1998.

[37] Chinga-Carrasco, G.; Lenes, M.; Johnsen, P. O.; Hult, E.-L. *Micron* 2009, *40*, 761-768.

[38] Tzaphlidou, M. *Micron* 2005, *36*, 593-601.

[39] Galván Josa, V.; Bertolino, S. R.; Riveros, J. A.; Castellano, G. *Micron* 2009, *40*, 794-799.

[40] Galván Josa, V.; Bertolino, S. R.; Laguens, A.; Riveros, J. A.; Castellano, G. *Microchem. J.* 2010, doi:10.1016/jmicroc.2010.03.010

[41] Galván, V.; Bertolino, S. R.; De La Fuente, G.; Riveros, J. A.; Castellano, G. *Acta Microsc.* 2009, *18* Supp. C, 385-386.

[42] Bertolino, S. R.; Galván, V.; Carreras, A.; Laguens, A.; De La Fuente, G.; Riveros, A. *X-Ray Spectrom.* 2009, *38*, 95-102.

[43] Cremonte, M. B.; Baldini, M.; Botto, I. L. *Intersec. en Antropol.* 2003, *4*, 3-16.

[44] González, A. R. *Rev. Inst. Antropol.* 1961, *II-III*, 205-252.

[45] Baldini, M.; Carbonari, J.; Cieza, G.; de Feo, M.; del Castillo, M.; Figini, A.; Rex González, A.; Huarte, R.; Togo, J. *Estudios Atacameños* 2002, *24*, 71-82.

[46] Laguens, A.; Bonnin, M. In *La Cultura de la Aguada y sus expresiones regionales* ; EUDELAR: La Rioja; 2005; pp 23-34.

[47] Pantorrilla, M.; Núñez Regueiro, V. *Intersecciones en Antropología* 2006, *7*, 235-245.

[48] Pérez Gollán, J. *La Cultura de la Aguada vista desde el Valle de Ambato* ; Publicaciones Arqueología CIFFyH: Córdoba, 1991; Vol. 46, pp 157-174.

[49] Assandri, S.; Avila, A.; Herrero, R.; Juez, S. *Introducción a la biogeografía y arqueología del Valle de Ambato (Provincia de Catamarca, Argentina)* ; Publicaciones Arqueología, CIFFyH: Córdoba, 1992; Vol. 46, pp 7-11.

[50] Pérez Gollán J. A.; Heredia, O. *Hacia un replanteo de la Cultura Aguada* ; Cuadernos; Inst. Nacional de Antropología: Buenos Aires, Argentina; 1975.

[51] Fabra, M. In IV Round Table *La Cultura de la Aguada y su Dispersión* ; Costa, M.; Llagostera, A.; Eds.; Inst. Investigaciones Arqueológicas y Museo, Universidad Católica del Norte: San Pedro de Atacama, 2006; pp 7-18.

[52] Laguens, A.; Juez, M. *Actas del XIII Congreso Nacional de Arqueología Argentina* , 1999, *II* 489-504.

[53] Laguens A. *Relaciones*; Soc. Arg. Antropología, 2005; Vol. *XXIX*, 137-162.

[54] Pérez Gollán, J.; Bonnin, M.; Laguens, A.; Assandri, S.; Federici, L.; Gudemos, M.; Hierling, J.; Juez, S. *Shincal* 2000, *6*, 115-124.

[55] Kusch M. F. In: *El Arte Rupestre en la Arqueología Contemporánea* ; Podesta M.; Llosas M. I. H.; Renard de Coquet S.; Eds.; FECIC: Buenos Aires, 1991; pp 14-24.

[56] Kusch M. F. *Shincal* 1996, *6*, 241-248.

[57] González A. R. *Arte Precolombino. Cultura La Aguada. Arqueología y sus diseños* ; Filme Ediciones Valero: Buenos Aires, 1998.

[58] Sorbier, L.; Rosenberg, E.; Merlet, C. *Microsc. Microanal.* 2004, *10*, 745-752.

[59] Gonzalez, R.; Woods, R. *Digital Image Processing* , 2nd ed.; Prentice-Hall: New Jersey, 2002.

[60] Savitzky, A.; Golay, M. J. E. *Anal. Chem.* 1964, *36*, 1627-1639.

[61] Bonnet, N. *Micron* 2004, *35*, 635-653.

[62] Chen, S. D.; Ramli, A. R. *Trans. Consumer Electron.* 2003, *49*, 1310-1319.

[63] Jafar, I.; Ying, H. *Integr. Comput.-Aid. Eng.* 2008, *15*, 131-147.

[64] Neycenssac, F. *Graph. Models Image Process.* 1993, *55*, 447-463.

[65] Rietveld H. M. *Acta Cryst.* 1967, *22*, 151-152.

[66] Rietveld, H. M. *J. Appl. Crystallogr.* 1969, *2*, 65-71.

[67] McCusker, L. B.; von Dreele, R. B.; Cox, D. E.; Löuer D.; Scardi P. *J. Appl. Cryst.* 1999, *32*, 36-50.

[68] Bish D. L.; Howard S. A. *J. Appl. Cryst.* 1988, *21*, 92-96.

[69] Ortiz, A.; Cumbrera F.; Sánchez Bajo F.; Guibertau F.; Xu H.; Padture, N. *J. Am. Ceram. Soc.* 2000, *83*, 2282-2286.

[70] De la Torres, A. G.; Bruqueand, S.; Aranda, M. A. G. *J. of Appl. Crystallog.* 2001, *34*, 196-202.

[71] Jones, R.; Bish, D. *Proc. Amer. Clay Min. Soc. Meeting* 1991, *28*, 84-94.

[72] Bish, D. L.; Post, J. E. *American Mineralogist* 1993, *78*, 932-940.

[73] Young, R. *The Rietveld Method*; International Union of Crystallography; Oxford University Press: Oxford, 1993.

[74] Bruker AXS: TOPAS V3.0: General profile and structure analysis software for powder diffraction data. Bruker AXS, Karlsruhe, Germany. 2004.

[75] Enzo S.; Parish W. *Adv. X-Ray Anal.* 1983, *27*, 37-44.

[76] March, A. Z. *Kristallogr.* 1932, *81*, 285-297.

[77] Dollase, W. A. *J. Appl. Cryst.* 1986, *19*, 267-272.

[78] Trincavelli, J.; Castellano, G.; Riveros, J. *X-Ray Spectrom.* 1998, *27*, 81-86.

[79] Packwood, R.; Brown, J. *X-Ray Spectrom.* 1981, *10*, 138-145.

[80] Riveros, J.; Castellano, G.; Trincavelli, J. *Mikrochim. Acta* 1992, *[Suppl.] 12*, 99-105.

[81] Reed, S. J. B. *Br. J. Appl. Phys.* 1965, *16*, 913-926.

[82] Newbury, D. E. *J. Res. Natl. Inst. Stand. Technol.* 2001, *107*, 567-603.

[83] Procop M.; Hübner W.; Wäsche R.; Nieland S.; Ehrmann O. *Microsc. Anal.* 2002, *336*, 17-18.

[84] Buxeda i Garrigós, I.; Cau Ontiveros, M. A. *Complutum* 1995 *6*, 293-309.

[85] Bruni, S.; Cariti, F.; Bagnasco Gianni, G.; Bonghi Jovino, M.; Artioli, G.; Russo, U. In *Archaeology and Clays*; I. C. Druc; Ed.; BAR International Series 942, 2001, pp 27.

[86] Grim, R. E. *Applied Clay Mineralogy*, McGraw Hill Book Co.: New York, 1962.

[87] Reed, S.J.B. *Electron Microprobe Analysis and Scanning Electron Microscopy in Geology*; Cambridge University Press: Cambridge, 1997.

[88] Perkins, S.T.; Cullen, D.E.; Chen, M.H.; Hubbell, J.H.; Rathkopf, J.; Scofield, J. *Tables and graphs of atomic subshells and relaxation data derived from LLNL evaluated atomic data library (EADL), Z=1-100*; Lawrence Livermore National Laboratory; Report UCRL-50400, 1991, Vol. 30.

In: Paints: Types, Components and Applications
Editor: Stephanie M. Sarrica

ISBN: 978-1-61761-813-0
© 2011 Nova Science Publishers, Inc.

Chapter 11

DESIGN AND MANUFACTURING OF ARTISTIC PAINT ROLLERS

Fusaomi Nagata[1], Yukihiro Kusumoto[2], Kaori Saito[3] and Takamasa Kusano[4]
[1]Tokyo University of Science, Yamaguchi, Japan
[2]Fukuoka Industrial Technology Center, Japan
[3]Masuo Fukumoto, Fukumoto Kogyo Co. Ltd., Japan
[4]SOLIC Co. Ltd., Japan
Keigo Watanabe, Okayama University, Japan

Abstract

In this chapter, a 3D design and machining system based on a 3-axis NC machine tool with a rotary unit is introduced to efficiently produce the artistic design of wooden paint rollers. The paint rollers are used to execute a relief wall just after painting. A simple post-processor is first proposed for the NC machine tool to transform a base tool path called cutter location data (CL data) to NC data, mapping the y-directional pick feed to the rotational angle of the rotary unit. Also, the post-processor has a novel function that elaborately adjusts feed rate values according to the curvature of each design so as not to chip the carved surface. The suitable feed rate values are generated by using a simple fuzzy reasoning method while checking edges and curvatures in a relief design. The post-processor allows the 3-axis NC machine tool with a rotary unit to easily carve an artistic relief design on a cylindrical wooden workpiece without an undesirable edge chipping. Experimental results show that wooden paint rollers with an artistic relief design can be successfully machined without any chipping. When a wooden paint roller is popular with the design, its mass production is further required. To cope with the mass production, rubber type paint rollers can be efficiently formed by using an aluminum mold with the same design of the wooden paint roller.

1. Introduction

Fukuoka prefecture is one of the largest furniture manufacturers in Japan. Interior city Ohkawa is located at the south part of Fukuoka. Although furniture is mainly made of various wooden materials, the production of furniture has been gradually decreased in two decades because of changes in consumer preference and industrial structure. From this fact, other promising wooden products except furniture are currently being considered to efficiently use wood resources in Fukuoka. As one example, wooden paint rollers attract our attention. In home-making industry, handy paint rollers with a simple pattern are generally used to transcribe its design to a wall just after painting. Interior planners and decorators want to use more artistically designed paint rollers to cope with various user needs, but the pattern designs are limited to several common ones. In order to efficiently provide user-oriented roller designs, a new 3D design and machining system should be considered for limited production with a wide variety of wooden paint rollers. However, it is not easy to perform its elaborate carving even when the latest woodworking machinery is used, because wooden paint rollers have a cylindrical shape. In furniture manufacturing industry, since 3-axis NC machine tools and 5-axis NC machine tools with a tilting head are used generally and widely, a new machining system for wooden paint rollers should be designed, keeping the use of such conventional machine tools to save investment in equipment.

5-axis NC machine tools have a major advantage that is not in 3-axis ones when machining a free-form surface efficiently. Up to now, 3D machining systems based on a 5-axis NC machine tool have been actively developed and studied in various manufacturing industries. Takeuchi et al. [1, 2] proposed a general method for generating NC data from the collision-free tool path of 5-axis control machining centers. The post-processor developed can convert CL data to actual NC data, taking account of the structure of machining centers, linearization, feed rate control and spindle rotation control. Xu et al. [3] presented a technique to avoid inefficient 5-axis machining practices by automatically creating and verifying a feasible tool-path prior to the actual metal cutting. Chen et al. [4] developed a hybrid 5-degrees-of-freedom parallel kinematic machine tool constructed by using the TRR-XY mechanism and its post-processing system. The effects of the cutter shapes and of the machine construction on the post-processing were investigated. Also, Lei et al. [5] developed a novel probe-ball measurement device which was designed to measure the overall position errors of a 5-axis machine tool and a method using double ballbar [6] was proposed to inspect the motion errors of the rotary axes out of five-axes. Furthermore, Cao et al. [7,8] proposed an offset approach of machining free form surface by using a 5-axis NC machine tool with a cylindrical cutter or toroidal cutter, in which Archimedes' helicoidal surface, pseudo-sphere and hyperbolic paraboloid surfaces were successfully machined.

When a 5-axis NC machine tool with a tilting head generally used in furniture manufacturing industry is applied to cylindrical machining or carving, a workpiece can not be made in a single pass, i.e. it needs to perform complicated repositionings several times as well as achieving a special zig. Such a 5-axis NC machine tool does not seem to be successfully used for the manufacturing of wooden paint rollers with a relief design at the present stage, due to the process and resulting workpiece setup time complexity.

Also, there are several researches tackling feed rate problems. For example, Timar et al. [9] developed an algorithm which computes the time-optimal feed rate variation along a curved path. Heo et al. [10] proposed a machining time estimation model using NC block distributions. An intelligent contour control strategy was investigated to improve the contour error of CNC machine tools by Tarng et al. [11], in which a cross-coupled fuzzy feed rate control scheme was presented to reduce the contour error. Zuperl et al. [12] discussed the application of fuzzy adaptive control strategy to the problem of cutting force control in high speed end-milling operations, in which it is designed to adaptively maximize the feed-rate subject to allowable cutting force on the tool. Also, Farouki et al. [13] reported variable feed rate CNC interpolators for constant material removal rates along Pythagorean hodograph (PH) curves. The proposed curvature-compensated feed rate scheme has important potential applications in ensuring part accuracy and in optimizing part programs consistent with a prescribed accuracy. Further, a new class of machine codes for the specification of PH curve tool paths, and associated feed rate functions were proposed in [14]. In there, experimental results from an implementation of real-time PH curve interpolators on an open-architecture CNC milling machine are described. However, it is not easy to apply such a method to conventional NC machine tools without using an open-architecture controller.

In addition to the above researches, there are already in commercial use machines that incorporate nonlinear interpolation (e.g. non-uniform rational B-spline curve interpolation called NURBS interpolation) into NC program to improve overall machining. Optimization programs are available to automatically adjust feed rate based on NURBS curves. However, it requires not only a special controller which can deal with NC program interpolated by NURBS curve, but also a corresponding CAM system which can generate an NURBS code such as G06. From the viewpoint of equipment cost, it is also important to consider how to effectively utilize conventional machine tools in wood product manufacturing. Moreover, advanced cylindrical grinding machines and improvements have been also proposed. For example, Frank et al. [15] developed a CNC cylindrical grinding machine for non-circular workpieces. Although parts with circular and non-circular cross-sections could be completely machined in one operation, the performance to circular cross-section with small edges was not mentioned. Therefore, the grinding machine would not be able to carve a wooden paint roller whose cylindrical surface has many small edges. Couey et al. [16] showed that force measurements were capable of providing useful feedback in precision grinding with excellent contact sensitivity, resolution, and detection of events occurring within a single revolution of the grinding wheel. Tawakoli et al. [17] developed a prototype internal grinding machine based on a new kinematics. The target was to defeat some problems of internal grinding such as poor enrichment of coolant lubricant, deflection of grinding tool and so on.

Summarizing the above, it can be seen that how to easily suppress undesirable edge chippings in machining cylindrical wooden workpieces with a relief design has not been discussed yet. In this chapter, a 3D design and machining system based on a 3-axis NC machine tool with a rotary unit is introduced efficiently and at a low price to produce wooden paint rollers with an artistic design. A key point is to use conventional and common 3-axis NC machine tools. In order to realize the machining system effectively, the following two points are considered in designing a corresponding post-processor. One is a basic function which efficiently generates suitable NC data to carve a relief design on a cylindrical wooden workpiece. The proposed post-processor transforms a base tool path called cutter location data (CL data) to NC data, mapping the y-directional pick feed to the rotational angle of the rotary unit [18]. The other is a systematic generation of feed rate values to suppress edge chippings on a relief design and to shorten the cycle time required for machining. The edge chipping is a particular problem when a wooden workpiece is machined by a router bit because wooden workpieces are considerably brittle compared with metallic materials. The edge chipping tends to appear with the increment of feed rate. Our proposed post-processor simply generates suitable feed rate values by using a simple fuzzy reasoning method based on an operator's experience while checking edges and curvatures calculated based on positions in CL data. The variation of curvature can be estimated by calculating the variation of the distance between two adjacent positions in CL data. An experimental result shows that wooden paint rollers with an artistic relief design can be successfully carved without a chipping.

2. Conventional 5-axis NC Machine Tool with a Tilting Head

In this section, a 3D machining system based on a 5-axis NC machine tool with a tilting head is evaluated as a conventional system [19]. The 5-axis NC machine tool is one of the most representative and popular machine tools in wood product manufacturing. The machining system consists of a 3D CAD/CAM with variable axis function and a post-processor. Figure 1 shows the 5-axis NC machine tool (HEIAN FF-151MC). The NC machine tool has a tilting head that can be simultaneously inclined and rotated. It should be noted however that corrected NC data are required to run the NC machine tool as computer simulation. The post process generally means to compute corrected NC data from CL data. Figure 2 shows the general process to calculate the corrected NC data for the 5-axis NC machine tool.

Examples of paint rollers with a relief design are given as shown in Fig. 3. We first draw the model of the relief design using a 3D CAD. Secondly, the CAM parameters such as pick-feed, path pattern (e.g., zigzag path), in/out tolerances and so on are set according to the requirement of actual machining. The main-processor of the CAM calculates cutter paths using the parameters. The cutter paths are called CL data. The i-th step $CL(i)$ in CL data is composed of position vector $p(i) = [x(i)\ y(i)\ z(i)]^T$ and normal direction vector $n(i) = [v_x(i)\ x_y(i)\ v_z(i)]^T$ as written by

$$CL(i) = [x(i)\ y(i)\ z(i)\ v_x(i)\ v_y(i)\ v_z(i)]^T \tag{1}$$

$$\{v_x(i)\}^2 + \{v_y(i)\}^2 + \{v_z(i)\}^2 = 1 \tag{2}$$

Finally, the post-processor generates the corrected NC data for the 5-axis NC machine tool with a tilting head, by only considering a tool length [19]. Figure 4 shows the tiltable main head which has a ball-end mill at the tip. The tool length $L_1 + L_2$ is defined as the distance from the center of swing to the tip of the end mill, which should be measured in advance. The post-processor transforms the CL data into NC data for 5-axis NC machine tools as shown in Fig. 1. The main head can be inclined and rotated within the range $\pm\,90$ and $\pm\,180$ degrees, respectively. The inclined and rotated axes are called the 4th (B) axis and 5th (C) axis, respectively. The i-th step in the corrected NC data is written by

$$NC(i) = [\tilde{x}(i)\ \tilde{y}(i)\ \tilde{z}(i)\ b(i)\ c(i)]^T \tag{3}$$

$$\tilde{j}(i) = j(i) + v_j(i)(L_1 + L_2), \quad j = x, y, z \tag{4}$$

where $b(i)$ and $c(i)$ are the head angles of the inclination and rotation as shown in Fig. 5, respectively. The CL data generated from the main-processor of the CAM are composed of

Figure 1. Conventional 5-axis NC machine tool with a tilting head.

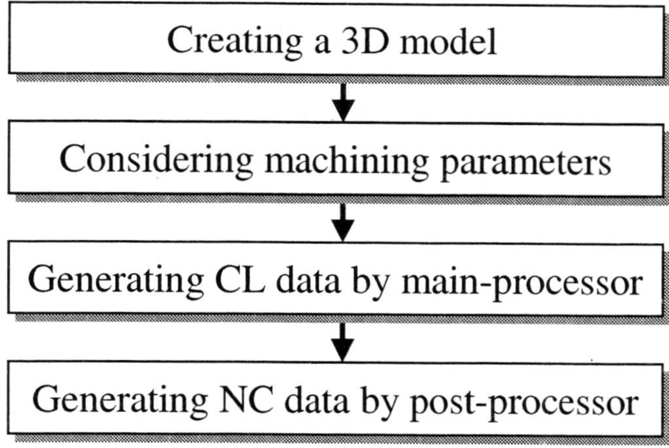

Figure 2. Main and post processes to compute corrected NC data for 5-axis NC machine tool.

sequential points on the model surface as shown in Fig. 6. If the NC data are transformed from the CL data without considering the tool length and are given to the NC machine tool, then the center of swing is obliged to directly follow the NC data. As can be expected, this would bring out a serious and dangerous interference between the main head and the workpiece. On the contrary, if the center of swing follows the corrected NC data given by (3) and (4), the tip of the ball-end mill can desirably move along the model surface as shown in Fig. 6.

Recently, such a 5-axis NC machine tool with a tilting head has become a center

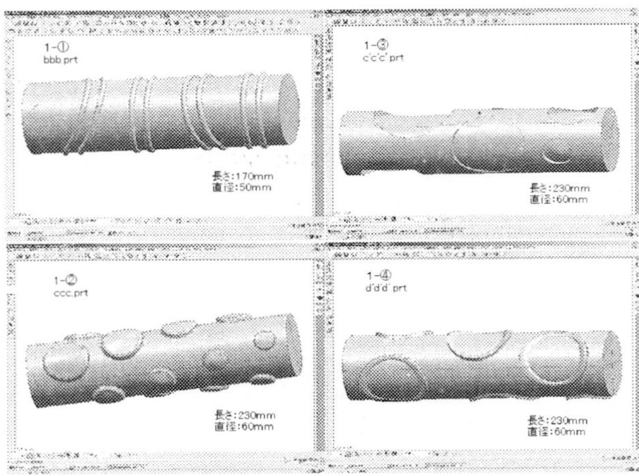

Figure 3. Artistic paint roller models designed by 3D CAD.

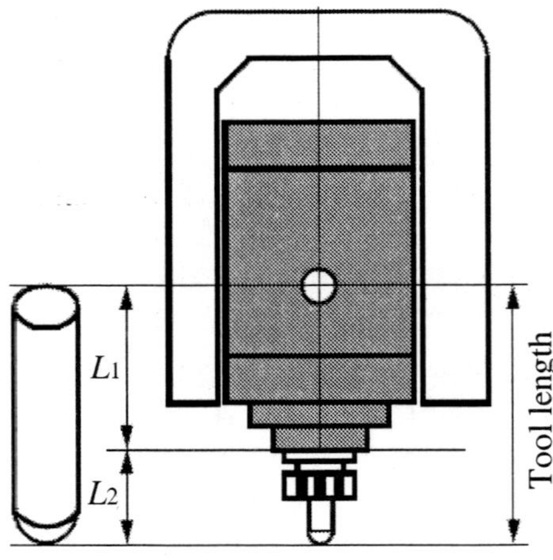

Figure 4. Definition of the tool length of a tilting head.

of attraction in wood product manufacturing. However, it has been recognized from a machining experiment that the 5-axis NC machine tool is not good at carving roller models as shown in Fig. 3. Although its high machining performance is expected to exceed the capability of standard NC machine tools, it has been hardly used for carving roller models yet. The main reason is that the roller models as shown in Fig. 3 cannot be machined in a single path of the cutter, i.e. models must be machined in several paths of the cutter, requiring extensive cutter repositioning; thus the 3D machining for artistic design paint rollers is complicated and time consuming. Furthermore, it is not easy to realize the modeling of a relief design on a cylindrical shape. To overcome these problems, a 3D machining system based on a 3-axis NC machine tool with a rotary unit is introduced in the next section.

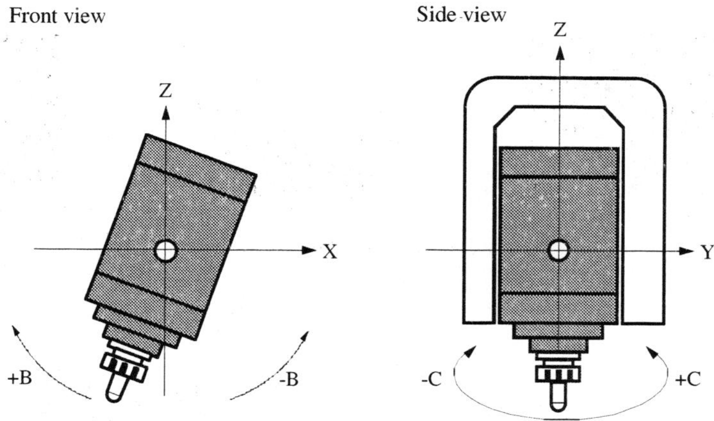

Figure 5. Inclination and rotation of main head.

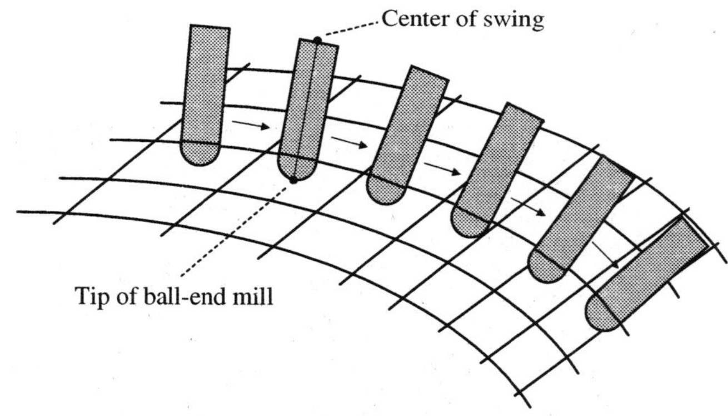

Figure 6. Motion of tip of ball-end mill generated from main-processor.

3. Intelligent Machining System for the Artistic Design of Wooden Paint Rollers

As described in the previous section, wooden paint rollers considered have a cylindrical shape with an artistic design, so that it is not easy to realize its elaborate carving even when the latest woodworking machinery such as a 5-axis NC machine tool with a tilting head is used. Generally, metallic cylindrical parts are precisely processed by an expensive CNC turning centre with milling capability. However, 3-axis NC machine tools and 5-axis NC machine tools with a tilting head have been mainly and widely used in furniture manufacturing industry, and therefore it is strongly required that a new woodworking machinery for wooden paint rollers should be designed by utilizing such existing NC machine tools and by considering equipment cost. In this section, a 3-axis NC machine tool with a rotary unit and its post-processor are proposed to efficiently produce artistic wooden paint rollers of many kinds of designs. The proposed machining system is realized easily and at low price only by adding a simple rotary unit into a conventional 3-axis NC machine tool. The post-processor provides two effective functions. One is a transformation technique from CL data without feed rate values to NC data, mapping the y-directional pick feeds to rotational angles of the rotary unit. The post-processor allows a well-known 3-axis NC machine tool to easily transcribe a relief design from the surface on a flat model to it on a cylindrical model. The other is an elaborate addition of feed rate values according to the curvature of each design to protect the cylindrical surface from undesirable edge chippings. The post-processor generates safe feed rate values by using a simple fuzzy reasoning method while checking edges and curvatures in a relief design, and appends them into NC data. The post-processed NC data mildly act on the fragile edges of wooden paint rollers.

3.1. Post-processor for a 3-axis NC machine tool with a rotary unit

Conventional paint rollers have generally no artistic designs at all and the designs are limited to flat or simple patterns even if they have. As mentioned in the previous section, unfortunately, it is not easy to carve a relief design on a cylindrical model even if the 5-axis NC machine tool with a tilting head would be used. First of all, we discuss a problem concerning the 3D machining of wooden cylindrical shape with a relief design. When the modeling of a roller is realized by using a 3D CAD, a base cylindrical shape is modeled in advance. Then a favorite relief design is drawn on the cylindrical model. However, the modeling of relief design on the cylindrical shape as shown in Fig. 3 is a complicated task, even when a 3D CAD software is used. Next, it is not easy to realize its 3D machining by using the 5-axis NC machine tool with a tilting head, in which the NC data generated from the CAM are composed of x-, y-, z-, b- and c- directional components. Thus, to easily provide many kinds of paint rollers with wide variety and low volume to home making industries, a machining system that can directly carve an artistic relief design on a cylindrical workpiece must be developed.

In this chapter, to solve these problems, a new 3D machining system is considered based on a 3-axis NC machine tool with a rotary unit, and a post-processor is proposed for the rotary unit. The overview of the proposed machining scheme is illustrated in Fig. 7. The post-processor allows such conventional woodworking machinery as the 3-axis NC machine tool to elaborately produce wooden paint rollers. An artistic design drawn on a flat model surface can be easily transcribed to a cylindrical model surface. In the remainder of this section, the system is described in detail. A 3-axis NC machine tool with x-, y- and z-axes must be first prepared to realize the proposed concept. As an example, an NC machine tool MDX-650A provided by Roland D.G. as shown in Fig. 8 is used for experiments. The NC machine tool equips with an auto tool changer ZAT-650 and a rotary unit ZCL-650A. The mechanical resolution of the rotary unit is about 0.0027 degrees. The NC machine tool has four degrees of freedom, i.e., three translations and one rotation. This chapter addresses how to easily make a wooden paint roller with an artistic relief design. The most important point is that proper NC data for the NC machine tool with a rotary unit can be generated in one step. To meet this end, the post-processor generates the NC data which transcribe the design on a flat model to it on a cylindrical model. By applying the post-processed NC data, the NC machine tool can directly carve an artistic relief design on a cylindrical workpiece.

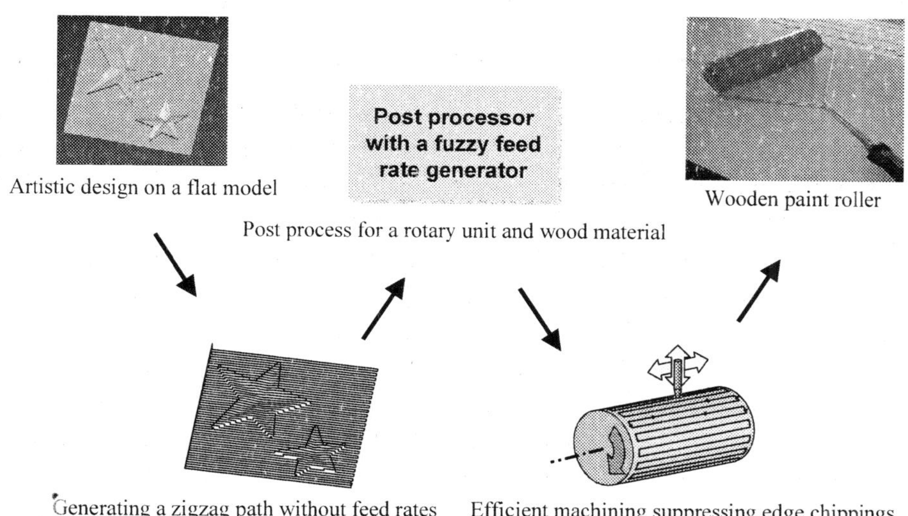

Figure 7. Overview of the proposed machining scheme for cylindrical wooden workpieces.

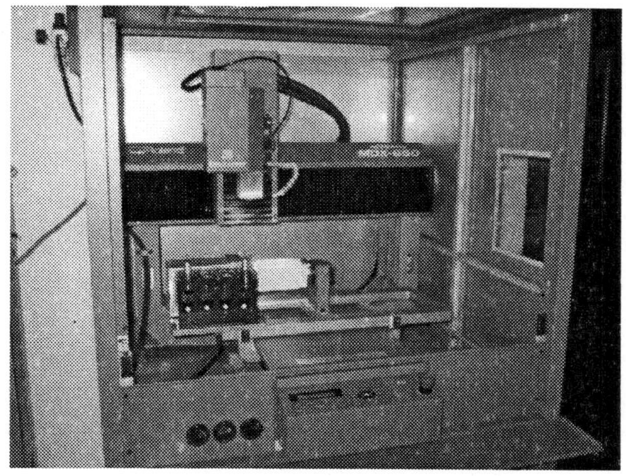

Figure 8. 3-axis NC machine tool with a rotary unit.

Next, we describe on the feature of the post-processor. A desired relief design is first modeled on a flat base model as shown in Fig. 9. CL data are secondly generated with a zigzag path as shown in Fig. 10. In this case, the coordinate system should be set so that the pick feed direction is parallel to the table slide direction of the NC machine tool, i.e., y-direction. The proposed post-processor transforms the CL data without gFE-DRAT/h statements into the corresponding NC data, mapping the y-directional position to the rotational angle of the rotary unit. As can be seen from the components of the NC data, when the rotary unit is active, the table slide motion in y-direction is inactive. The post-processor first checks all steps in the CL data, and extracts the minimum value y_{min} and the maximum value $y_{\_max}$ in y-direction. The angle $a(i)$ for the rotary unit at the i-th step is calculated from

$$a(i) = \frac{360 \times \{y(i) - y_{\_min}\}}{y_{\_length}} \tag{5}$$

where $y_{\_length}$ is the length in y-direction, which is easily obtained by $y_{\_max} - y_{\_min}$. The CL data $p(i) = [x(i)\ y(i)\ z(i)]^T$ at the i-th step is transformed into the NC data composed of $[x(i)\ a(i)\ z(i)]^T$ by using (5). The length in y-direction is transformed into the circumference of the roller model. It is expected that the relief design shown in Fig. 9 is desirably carved on the surface of a cylindrical workpiece. Thus, the proposed system provides a function that easily transcribes an artistic design from the surface on a flat model to it on a cylindrical wooden workpiece fixed to the rotary unit.

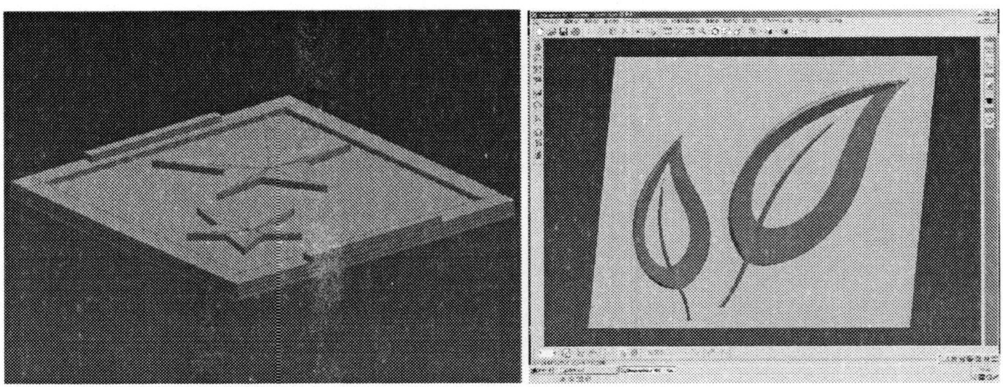

Figure 9. Artistic relief designs illustrated by a 3D CAD.

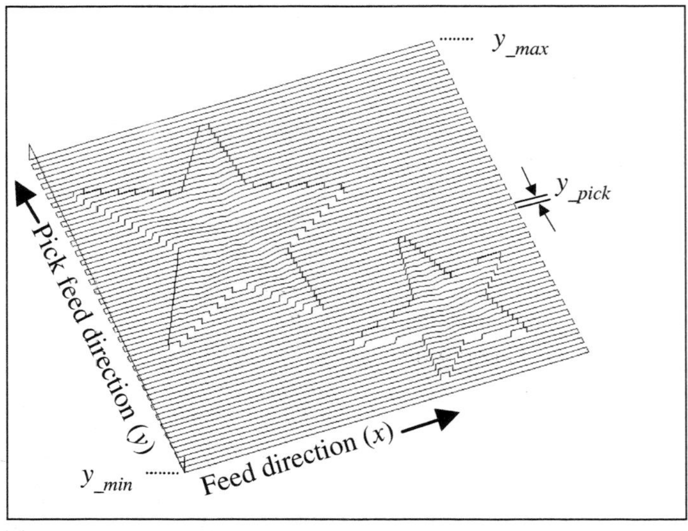

Figure 10. An example of zigzag path on a flat model ($y_{\_length} = y_{\_max} - y_{\_min}$).

3.2. Feed rate generation using fuzzy reasoning

3.2.1. Feed rate generation according to curvature

It is generally known that feed rate is one of the most important parameters to smoothly control NC machine tools and to reduce the total machining time. Wooden materials are essentially brittle in machining, so that the regulation of feed rate is especially significant to protect the surface from an edge chipping as shown in Fig. 11. The key factors which affect the feed rate to be set are mainly the complexity of the relief design and the kind of the wooden material. The feed rate must be adjusted not only to avoid the edge chipping on the cylindrical wooden workpiece, but also to shorten the machining time. When a common 3D CAD/CAM is used, the feed rate values are given with a gFEDRAT/h statement in CL data through main process of CAM, and F-codes such as F3000.0 (i.e., 3000 mm/min.) corresponding to the feed rate values are written in NC data through post process. Thus, the feed rate value can not be considered and generated in the post process, which is very inconvenient and a drawback for the case of using a common 3D CAD/CAM. Feed rates should be recalculated rapidly according to an actual machining situation. Also, if the cutter path has a large curvature or small edge then undesirable material chipping becomes easy to occur. This means that the machining accuracy tends to go down and therefore we can not obtain the faithful shape as the model designed by a 3D CAD. Especially, when a wooden paint roller with a relief design is machined, the problem of edge chipping can not be avoided. Therefore, the feed rate should be slowed down suitably so that a small artistic design is not damaged by the edge chipping. Note, however, that conventional post-processors do not possess a function to systematically and elaborately decrease feed rate values according to linear interpolated positions given by gGOTO/h statements in CL data. In other words, the operator must return to the main-process of CAM to correct the feed rate values.

Figure 11. Undesirable edge chipping

The proposed post-processor has a function that can automatically generate feed rate values, e.g. F1000.0, from CL data without using feed rate statements according to the curvature of each model not only to keep out the edge chipping but also to shorten the machining time. When the post-processor is used, the operator sets only two parameters, which are the minimum and maximum values of the feed rate. Of course, the post-processor is easily applied to NC machine tools that do not have an open architecture controller. Generally, the main-processor of CAM calculates the CL data $p(i) = [x(i)\ y(i)\ z(i)]^T$ with a linear approximation so that a workpiece can be machined within the tolerance of a designed model. Therefore, the larger the curvature is, the higher its point density is. Accordingly, considering the curvature results in acquiring the distance $d(i) = \|p(i+1) - p(i)\|$ between two adjacent steps in CL data and its increment $\Delta d(i) = d(i+1) - d(i)$. Figure 12 shows an example for the relation between the distance and the point density with respect to CL data.

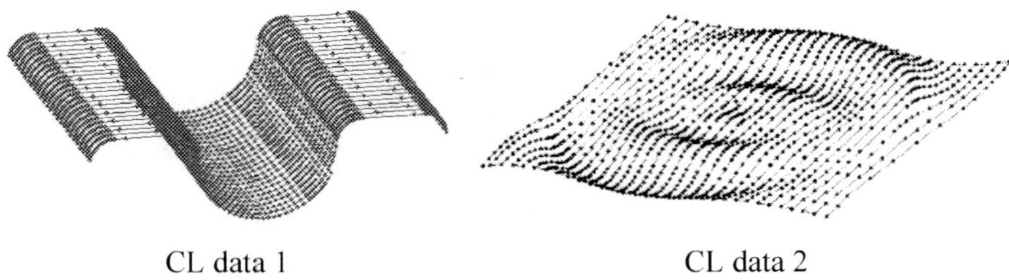

Figure 12. Examples of relation between the distance and the point density with respect to CL data.

3.2.2. Fuzzy feed rate generator

In this section, we propose a fuzzy feed rate generator that generates suitable feed rate values according to $d(i)$ and $\Delta d(i)$. The fuzzy feed rate generator consists of two simple fuzzy reasoning parts whose consequent parts have constant values. When the current position $X(k) = [X(k)\ Y(k)\ Z(k)]^T$ of the end mill at the discrete time k is $X(k) \in [p(i), p(i+1)]$, the fuzzy rules are described by

Rule 1: IF $d(i)$ is $\tilde{A}_1$, THEN $F_{\tilde{A}}(i) = c_1^A$
Rule 2: IF $d(i)$ is $\tilde{A}_2$, THEN $F_{\tilde{A}}(i) = c_2^A$

$\vdots$

Rule L: IF $d(i)$ is $\tilde{A}_L$, THEN $F_{\tilde{A}}(i) = c_L^A$

and

Rule 1: IF $\Delta d(i)$ is $\tilde{B}_1$, THEN $F_{\tilde{B}}(i) = c_1^B$
Rule 2: IF $\Delta d(i)$ is $\tilde{B}_2$, THEN $F_{\tilde{B}}(i) = c_2^B$

$\vdots$

Rule L: IF $\Delta d(i)$ is $\tilde{B}_L$, THEN $F_{\tilde{B}}(i) = c_L^B$

where $\tilde{A}_j (j = 1, ..., L)$ and $\tilde{B}_j$ are the j-th antecedent fuzzy sets for two fuzzy inputs $d(i)$ and $\Delta d(i)$; c_j^A and c_j^B are the consequent constants at the j-th rule for the feed rate $F_{\tilde{A}}(i)$ and its compensation $F_{\tilde{B}}(i)$; L is the total number of fuzzy rules. The confidence of each antecedent part at the j-th rule is obtained by

$$\omega_j^A = \mu_{\tilde{A}j}\{d(i)\} \tag{6}$$

$$\omega_j^B = \mu_{\tilde{B}j}\{\Delta d(i)\} \tag{7}$$

where $\mu_X(\bullet)$ denotes the confidence of a fuzzy set labeled by X. Therefore, the fuzzy reasoning results for the feed rate $F(i)$ and its compensation $\Delta F(i)$ are respectively calculated by

$$F(i) = \frac{\sum_{j=1}^{L} \omega_j^A c_j^A}{\sum_{k=1}^{L} \omega_k^A} \tag{8}$$

$$\Delta F(i) = \frac{\sum_{j=1}^{L} \omega_j^B c_j^B}{\sum_{k=1}^{L} \omega_k^B} \tag{9}$$

The resultant fuzzy feed rate $\tilde{F}(i)$ is realized in the form

$$\tilde{F}(i) = F(i) + \Delta F(i) \tag{10}$$

Note that the fuzzy set used is the following Gaussian membership function

$$\mu_X(x) = \exp\{\log(0.5)(x - p)^2 q^2\} \tag{11}$$

where p is the center of membership function and q is the reciprocal value of standard deviation. Figures 13 (a) and (b) show the antecedent membership functions designed for $d(i)$ and $\Delta d(i)$, respectively. The reciprocal values of the standard deviations are 0.2 and 0.1, respectively. These antecedent membership functions are designed by analyzing $d(i)$ and $\Delta d(i)$ included in several CL data files for machining paint rollers. An example of corresponding constant values in consequent parts is tabulated in Table 1, in which $F_{\max}$ and $F_{\min}$ are the maximum and minimum values (mm/min.) of feed rate; F_{base} denotes $F_{\max} - F_{\min}$. Consequent constant values determined with $F_{\max}$ and $F_{\min}$ must be suitably tuned according to each wooden material, e.g., based on the experience of a skilled NC operator. It is better that the parameters $F_{\max}$ and $F_{\min}$ are given according to the kind of wooden material through a preliminary experiment, while checking the occurrence of an edge chipping and considering the efficiency. It should be noted also that the allowable values of the parameters tend to depend on the complexity of relief design, the depth of material removal, the kind of wooden materials and the direction of the woodgrain. As can be seen from the table, the key parameters that affect the feed rate are only $F_{\max}$ and $F_{\min}$. The operator sets the two parameters according to the relief design and the target wood material. Furthermore, note that the fuzzy reasoning part yields not only larger values than

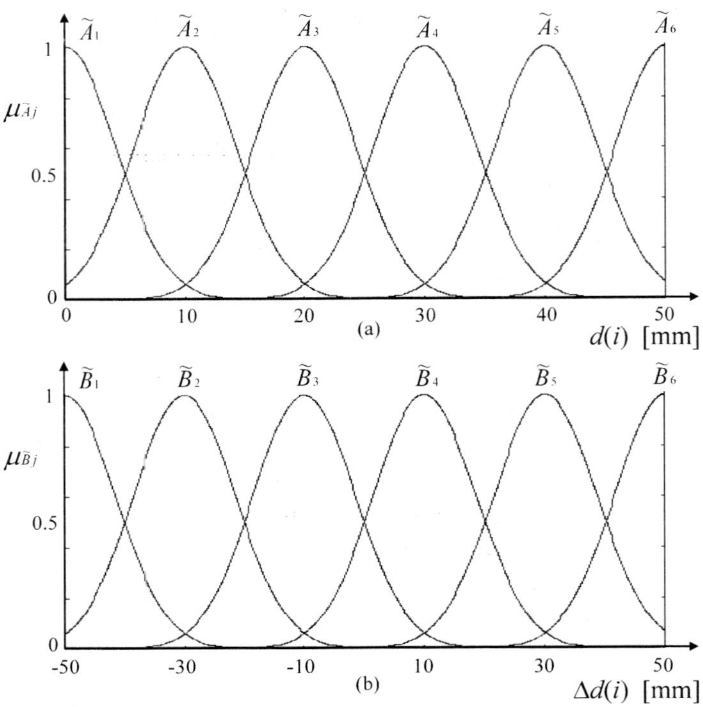

Figure 13. Antecedent membership functions for $d(i)$ and $\Delta d(i)$.

$F_{\max}$ but also smaller values than $F_{\min}$ by using the combination of $d(i)$ and $\Delta d(i)$. Our proposed post-processor performs the post process for the linear approximated CL data without gFEDRAT/h statements and generates suitable feed rate values, so that conventional NC machine tools can be effectively utilized. Also, it need not go back to the main process to recalculate feed rate values. It is expected that the post-processor allows a 3-axis NC machine tool with a rotary unit to easily carve an artistic relief design on a cylindrical wooden workpiece, without giving any undesirable edge chipping.

Table 1. Consequent constants of fuzzy reasoning part for $d(i)$ and $\Delta d(i)$.

c_1^A	c_2^A	c_3^A	c_4^A	c_5^A	c_6^A
$F_{\min} + 0.1F_{base}$	$F_{\min} + 0.2F_{base}$	$F_{\min} + 0.4F_{base}$	$F_{\min} + 0.6F_{base}$	$F_{\min} + 0.8F_{base}$	$F_{\min} + F_{base}$

c_1^B	c_2^B	c_3^B	c_4^B	c_5^B	c_6^B
$-0.5F_{\min}$	$-0.2F_{\min}$	$-0.1F_{\min}$	$0.1F_{\min}$	$0.2F_{\min}$	$0.5F_{\min}$

4. Experiments

In the previous section, a 3-axis NC machine tool with a rotary unit and its post-processor were introduced to efficiently machine cylindrical wooden workpices. The machined workpiece can be used as an artistic design paint roller, which is very useful and convenient to directly transcribe a relief design to a wall just after painting.

In this section, an actual machining experiment of cylindrical wooden workpiece is conducted by using the proposed system. Although up to now there have been few studies to investigate the optimal machining conditions with respect to the machined curved surface of wood, Fujino et al. [20] examined the influences of machining conditions such as feed rate and feed direction to the grain by using two wood species, in which constant feed rate values under 2000 mm/min. were evaluated. The roughness of machined surfaces was measured on 3D profiles obtained by laser scanning, where the surface roughness was shown to be improved as the feed rate was decreased. It is expected, from the above result, that the decrease of feed rate would be effective also to suppress any edge chipping. Wood materials are essentially brittle compared to metallic materials, so that the increase of feed rate brings out any undesirable edge chipping. Such a tendency was confirmed in preliminarily conducted machining test, in which a feed rate override, i.e., a variable manual control function that allows the NC machine tool to increase or decrease programmed feed rates was used. Figure 14 shows an example of machining scene of a paint roller without giving any undesirable edge chipping. In this case, the feed rate values generated from the fuzzy feed rate generator described in the previous section are shown in Fig. 15. The maximum and minimum values $F_{\max}$ and $F_{\min}$ of feed rates were determined to be 2000 mm/min. and 600 mm/min. respectively through a preliminary machining test. The kind of the wooden material used is a glued laminated wood. The maximum cutting depth of material removal is 3 mm. In the case that this wooden material was machined with the design as shown in Fig. 9 giving the maximum cutting depth of 3 mm, an undesirable edge chipping occurred when feed rate values higher than about 800 mm/min. were given. From this fact, 600 mm/min. was set to $F_{\min}$ as a safe value. Also, although a higher value than 2000 mm/min., e.g. 3000 mm/min. was able to be given to $F_{\max}$ within a long straight path, a safety margin of 1000 mm/min. for an edge chipping gave a sufficient reduction in machining time.

It is observed, from the results shown in Fig. 15, that the feed rate $\tilde{F}(i)$ is varied according to the curvature of the model surface. Note that the feed rate of 600 mm/min. appearing periodically is forcibly given every pick feed motion, when the rotary unit is rotated with a small angle (e.g., 0.56 degrees). The amount of the small angle depends on the ratio of y_length to y_pick as shown in Fig. 10. NC data were given to the NC machine tool MDX-650A through a standard RS232C interface by using a DNC (direct numerical control) software. Generally DNC systems with a RS232C interface do not provide a real time communication but an asynchronous serial communication using XON/XOFF flow control, so that the actual feed rate variation for time can not be obtained from the NC machine tool. Instead, however, the relation among the feed rate value $\tilde{F}(i)$ given to the NC machine tool, the distance $d(i)$ and its increment $\Delta d(i)$ is plotted using the step number as the abscissa, as shown in Fig. 15. The relation between $\tilde{F}(i)$ and the point density in CL data can be observed from the result. The curvature can be estimated from the point density because the main-processor generates points written with gGOTO/h statement in CL data accord-

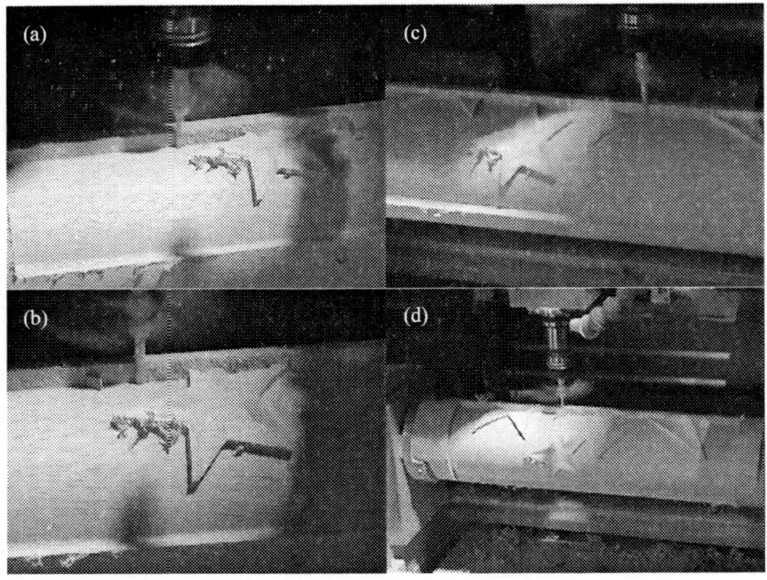

Figure 14. Carving scene of a wooden paint roller with an artistic design.

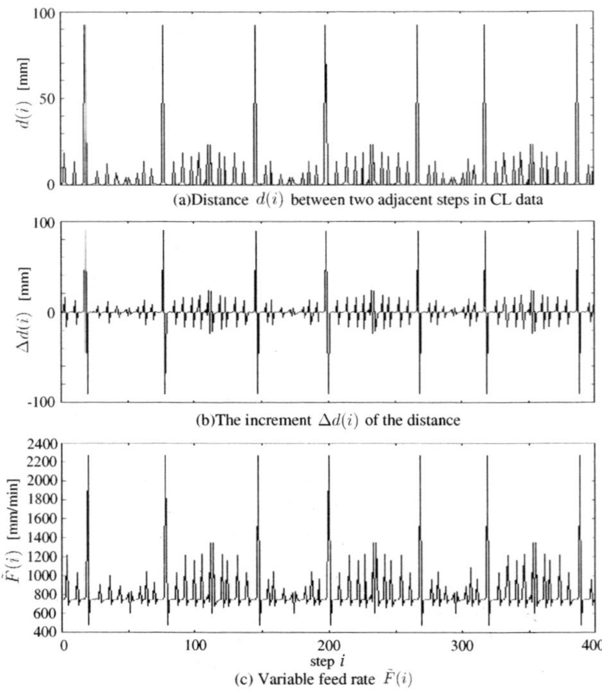

Figure 15. An example of the feed rate values given to NC machine tool, in which the total number of steps is 400, $F_{\max}$ and $F_{\min}$ are set to 2000 mm/min. and 600 mm/min., respectively.

ing to the curvature and the tolerance. In the case of Fig. 15, the total machining time was reduced about 20% compared with the case of using a constant feed rate value of 600 mm/min. As can be seen, the proposed fuzzy feed rate generator provides a more intuitive and finely tunable feed rate function for the post process; however, the quantitative performance evaluation would depend on the combinatorial condition between the complexity of the relief design and the kind of the wooden material. Figure 16 shows the paint rollers with an artistic design carved by the proposed system.

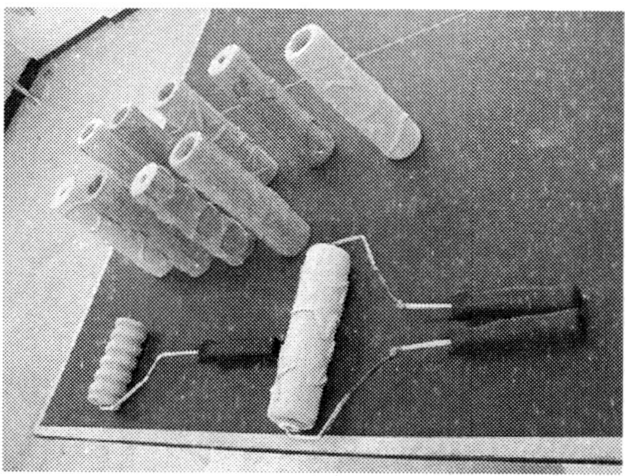

Figure 16. Artistic design wooden paint rollers carved by the proposed system.

5. Mass Production

When a wooden paint roller is popular with the design, its mass production is further required. To cope with the mass production, rubber type paint rollers can be efficiently formed by using an aluminum mold with the same design of the wooden paint roller. Figure 17 shows the machining scene of a part of an aluminum mold, where the aluminum workpiece is fixed with the rotary unit and the inner surface is machined based on the relief design. Figure 19 shows a rubber type paint roller formed with the mold shown in Fig. 18. As can be seen, the aluminum mold consists of four parts in consideration of the easiness of machining.

6. Conclusions

In this chapter, an intelligent machining system based on a 3-axis NC machine tool with a rotary unit has been introduced to efficiently and easily produce many kinds of artistic designed wooden paint rollers. A post-processor with the fuzzy feed rate generator has been also presented to effectively run the machining system. The post-processor allows the NC machine tool to easily transcribe a relief design from the surface on a flat model to it on a cylindrical workpiece. The fuzzy feed rate generator generates suitable feed rate values from CL data without using a feed rate statement according to the curvature of each relief

Figure 17. Machining scene of a part of an aluminum mold.

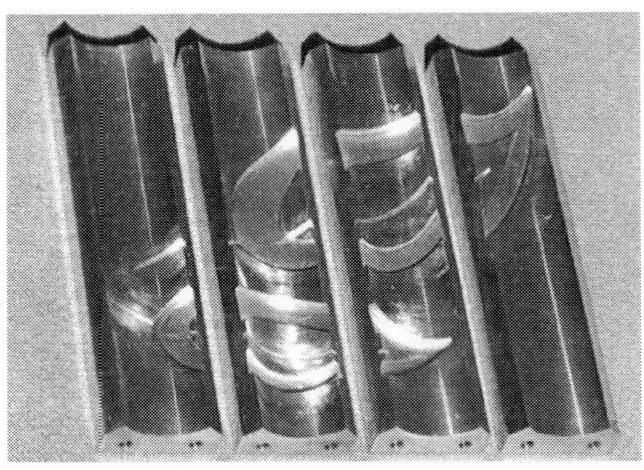

Figure 18. Aluminum mold for forming rubber paint rollers.

design. By the fuzzy feed rate generator, it has been confirmed that not only the edge chippings on the cylindrical surface can be excluded but also the total machining time can be reduced about 20%. Experimental results, thus, have shown that artistic designed wooden paint rollers can be successfully carved by using the proposed machining system. When a wooden paint roller is popular with the design, its mass production is further required. To cope with the mass production, rubber type paint rollers can be efficiently formed by using an aluminum mold with the same design of the wooden paint roller.

In future works, we plan to conduct similar machining experiments using other kinds of

Figure 19. A rubber type paint roller formed with the mold shown in Fig. 18.

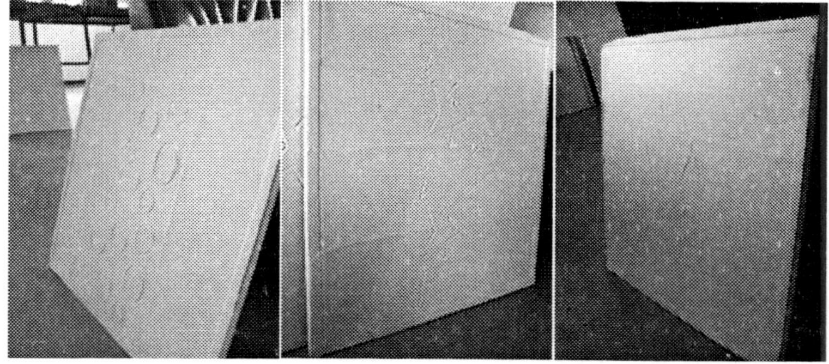

Figure 20. Examples of attractive wall painted by using paint rollers.

wooden materials except for the glued laminated wood and different relief designs. By evaluating the performances and taking the results in fuzzy rules, the fuzzy feed rate generator will be more flexible to the kinds of both wooden materials and relief designs for artistic paint rollers.

References

[1] Takeuchi Y., Watanabe T., *Journal of the Japan Society for Precision Engineering*, 1992, vol. 58, no. 9, pp. 1586–1592, Study on post-processor for 5-axis control machining, (in Japanese).

[2] Takeuchi Y., Wada K., Hisaki T., Yokoyama M., *Journal of the Japan Society for Precision Engineering*, 1994, vol. 60, no. 1, pp. 75–79, Study on post-processor for 5-axis control machining centers —In case of spindle-tilting type and table/spindle-tilting type, (in Japanese).

[3] Xu X. J., Bradley C., Zhang Y. F., Loh H. T., Wong Y. S., *International Journal of Production Research*, 2002, vol. 40, no. 14, pp. 3253–3274, Tool-path generation for five-axis machining of free-form surfaces based on accessibility analysis.

[4] Chen S. L., Chang T. H., Inasaki I., Liu Y. C., *The International Journal of Advanced Manufacturing Technology*, 2002, vol. 20, no. 4, pp. 259–269, Post-processor development of a hybrid TRR-XY parallel kinematic machine tool.

[5] Lei W. T., Hsu Y. Y., *Machine Tool & Manufacture*, 2002, vol. 42, pp. 1153–1162, Accuracy test of five-axis CNC machine tool with 3D probe-ball. Part I: design and modeling.

[6] Lei W. T., Sung M. P., Liu W. L., Chuang Y. C., *Machine Tool & Manufacture*, 2006, vol. 47, pp. 273-285, Double ballbar test for the rotary axes of five-axis CNC machine tools.

[7] Cao L. X., Gong H., Liu J., *Journal of Materials Processing Technology*, 2006, vol. 174, pp. 298–304, The offset approach of machining free from surface. Part 1: Cylindrical cutter in five-axis NC machine tools.

[8] Cao L. X., Gong H., Liu J., *Journal of Materials Processing Technology*, 2007, vol. 184, pp. 6–11, The offset approach of machining free from surface. Part 2: Toroidal cutter in 5-axis NC machine tools.

[9] Timar S. D., Farouki R. T., Smith T. S., Boyadjieff C. L., *Robotics and Computer-Integrated Manufacturing*, 2005, vol. 21, no. 1, pp. 37–53, Algorithms for time optimal control of CNC machines along curved tool paths.

[10] Heo E. Y., Kim D. W., Kim B. H., Chen F. F., *Robotics and Computer-Integrated Manufacturing*, 2006, vol. 22, nos. 5/6, pp. 437–446, Estimation of NC machining time using NC block distribution for sculptured surface machining.

[11] Tarng Y. S., Chuang H. Y., Hsu W. T., *International Journal of Machine Tools and Manufacture*, 1999, vol. 39, pp. 1673–1692, Intelligent cross-coupled fuzzy federate controller design for CNC machine tools based on genetic algorithms.

[12] Zuperl U., Cus F., Milfelner M., *Journal of Materials Processing Technology*, 2005, vols. 164/165, pp. 1472–1478, Fuzzy control strategy for an adaptive force control in end-milling.

[13] Farouki R. T., Manjunathaiah J., Nicholas D., Yuan G. -F., Jee S., *Computer-Aided Design*, 1998, vol. 30, no. 8, pp. 631–640, Variable-feedrate CNC interpolators for constant material removal rates along Pythagorean-hodograph curves.

[14] Farouki R. T., Manjunathaiah J., Yuan G. -F., *Machine Tools & Manufacture*, 1999, vol. 39, no. 1, pp. 123–142, G codes for the specification of Pythagorean-hodograph tool paths and associated feedrate functions on open-architecture CNC machines.

[15] Frank A, Schmid A., *Robotics and Computer-Integrated Manufacturing*, 1988, vol. 4, nos. 1/2, pp. 211–218, Grinding of non-circular contours on CNC cylindrical grinding machines.

[16] Couey J. A., Marsh E. R., Knapp B. R., Vallance R. R., *Precision Engineering*, 2005, vol. 29, pp. 307–314, Monitoring force in precision cylindrical grinding.

[17] Tawakoli T., Rasifard A., Rabiey M., *Machine Tools & Manufacture*, 2007, vol. 47, pp. 729–733, High-efficiency internal cylindrical grinding with a new kinematic.

[18] Nagata F, Kusumoto Y, Hasebe K, Saito K, Fukumoto M, Watanabe K., *Procs. of 2005 International Conference on Control, Automation and Systems*, 2005, pp. 438-443, Post-processor using a fuzzy feed rate generator for multi-axis NC machine tools with a rotary unit.

[19] Nagata F., Watanabe K., *Journal of the Japan Society for Precision Engineering*, 1996, vol. 62, no. 8, pp. 1203–1207, Development of a post-processor module of 5-axis control NC machine tool with tilting-head for woody furniture (in Japanese).

[20] Fujino K, Sawada Y, Fujii Y, Okumura S., *Procs. of the 16th International Wood Machining Seminar*, 2003, Part 2, pp. 532–538, Machining of curved surface of wood by ball end mill – Effect of rake angle and feed speed on machined surface.

INDEX

SiO2, 143, 144, 149

skin, 32, 162, 167

Slovakia, 8

smart materials, 146

smoothing, 176, 177, 181, 182, 184, 186

social inequalities, 173

society, ix, 116, 141, 147, 173

softener, 59, 60

software, x, 117, 119, 125, 169, 178, 181, 183, 197, 202, 210, 220, 227

solid state, 154

solubility, 13

solution, 12, 24, 27, 28, 30, 31, 32, 53, 60, 62, 63, 65, 68, 81, 149, 154, 155, 162, 163, 165

solvents, ix, 31, 58, 59, 62, 114, 127, 129, 130, 133, 134, 142

South Korea, 161

sowing, 163

Spain, 5

specialists, 88

speciation, 10

species, 2, 12, 145, 177, 196

specific heat, 98

Spectroanalytical techniques, vii, 1

spectroscopic techniques, 7

spectroscopy, 11, 13, 171

spindle, 214, 231

sponge, vii, viii, 23, 27, 29, 30, 32, 33, 34, 36, 37, 38, 44, 45, 46, 47, 48, 49, 50, 51, 52, 57, 66, 67, 68, 69, 70, 71, 72, 73, 74, 75, 76, 77, 78, 79

stability, 13, 132, 135, 142, 144, 145, 148, 156

stabilization, 132, 134, 142, 148

stabilizers, 134, 145

stakeholders, 115

standard deviation, 99, 121, 225

Staphylococcus aureus, x, 161, 162, 163, 165

starch, 122

state, viii, 6, 12, 57, 116, 132, 134, 171

statistics, 16, 98, 178, 181, 199

steel, ix, x, 12, 13, 96, 97, 98, 127, 130, 136, 148, 149, 151, 157, 159

stem cells, 166

still life, 9

stoichiometry, 203, 204

storage, 91, 132, 133, 174

stress, 152, 154, 177

stretching, 176

structure, 6, 9, 14, 58, 96, 115, 116, 117, 126, 131, 132, 176, 188, 210, 213, 214

structuring, 119

style, xi, 5, 169, 173, 174, 193, 205

styrene, 14, 130

subgroups, 10

substitutions, 205

substrate, vii, ix, 2, 127, 133, 136, 145, 149, 151, 172

substrates, 11, 12, 133, 135

sulfate, 30, 131

sulfur, 165

sulphur, 182, 202, 203

supplier, 115, 117, 121

supply chain, 116

surface area, 100, 131

surface boundary layer model, viii, 85, 91, 110

surface chemistry, 5

surface coating, vii, ix, 127

surface heat budget model, viii, 85, 100, 102, 105

surface modification, 146

surface properties, 53, 54, 81, 135, 142

surface temperature, viii, 14, 85, 91, 92, 93, 100, 102, 109, 110, 111

surface tension, 28, 62, 63, 129

surface treatment, x, 31, 121, 125, 136, 169, 188

surfactant, 24, 28, 29, 30, 31, 32, 60, 61, 62, 63, 64, 66, 67, 82, 130, 137

surfactants, vii, viii, 23, 26, 27, 28, 31, 32, 53, 57, 58, 59, 61, 62, 63, 65, 67, 68, 81, 82, 83, 130, 134, 147

sustainability, vii, ix, 113, 114, 115

sustainable development, ix, 113, 115

Switzerland, 52, 59, 80

symmetry, 171, 177

synergistic effect, 31

synthesis, 167

synthesized silver nanoparticles, x, 161

T

talc, 14, 131

techniques, vii, ix, x, 1, 2, 3, 4, 5, 6, 7, 9, 10, 11, 12, 13, 15, 25, 64, 113, 114, 115, 116, 118, 120, 122, 133, 137, 169, 170, 171, 172, 174, 205, 206

technological advancement, 86

technological progress, 116

technologies, vii, viii, 85, 86, 88, 90, 110, 114, 115, 135, 142, 147, 148

technologies of improvement, viii, 85

technology, viii, ix, x, xi, 10, 23, 25, 26, 57, 64, 65, 85, 86, 88, 89, 91, 106, 108, 110, 111, 115, 116, 117, 118, 123, 133, 137, 141, 142, 143, 146, 148, 153, 162, 170, 173, 206

temperature, viii, x, 10, 11, 23, 24, 30, 31, 32, 58, 62, 63, 65, 68, 85, 86, 87, 88, 91, 92, 93, 96, 97, 99, 100, 102, 103, 109, 110, 111, 130, 133, 134, 135, 145, 146, 151, 155, 156, 157, 158, 159, 163, 187

tensile strength, 134

tension, 130, 137